Phyto-nematodes in Crops

NIPA® GENX ELECTRONIC RESOURCES & SOLUTIONS P. LTD.
New Delhi-110 034

www.nipaers.com
Browse, Search, Read & Buy...
Print Books
eChapters
eBooks
Publishers
Forthcoming
Subject Catalogues
New Titles
Online Resources on:
✓ Current Affairs
✓ Reasoning, Logic & Aptitude
✓ Competitive Examinations
✓ English Language Lab
✓ Writing & Pronunciation Tools
✓ Personality Development
✓ Online Programmes for Professionals
✓ Interview Preparation
Pay Using
PhonePe
Google Pay
Paytm
PayPal
UPI
CC Avenue

Phyto-nematodes in Crops: Their Identification, Treatment and Management

G.L.SHARMA
Professor
Department of Nematology S.K.N. College of Agriculture (Swami Keshwa and Rajasthan Agricultural University) Jobner- 303 329 (Jaipur), Rajasthan, India

NIPA® GENX ELECTRONIC RESOURCES & SOLUTIONS P. LTD.
New Delhi-110 034

NIPA® GENX ELECTRONIC
RESOURCES & SOLUTIONS P. LTD.

101,103, Vikas Surya Plaza, CU Block
L.S.C. Market, Pitam Pura, New Delhi-110 034
Ph : +91 11 27341616, 27341717, 27341718
E-mail: newindiapublishingagency@gmail.com
web: www.nipabooks.com

For customer assistance, please contact
Phone: + 91-11-27 34 17 17
Fax: + 91-11- 27 34 16 16
E-Mail: feedbacks@nipabooks.com

ISBN: 978-81-19254-12-5

Composed and Designed by NIPA®.

Preface

Since last four decades, after the discovery of first phytonematode from India, a continuous survey and surveillance processes remained continue, because of causing immense crop losses to various agricultural/ horticultural/plantation crops. Based on FAO data (1984) the crop losses caused by these on world wide scale seems to about 12.3%. In many crops , where either monocropping remained continue by the cultivators or the prevailing ecosystem suited to the phytonematodes, then the range of crop losses may even exceeds more than 60-70 % and which some times causes menace condition during crop cultivation. After having such data, various phytonematode management strategies; viz. cultural, chemical, biotic, breeding for disease resistance and Integrated Nematode Management were initiated, a lot of compilation of such work has been done and various packages of practices have been developed by the nematologists for economic management of phytonematodes for the use of cultivators.

This book deals with total 11 chapters. Of which, some are related to the thrusts caused by the phytonematodes and their managements in cereals, forage, pulses, seed spices, horticultural and plantation crops. In some of these, the protective cultivation and interaction between nematode and phytopathogens have also been dealt. Coverage of this book is with recent literature pertinent to the related topics and their compilation also made by subject learned scientist. It is to hope that this book will be serving as a good reference book to the subject related researchers and scientists.

G.L.SHARMA

Contributors

A.K Singh
Unit of Nematology
Directorate of Wheat Research
Karnal-132 001, India

A.K. Sinha
Department of Nematology
Faculty of Agriculture
Assam Agricultural University
Jorhat-785 013, Assam, India

A.N.Srivastava
Division of Nematology
Indian Agricultural Research Institute
New Delhi-110 012, India

B. Mahanta
Department of Nematology
Assam Agricultural University
Jorhat-785 013, Assam, India

G. Chawla
Division of Nematology
Indian Agricultural Research Institute
New Delhi-110 012, India

G.L.Sharma
Department of Nematology
S.K.N.College of Agriculture
Jobner-303 329, India

Hari Chand
Department of Plant Pathology
CCS Haryana Agricultural University
Hisar- 125 004, India

I.J. Paruthi
Department of Nematology
CCS Haryana Agricultural University
Hisar- 125 004, India

K.K. Walia
Department of Nematology
CCS Haryana Agricultural University
Hisar-125 004, India

M.S.Sheela
Department of Nematology
Kerala Agricultural University
Kerala, India

Pankaj
Division of Nematology
Indian Agricultural Research
Institute, New Delhi-110 012, India

P.P. Neog
Department of Nematology
Assam Agricultural University
Jorhat-785013, India

Mrs. P.Sharma
Division of Plant Pathology
IARI, New Delhi-110 012, India

R.K. Chugh
Centre for Plant Biotechnology
HSCS&T Department of Science
and Technology, CCS HAU
New Campus, Hisar-125 004, India

R.K. Walia
Department of Nematology
CCS Haryana Agricultural University
Hisar-125 004, India

R.K.Jain
Indian Grassland and Fodder
Research Institute
Regional Research Station
Avikanagar-304 501, Rajasthan
India

R.K.Jaiswal
Division of Nematology
Indian Agricultural Research Institute
New Delhi-110 012, India

R.P.Nagar
Indian Grassland and Fodder
Research Institute
Regional Research Station
Avikanagar-304 501, Rajasthan, India

Rajan
Plant Quarantine Division
National Bureau of Plant Genetic
Resources, New Delhi-110 012, India

Rashid Parvez
Indian Institute of Spices Research
Calicut, Kerala, India

S.S.Ali
Emeritus Scientist
Indian Institute of Pulses Research
Kanpur, India

Sudershan Ganguly
Division of Nematology
Indian Agricultural Research Institute
New Delhi-110 012, India

Mrs. V.L.Majumdar
Department of Plant Pathology
A.R.S.Durgapura, Jaipur-302 018, India

□□□

Contents

Phyto-nematodes in Crops: Their Identification, Treatment and Management, 2011
G.L. Sharma (editor), pp. 1-8
New India Publishing Agency, New Delhi (India)
E-mail : info@nipabooks.com; Website : www.nipabooks.com

Chapter 1

Cereal Cyst Nematode *(Heterodera avenae)*, A Menace causing Phytonematode in Cereals in Light Soils

G.L.SHARMA, P. SHARMA AND V.L.MAJUMDAR

Introduction

The cereal cyst nematode, (*Heterodera avenae* Woll.) a causal organism of 'molya' disease in wheat and barley, since after its first report (by Wollenweber, 1924, Vasudeva, 1958 and Prasad *et al.*, 1959), it has been timely reviewed about its spread and losses caused in the various parts of wheat and barley growing areas of the world. It has been reported to occur in Australia, Canada, Israel, South Africa, Japan, in most of the European countries, North America, India, Morocco, Tunisia, Pakistan, Libya, Turkey and Australia (Prasad *et al.*, 1959, Smiley *et al.*, 1994, Rumpen horst *et al.*,1996, Krall *et al.*, 1999, Nicol, 2002 and Rao *et al.*, 2002). Its economic importance has been assessed by mapping the yield losses caused in cereals i.e. ranged from 15-20% in Pakistan, 40-92% in Sauthi Arabia, 23-50% in Australia (Nicol, 2002) and 40-50% in India (Mathur 1969, Mathur *et al.*, 1980, Handa *et al.*, 1985, Mathur, *et al.*, 1987). This phyto-nematode has alarmed the cultivators of the wheat growing areas of the world due to its intensity of causing crop yield losses. Rajvanshi and Sharma (2002) also found a varied declination of yield status in barley crop due to change in the population of *Heterodera avenae.*

Based on world wide report, as the population is increasing, the wheat consumption also found to increase along with an enhancement of its area of cultivation.

In India total wheat production varies from 70-80 million tons since last one decade to meet out the food demand of about one Arab person.In Rajasthan state, the total wheat cultivated area is 1.97 million ha (as per record of 2003-04) with an average yield of 3.13 tons/ha (Mathur *et al.,* 1980) having a total production of 5.99.M.Tons.The total cereal cyst nematode (CCN) infested area of wheat is 0.15 Million ha, which instead of producing 0.43 M tons,only yielding 0.22 M tons (2001-02) and grain loss in terms of money amounts about Rs 126 millions in Rajasthan.

Survey of the Cereal Cyst Nematode

As per the mapping reports of CCN in India, all wheat growing states are having its presence

States		Area Infested with CCN
Jammu & Kashmir	-	Leh region in Laddakh
Himachal Pradesh	-	Una and Kangra
Punjab	-	Bhatinda, Faridkot, Hoshiyarpur, Jalandhar, Kapurthala, Ludhiana, Patiala and Sangrur
Hariyana	-	Ambala, Bhiwani, Gurgaon, Mohindergarh, Rohtak, Sirsa and Sonipat.
Rajasthan	-	Alwar, Ajmer, Bhilwara, Dausa, Hanumangarh, Jaipur, Jhunjhunu, Karauli, Nagaur Pali, Rajsamund, Sawai- Madhopur, Sikar, Sirohi, Tonk and Udaipur etc.
Madhya Pradesh	-	Tikamgarh
Uttar Pradesh	-	Aligarh, Badaun, Gaziababd and Jhansi
Delhi	-	Najabgarh block

Besides it, the reports about the presence of CCN is still continuously receiving from the wheat growing areas of the country. About 40 districts of 8 states have shown varied infestation of CCN (mild, moderate and severe) on wheat and barley crops. In last few years the barley crop cultivation area is also day to day increasing due exploration its use in malt industry, food and fodder etc. in the states of Rajasthan and Karnataka.

Since the menace caused in wheat production has been recorded due to certain fungi (Rusts) and insects (Termites), but after reviewing the loss estimates caused by phytonematodes, mainly by cereal cyst nematode (*Heterodera avenae*), it has been recognized as one of the emerging pest to decrease the yield production of wheat/barley in India. After the Green Revolution in wheat, the evolution of triple dwarf gene based cultivars have solved the food problem of the country, but due outreaching data of human population (estimated to be approximately 1.50 Arab up to the year 2030), the presently available wheat amount shall be insufficient to cater the food requirement. Since after incorporating all optimum agronomical /plant breeding/soil chemistry recommendations, the yield of wheat in India is still roaming about 70-80 millions tons. Average losses recorded world wide by various pests viz. fungi, bacteria, insects and phytonematodes have been estimated up to 18-20%. Therefore, if these losses can be managed by employing certain economic management strategies, a marked increase in the wheat production could be achieved, which in future can solve the food problem to some extent.

In the light soils, looking towards the extent of losses caused by *Heterodera avenae* in wheat growing areas, this pest is considered to be a menace causing one. With this view the further detailed studies carried on its various aspects have been included in this chapter.

Management

Based on research work conducted earlier, the damage caused by CCN may be affected by a number of biotic and abiotic factors. This nematode used to cause greater damage instances where plant growth is stressed, either by poor soil nutrition, temperature or water (Barker and Noe, 1987) or where pathogen pressures occur (Taheri *et al.*, 1994).

Cultural Practices : The mono-cropping of wheat and barley have been noticed to cause greater damages. The reduction in the incidence of CCN has variously been practiced (Mathur *et al.*, 1998) viz. such as cultural practices [crop rotation(Mathur and Handa, 1984),Host efficiency (Handa *et al.*, 1985), summer plouhging and solarisation (Haq and Prasad,1981, Handa,1983, Stapleton and Devay,1983, Handa *et al.*,1985, Mathur *et al.*, 1985, Mathur *et al.*, 1983 and 1987), effect of organic and inorganic fertilizers (Mathur, 1969, Sakhuja *et al.*, 1978, Mathur and Handa, 1984), mixed cropping (Handa *et al.*, 1980), soil amendments using oil cakes, Neem kernel powder, decomposed leaves of Neem, Datura and Aak as well as Carrot and Turmeric powders (Mukhopadhyaya *et al.*, 1972, Mishra, 1974, Dhawan and Kaushal, 1988, Rajvanshi and Vishnoi, 1998, Rajvanshi *et al.*, 2002), intercropping of carrot, fenugreek, mustard and reddish with wheat and barley (Rajvanshi *et al.*, 2002)].

Breeding for Disease Resistance

CCN resistance varieties for barley- RD-387(RAJ KIRAN) (published by Handa, 1992), RD-2052, RD-2035 and RD-2508 released from A.R.S. Durgapura, Jaipur; CCN resistant wheat line-CCNRV1 was developed (Mathur *et al.*, 1998, Sharma and Sharma, 2000 and Sharma *et al.*, 2007) and released as a cultivar- RAJ MR-1 (pub. Mathur *et al.*, 2004), C-164 (from CCHAU, Hisar) and DWR-171 (from SAU, Dharwar, Maharastra).

Integrated Nematode Managements

Its adoption has also reduced the CCN population in wheat and barley crops viz. (summer ploughing + Chemicals (Mathur *et al.*, 1984), use of bio-agents- *Trichoderma viride* (Pankaj *et al.*, 2002, Rajvanshi, 2003), *Beuvaria bassiana and Metarhizium anisoplae* (Rajvanshi and Sharma, 2005 and Sharma and Gupta, 2010) have shown marked increase in grain yield and reduced the CCN population.

Use of Agro-Chemicals

Among pesticides, the halogenated hydrocarbons (DD soil fumigants and DBCP), carbamates (Aldicab 10 G, Carbofuran 3G, Oxamil 5 & 10 G) and organophosphates (Thimet 10G Diazinon 10G and Solvirex 10G) have also showed marked reduction in CCN population when used against wheat and barley crops (Handa *et al.*, 1980, Handa and Mathur, 1982, Handa *et al.*,1985,Mathur *et al.*,1986, Swarup, 1986 and Sharma and Rajvanshi,2004). But due to having health hazardous effect and costlier in manufacture most of these chemicals have been ceased or eradicated from market.

Besides the granular formulations, the seed dressing powders (25 SP) and emulsified concentrates (30 EC) of Carbosulfon were also found very effective and economic to manage the CCN in wheat and barley crops (Rajvanshi and Bishnoi, 1995, Rajvanshi and Sharma, 2000).

Sharma and Rajvanshi, (2004) also concluded that a combination of chemicals and bio-pesticides (Achook- a neem formulation) reduced the CCN population level and increased the grain yield of wheat.

Conclusion

After reviewing the status of the Cereal Cyst Nematode (*Heterodera avenae*) in cereals, it is noted that in wheat sown light soil areas of the world, this phyto-nematode seems to be a menace causing pest because the crop yield losses in wheat/barley have been estimated 40-50% which sometimes reaches even up to 70% at high CCN inoculum level in the field.

Among various management strategies- the cultural practices viz. deep summer ploughing and crop rotation have been able to minimize the pest population as well documented as a feasible way by cultivators. The use of chemicals (granular nematicides) although drastically reduced the nemic population but being costlier preposition its use seems to be beyond the reach of cultivators. The use of seed dressers and emulsified concentrates gave significant reduction in CCN population and also proved economic. Looking towards the use of bio-agents- some fungi viz. *Trichoderma, Metarhizium and Beuvaria* spp. gave encouraging results as alone or with combination of reduced doses of chemicals or organic amendments in soil using neem kernel powder. But rearing these bio-agents and their acclimatization to the field in which they used to be applied, still needs further research and vigorous field experimentation.

References

Barker, K.R. and J.P. Noe, (1987). Establishing and using threshold population levels.In: *Vistas on Nematology*, Society of Nematologists, Maryland, USA. pp-75-81.

Dhawan, S.C.and Kaushal,K.K (1988). Effect of neem emulsion on the hatching of cereal cyst nematode, *Heterodera avenae. Ind. J. Nematol.*, 18(2):364.

Handa, D.K. (1983). *Studies on molya disease (Heterodera avenae) of wheat and barley with special emphasis on its control.* Ph. D. Thesis, University of Rajasthan, Jaipur.

Handa, D.K., B.N.Mathur, G.L.Sharma and B.D.Yadav (1980). Relative loss in between tall and dwarf varieties of wheat and barley due to molya disease. *Ind. J. Mycol. & Plant Pathol.* 10: vii.

Handa, D.K., B.N. Mathur, R.L.Mathur, B.D.Yadav and G.L.Sharma (1980). Evaluation of barley varieties for resistance to molya disease incited by *Heterodera avenae. Ind. J.Mycol. & Plant Pathol.* 10: ix.

Handa,D.K.,B.N.Mathur,R.L.Mathur,B.D.Yadav and G.L.Sharma(1985).Control of molya disease caused by *Heterodera avenae* in wheat and barley. I.Through seed treatment.In:IV All India Nematology Symposium,Udaipur (May 17-18).

Handa, D.K. (1982). Management of cereal cyst nematode *(Heterodera avenae)* in barley. In: *Recent advances in Nematology,* pp-83-99.

Handa,D.K.,R.L.Mathur,B.N.Mathur, G.L.Sharma and B.D.Yadav (1985).Host efficiency of tall and dwarf wheat and barley crops to cereal cyst nematode *(Heterodera avenae).Ind. J.Nematol.,*15(1):39-42.

Handa, D.K., R.L.Mathur, B.N.Mathur and B.D.Yadav (1985). Estimation of losses in barley due to cereal cyst nematode in sandy and sandy loam soils.*Ind. J.Nematol.* 15(2):163-166.

Haq, M.M. and D. Prasad (1981). Effect of agronomical practices and D-D, its application on nematode population and crop yield. (Abstract) *Ind. J. Nematol.* 11:124.

Krall, E., D. Sturhan and J.Muur. (1999). Turaviljadel jakorsheeintel parasiteerivad tsuste mood ustaved nematoodid Eestis, *Teadustoode Kogunik, 203:44-48.*

Mathur, B.N., D.K.Handa, and G.Swarup. (1987). Effect of deep summer ploughings on the cereal cyst nematode, *Heterodera avenae* and yield of wheat in Rajasthan, India. *Ind. J. Nematol.,* 17: 292-295.

Mathur, B.N.(1969). *Studies on cereal root eelworm Heterodera avenae with special reference to molya disease of wheat and barley in Rajasthan. Ph.D. Thesis,* University of Rajasthan, Jaipur.

Mathur, B.N., D.K.Handa, G.Swarup, C.L.Sethi, G.L.Sharma and B.D.Yadav (1986). On the loss estimation and chemical control of molya disease of wheat caused by *Heterodera avenae* in India. *Ind. J. Nematol.,*16(2):152154.

Mathur, B.N and Handa, D.K. (1984). Cultural practices and their effects on cereal cyst nematode and yield of wheat in Rajasthan, India. First Internatl. *Congress of Nematology, Guelf, Ontario, Canada, Aug.-5-10,pp62.*

Mathur,B.N.,D.K.Handa,G.L.Sharma and B.D.Yadav(1984).Control of molya disease of wheat caused by *Heterodera avenae* by post planting application of chemicals.Presented in All India Symposium on Mycology& Plant Pathology held at Solan from 23-25 May.

Mathur, B.N., D.K.Handa and G.Swarup (1987). Effect of deep summer ploughings on cereal cyst nematode, *Heterodera avenae* and yield of wheat in Rajasthan, India. *Ind. J.Nematol,* 17(2):292-295.

Mathur, B.N., V.K.Bhatnagar, S.N.Sharma and G.L.Sharma (1998). Evaluation of CCN resistant variety for higher yields. *Proc. of III rd Internatl. Symp. of Afro-Asian Society of Nematologists, Coimbatore, India.pp-45.*

Mathur,B.N.,S.N.Sharma,V.K.Bhatnagar,G.L.Sharma, R.S.Sain, S.M. Bhatnagar, H.Singh A.Mishra, C.P.Nagpal and R.L.Midha (2004). Raj Molya Rodhak-1, first commercial unique wheat variety resistant to cereal cyst nematode in India. *Wheat Information Service,* 98: 25-27.

Mishra, S.D. and S.K.Prasad (1974). Effect of soil amendment on nematodes and crop yield. *Ind. J.Nematol.* 4: 1-19.

Mukhopadhyaya, M.C., M.R.Dalal, S.Saren and S.S.Kharab (1972). Studies on molya disease of wheat and barley. *Ind.J.Nematol.* 2: 11-20.

Pankaj, S.D.Mishra and G.L.Sharma (2002). Integrated management of Heterodera avenae infecting wheat. In: *Proceedings of Symposium on "Bio-diversity and management of nematodes in cropping systems for sustainable agriculture" organized by ICAR & NSI at ARS, Durgapura, Jaipur* (11-1 3 Nov., 2002) pp-265.

Rajvanshi, I.and S.P.Bishnoi (1998). Effect of decomposed leaves on cereal cyst nematode of barley. *Third Internatl. Sympo. of Afro-Asian Society of Nematologists, April (16-19), Coimbatore, India.*

Rajnanshi, I. and G.L.Sharma (2000). Management of cereal cyst nematode, *Heterodera avenae* in barley through chemicals.*Current Nematology,*11: 35-37.

Rajvanshi, I. B.N.Mathur and G.L.Sharma (2002). Effect of intercropping on incidence of *Heterodera avenae* in wheat and barley crops. *Annals of Plant Protection,*10(2): 406-40

Rajvanshi, I. and G.L.Sharma (2002). Declination of yield status of barley due to *Heterodera avenae* in light soils of Rajasthan State. *Current Nematology,* 13(1,2): 27-29.

Rajvanshi, I. and S, N, Sharma (2003). Breeding for resistance to cereal cyst nematode, *Heterodera avenae* in barley. *Current Nematology,*14(1,2):99101.

Rajvanshi, I. and G.L.Sharma (2005). Bio-control of cereal cyst nematode, *Heterodera avenae* infecting wheat. In: *Proceedings of National Symposium of NSI, New Delhi* (9-10 Dec.2005) pp-65.

Rampenhorst, H.J., I.H. Elekcioglu, D.Sturhan, G.ozturk and S. Enneli (1996). The cereal cyst nematode, *Heterodera filipjevi* (Madzhidov) in Turkey. *Nematologia Mediterrania,* 24: 135-138.

Rao, U., K.K.Kaushal, K.Singh, V.Singh and A.K.Gaguli (2002). Genetic diversity in the Indian populations of cereal cyst nematode, *Heterodera avenae* evidenced by molecular markers. *Second Intl. Meeting on Wheat Technologies for warmer areas, MACS-Agarker Institute, Pune,India.* Pp49-50.

Sakhuja,P.K.,I.Singh and S.K.Sharma(1978).Effect of nitrogen level alone and in combination with some nematicides on population of cereal cyst nematode,*Heterodera avenae* and yield of wheat.*Ind.J.Nematol.*8(3): 159-162.

Sharma, G.L. and S.N.Sharma (2000). Evaluation of cereal cyst nematode *(Heterodera avenae)* resistant wheat variety in Rajasthan, India. *Wheat Information Service*, 90: 4951.

Sharma, G.L.and K.N.Gupta (2010).Bio-management of cereal cyst nematode, *Heterodera avenae in cereals. 13th All India Congr. Of Agri. Scientists & Farmers held at Allahabad (Feb.2010), organized by Bio-ved Research Society, Allahabad.*

Sharma, S.N.,R.S.Sain, B.N.Mathur, G.L.Sharma,V.K.Bhatnagar,H.Singh and R.L.Midha(2007). Development and validation of first cereal cyst nematode *(Heterodera avenae)* resistant wheat CCNRV1 for North India. *SABRAO J.of Breeding & Genetics*,39(1):1-16.

Smiley, R.W.,R.E.Ingham, W.Uddin and G.H.Cook (1994).Crop sequences for managing cereal cyst nematode-fungal pathogen of winter wheat, *Plant Disease* 78:1142-1149.

Stepen, J.J. and J.E.Devay (1983). Response of phytoparasitic and free living nematodes to soil solarisation and 1, 3-di chloropropane in California, *Phytopatholohy,* 73:1426-1436.

Taheri,A.,G.J.Hollamby and V.A.Vanspone(1994).Interaction between root lesion nematode, *Pratylenchus neglectus* (Rench, 1924) Chit wood & Oteifa, 1952 and root rotting fungi of wheat. *New Zealand Crop Hortic. Science,* 22: 181-185.

Vasudeva, R.S. (1958). *Scientific Report, IARI,* New Delhi.

Wollenweber, H. (1924). Zur Kenntis Der Kartoffel- *Heterodera (I).Landw.Ztg.* No.1.

Phyto-nematodes in Crops: Their Identification, Treatment and Management, 2011
G.L. Sharma (editor), pp. 9-23
New India Publishing Agency, New Delhi (India)
E-mail : info@nipabooks.com; Website : www.nipabooks.com

Chapter 2

Heterodera and Anguina - Persistent Threat to Wheat Cultivation in India

A.K. SINGH

Preamble

In India, agriculture plays an important role as it accounts for about 18% of the GDP and supports nearly 58% of the total working population for employment. Food grain production increased significantly from a low level of 50 million Tons during 1950-51 to a record of 230.7 million Tons in 2007-08. The demand for food is growing rapidly due to increasing population and rising income levels, hence we need to produce about 320 million Tons of foodgrains by the year 2025. It implies more pressure on our exhisting land, soil and water resources which are already in short supply, degraded, fragile and this similar situation is being faced by mainly developing countries today. Plant parasitic nematodes are of great economic significance. Since most of them live in the soil hence represent one of the most difficult pest problems in terms of identification, demonstration and management. Their influence on crops usually get underestimated by planners, farmers, development workers, crop scientists as well as pest management consultants, though it has been estimated that some 10%

of the world crop produce is lost due to nematode damage. Although many nematodes have been found associated with small grain cereals, but only a few of them are considered economically important. These include CCN (*Heterodera spp*), Seed gall nematode (*Anguina tritici*), Root lesion nematode (*Pratylenchus spp*), Root Knot Nematode (*Meloidogyne graminicola*), Stunt nematode (*Tylenchorhynchus vulgaris*) etc.

Of late, researches have delineated the fact that large number of nemic population regularly being encountered in the rhizosphere of wheat crop, but only two genera, namely *Heterodera spp* and *Anguina tritici* are economically important; other four species namely, *Meloidogyne javanica, M incognita, Pratylenchus thornei* and *Tylenchorhynchus vulgaris* are posing potential threat while role of the rest of the species has been investigated but nothing conclusive in pathogenic term has come out so far but may be having some role in nutrient recycling and hyperparasitism.

M graminicola is coming on wheat in other parts of the country also. *Pratylenchus* and *Tylenchorhynchus vulgaris* are of the potential importance while role of the rest of the species has not been investigated so far in our context. Though Molya was first recorded in 1956 when wheat sickness was spotted in the Sikar district of Rajasthan and in 1958 its association with *Heterodera avenae* was finally established. The survey conducted during 1958-62 revealed that its infestation has gone to each direction in Rajasthan and most specifically it was recorded in places like Churu, Jhunjhunu, Jaipur and Ajmer. Additionally, Neem Ka Thana was also found infested. Subsequent surveys revealed infestation in Narnaul (Haryana), Beawar, Udaipur and Chittorgarh (all in Rajasthan). The infestation sites in Punjab were reported only in late 1960s with Hoshiarpur and Ludhiana as the only two sites. To tackle this emerging problem, the All India Coordinated Wheat and Barley Improvement Project (AICW&BIP) came in to existence in 1965. This was one of the first organized projects initiated by the Indian Council of Agricultural Research which laid the foundation of organized scientific and coordinated research under its purview in the 1960s, when a disease problem called locally as Molya in Rajasthan was found to be caused by a nematode parasite, *Heterodera avenae*. Initially, the work started at the Indian Agricultural Research Institute, New Delhi as a voluntary centre to undertake research on the plant parasitic nematodes associated with wheat crop. Subsequently, centres at Jaipur and Udaipur in Rajasthan state, Hisar in Haryana State, and Ludhiana in Punjab state were established as part of the coordinated project to carry out nematological research on wheat and barley. At present, 5 voluntary centres namely, NDUAT, Faizabad, RAU Pusa, Samastipur, IARI New Delhi and SKAUST Jammu are actively participating in this programme. The main stress now is on Molya (CCN), Ear Cockle (ECN), other plant parasitic nematodes in different wheat based cropping system and their integrated management. More than 25 species of plant parasitic nematodes have been recorded on wheat in India and of these six are endoparasitic and the rest are ectoparasitic.

Losses : As per trials report in 1960s, CCN (*H avenae*) caused losses worth Rs 40 million and Rs 30 million in wheat and barley, respectively in Rajasthan. Little is known about the economic importance of the species *H filipjevi* but it is assumed that it can cause damage up to 20%. CCN, *Heterodera avenae* is a root pathogen of cereal especially wheat and barley that can cause sever losses in intolerant wheat cultivars. The sedentary soil borne CCN is an important biotic constraint limiting the productivity and production of wheat and barley crops in sandy soil of semi arid areas of Rajasthan and Haryana. *Anguina* causes less than 1% average loss of wheat amounting to Rs 750 million at current price.

A. Climate change, tillage options and its impact

Climate change could strongly affect the wheat crop that accounts for 21% of food and 200 million hectares of farmland worldwide. Climate change may inflict in germplasm adaptation, system management and mitigation on wheat in some of the most important wheat growing areas. Future climate scenario suggests that global warming may be beneficial for the wheat crop in some regions, but could reduce productivity in zones where optimal temperature already exists. Farmers are accepting conservation tillage due to economics of time and other resources. The loss of many nematicides due to environmental concerns and high costs of pesticides registration has focussed considerable attention on the development of cropping schemes that utilize biocontrol agents, green manure crops, resistant cultivars and non host crops for managing nematodes in both agronomic and horticultural crops. The progress made by many projects in the discovery, identification, culture and evaluation of field efficacy of several strains of biocontrol agents contributed to the development of urgently needed biocontrol agent that will be useful over large acreage. The use of antagonistic cover and rotational crop and compost application were demonstrated to the beneficial in reducing population of RKN and RLN. Host resistance to plant parasitic is one of the most inexpensive, environmentally friendly and feasible methods for managing plant parasitic nematodes in crop plants. Integration of host resistance with biocontrol agent, antagonistic cover, rotational crops and compost application will contribute to the development and implementation of environmentally safe, alternative management strategies for plant parasitic nematodes. Reaction of cover, rotational and resistance crops to plant parasitic nematodes has been characterized. Experiments to evaluate cover crops, compost and green manure treatments, and rotational crops for managing nematodes have been conducted. Plant parasitic nematodes play great roles both harmful and beneficial in agriculture. Both are important as their actions have economic propositions.

B. Resistance and population dynamics - the soil biology way

One of the most efficient methods of controlling *H avenae* is with grass free rotation using non host crops. In long term experiments, non host or resistant cereals keep population below damaging thresholds. Clean fallow and deep summer ploughing reduce the population density but are not always environmentally sound. Cultivars resistance is considered one of the best methods of nematode control and has been found to be successful in several countries such as Australia, Sweden, and France on a farm scale. However, it has been observed that the use of resistance, especially derived from single dominant gene, may cause disequilibrium in the biological communities and possibly ecological replacement with other nematodes, such as *Pratylenchus*. Another potential concern is the breakdown of resistance sources with repeated use. Since 1998-99 soil biology/biodynamic aspects of important plant parasitic nematodes have been studied under different prevailing wheat based cropping systems in details under AICW&BIP.

The cropping systems which were considered for this study were Rice-Wheat (at Ludhiana, Karnal, Delhi, Pusa-Bihar, and Faizabad), Cotton-Wheat (at Hisar and Ludhiana), Maize-Wheat (at Pusa-Bihar and Faizabad), Soybean-Wheat (at Ludhiana and Delhi), Groundnut-Wheat(at Durgapura and Anand), and Bajra-Wheat (at Durgapura and Anand) throughout the wheat growing zones of the India. The findings revealed that *Tylenchorhynchus* and *Hoplolaimus* were present in all four cropping systems at Ludhiana. *Hirschmanniella* was present in rice under rice-wheat only whereas *Heterodera zeae* was present in maize under Maize-wheat system. Under Bajra-Wheat, *Tylenchorynchus* and *Hoplolaimus* increased when wheat was sown after Bajra. It was also recorded that under Bajra cysts of *Heterodera avenae* did not hatch.

At Karnal, R-W system as such had encouraged the growth of *Hoplolaimus spp* and *Tylenchorhynchus spp* in rice. *Hirshmanniella* was common in all the cropping system but its population was high during rice season.

In Himachal Pradesh, Rice was affected by *Helicotylenchus diystera, Tylenchorhynchus nudus, Hirschmanniella gracilis, Pratylenchus zeae* and *Macoposthonia zenoplax* in order of dominance. *Hirschmanniella oryzae* was present in rice and *H.zeae* in maize.At Pusa, a steep decline in population of *Meloidogyne,* Reniform nematode and lesion nematode were noticed but stunt, lance and spiral nematodes increased in wheat.In Haryana, *Hirschmnniella* and *Meloidogyne graminicola* were more in rice compared to wheat. The population of *Tylenchorhynchus* and *Helicotylenchus* have shown increasing trend in wheat and peak was achieved during January to March months. *Tylenchorhynchus* was higher during wheat season. In Soybean-Wheat population buildup of all important nematodes was high. *Tylenchorhynchus* was high during maize as compared to wheat but population of *Hoplolaimus* was same during both the seasons.In Cotton-Wheat, *Helicotylenchus, Hoplolaimus, Pratylenchus* and *Tylenchorhynchus* are the important ones. In Cotton-Wheat, population of *Hoplolaimus* was maximum at cotton harvest whereas *Tylenchorhynchus* and *Helicotylenchus* were maximum during mid season of wheat at Hisar location.In Ludhiana, in cotton, population of *Tylenchorhynchus* was very high (1600 nem/250 ml soil). Population of *Pratylenchus spp* got decreased in wheat in Groundnut-Wheat system but for *Tylenchorhynchus* it was reverse. Cysts of *Heterodera avenae* remained dormant and unhatched in groundnut.

C. Seed gall nematode –national scenario

Ear cockle nematode (seed gall nematode), an important pest of wheat has potential to cause heavy losses in wheat.In the recent years, the disease has been reported to assume high incidence (1-10%)in some parts of the states like Bihar, Uttar Pradesh, and Madhya Pradesh (Vasudeva and Hingorani, 1952;Paruthi and Bhatti, 1985;Paruthi & Gupta, 1987;Nath & Pathak, 1993).The grain samples were obtained from mandies during the last six crop seasons to establish the severity and infestation so as to address trade related issues like sanitary, phytosanitary and market pathology.To address these issues, a total of 21907 grain samples were collected during the period 1998-99 to 2003-2004.In 1998-99, infestation with galls ranged between 0.01 to 50%, in 1999-2000, between 0.01-0.35%, in 2000-01 between 0.012 and 0.7%, in 2001-02 between 0.05 and 1.6%, during 2002-03 up to 0.24% whereas in 2003-2004, it ranged between 0.05 and 8%.Out of 7140 samples analyzed from Punjab none of the samples was found infested during last 3 crop seasons (2002, 2003 and 2004)indicating the areas to be disease free. In Haryana, 3 samples (2 from Gurgaon and 1 from Faridabad) were found infested during 2002-03. In the year 2003-04 grain infestation to the extent of 8% was noticed at Jatmalpur in Darbhanga district of Bihar whereas, at rest of the places in Bihar, it was up to 2.5%, besides other places.Contrary to it, in 2003-2004 only 2 samples from Nuh and Palwal in Haryana contained infestation in the range of 0.05% to 0.10%.This indicates nematode is still prevalent in the zones.The recurrence of disease is due to seed contamination and the disease incidence depends upon the degree of admixture at the time of sowing.It can be effectively eradicated from the country by using clean seed obtained from seed certification agencies. Hence, seed industry can play a decisive role by developing entrepreneur-ship in providing gall free seed to the farmers to contain the disease.

D. Breeding materials and their utilization in development of CCN resistant cultivars

Breeding for resistance or tolerance to *Heterodera spp* is one of the most economical and effective ways to help farms to cope up with the problem, but it is not the only way. The scientists are using every tool at their disposal to identify resistant wheat and barley lines and provide them to breeding programmes. Wild relatives of wheat have resistance to some nematodes. Breeders from Australia, France, Spain and CIMMYT have used conventional approaches to put resistance in to bread wheat. The resistance of these wheats is being studied throughout the world. Researchers in collaboration with research institutes in located in Australia and else where are also using molecular markers to identify and select resistant wheat early in breeding cycle hence reducing the need for expensive field trials. To cite an example, it can be mentioned that in Australia, CCN losses have declined significantly in last 16-17 years due to more effective rotations and the widespread adoption of well adapted cereal cultivars with nematode resistance. Many developing countries unfortunately have limited resources and /or expertise to establish this information, and current control methods are based on the understanding the response of local cultivar to the pathogen.However, identification of consistent sources of resistance in wheat has not been possible in spite of the fact that thousands of genotypes have been screened in India alone (Dhawan and Nagesh, 1987).Therefore, there is need to search sources of resistance to this nematode among wild relatives of wheat.One of the most cost effective and sustainable methods of control is with the use of host resistance.To date, more than 9 single dominant genes known as Cre gene family has been made known worldwide with many of these have originated from wild relatives of wheat.In order for cultivar resistance to be effective and desirable, a sufficient understanding of the number of species and pathotypes within species is essential.Although useful, a pathotype scheme for a species complex based on interaction with these cereal genera will not easily describe extensive variation in virulence. Furthermore, to date, these are few molecular or other diagnostic methods that can provide consistent and reliable pathotype and pathogenically differentiation.Anyhow, the traditional approach is still employed as being convenient and reliable. We need to look at reduced tillage practices, crop rotations and different ways of improving soil fertility.

E. Alternative approaches in managing CCN

a) *Crop Rotation*: To combat *Heterodera avenae* menace, growing non host crops after host crop to bring down the nematode population to a level which may prevent the significant damage to cereal crops.Attempts were made to manage the population through fallowing/growing non host crops because CCN being host specific to cereals.One study indicated that nematode population decreased by 70% with continued rotation with non host crops like mustard, carrot, fenugreek, onion and gram or by following fallowing and other means, like 2 years rotations of non host crops, the yield increased by 87%.

b) *Manure*: Organic manure influence the development of *Heterodera spp* as FYM, compost and saw dust application resulted in improved growth and discouraged the multiplication of CCN.

c) *Irrigation*: Higher multiplication of nematodes in well irrigated fields of wheat and barley was observed as compared to soil with poor soil moisture.

d) *Summer ploughing* : Soil temperature during may-June generally remains more than 30 degree Celsius and Sgoes up to 48 degree celsius.By ploughing, the soil is loosened and freshly formed cysts are disturbed and with the turning of the loose soil they also come up in the soil zone.The moisture of soil soil evaporates and cysts are subjected to dry heat.The nematode is sensitive to desiccation, hence its contents start drying and ultimately perished.

e) *Agrochemicals*: Earlier the use of soil fumigants like DD @ 300 litre/ha and DBCP @ 45 litre/ha were used.Later granular nematicides like Aldicarb 10G and carbofuran 3G @ 1.5 Kg a.i./ ha, respectively, were used which gave good protection of molya disease and also gave better plant growth and increased grain yield(Handa et.al., 1980;Mathur and Handa, 1984).Alternative to exclusive chemical management strategy of Molya disease(*Heterodera avenae*), concerted efforts were made by evaluating different organic, inorganic, resistant source and biological agents in combination during the wheat crop seasons of 1997-2005(8 years)in order to keep cereal cyst nematode population below threshold level in endemic and hot spot areas. Though other treatments, like CCNRV-1, carbofuran in isolation reduced the number of cysts per plant but combination involving *Trichoderma spp*, FYM, and reduced amount of carbofuran proved better at Durgapura. At Karnal, *Trichoderma viride* in combination with VAM fungi gave substantial growth and discouraged the cyst number to 5.40 per plant compared to 20.90 in check. When this exercise was repeated at Durgapura, *T.viride* + FYM+Carbofuran 3G @ 1.0 Kg a.i./h had yielded at par to that of carbofuran @ 1.5 Kg a.i./ha in getting 40.76 q/h and 41.94 q/ h yield and reduced the average no.of cysts/plant up to 2.9, and 3.0, respectively, over the untreated control (yield -24.24 q/h and 18.5 cysts/plant). However, the treatment involving *T. viride* + FYM also gave significantly higher yield of 36.39 q/h and reduced number of cysts/plant to 6.8 over 24.24 q/h and 18.5 cysts/plant in control.At Hisar, all the treatments inhibited the nematode multiplication as compared to control but grain yield and shoot biomass increased in FYM @ 20 Tonnes/h;Posse ST @ 0.75 Kg a.i./h + FYM @ 20 Tonnes/h;seed treatment with *T.viride*+FYM @ 10 Tonnes/h;seed treatment with *T. viride* + FYM @ 10 Tonnes/h + Carbofuran 3G @ 0.75 a.i./h and seed treatment with *Gliocladium virens*.Carbofuran 3G @ 1.5 Kg a.i./h and seed treatment with *Trichoderma harzianum* increased yield but not biomass.At Karnal, among treatments, *T.viride* @ 4g/kg seed + Carbofuran 3G@ 1 Kg a.i./h was the best after Carbofuran @ 1.5 Kg a.i./h.In subsequent year at Durgapura, the treatments of Carbofuran 3G @ 1.5 Kg a.i./ha and compost + *T. viride* + half dose of Carbofuran 3G @ 0.75 Kg a.i./ha gave the highest yield of 44.37 q/ha and 43.75 q/ha, respectively, and also reduced the cyst population significantly to 2.34 and 2.68 cysts/plant, respectively, over untreated control(yield 20.94 q/ha and 5.13 cysts/plant).Besides this, the treatment of compost + half dose Carbofuran, Neem Seed Powder and CCN resistant line (CCNRV-1) also gave significantly higher yield- 40.52.38.02 and 40.83 q/ha and also reduced the average CCN count up to 2.73, 2.85, and 0.70, respectively, over untreated control(yield 20.94 q/ha and 5.13 cysts/ plant).It was observed that *T.viride,* when used alone as seed treatment, did not produce significantly higher yield but gave slight reduction in the cyst counts.At Hisar, result showed that all the treatments except seed treatment with *Azotobacter chroococcum* @ 4% proved effective in reducing the cyst population significantly as compared to control. Seed treatment either with *A. chroococcum* or *T. viride* along with soil application of carbofuran 3G @ 0.75 kg a.i. /ha was as effective as 1.5 kg a.i./ha soil application of carbofuran 3G. These findings suggest that *T. viride* in combination with compost and half dose of carbofuran can reduce substantially the cyst population and improve the health of soil.

F. Wheat seed galls and their importance in wheat export

Wheat galls (the inciter of this disease is *Anguina tritici,* a nematode) though cause insignificant damage but in modern agriculture ability to export wheat grains in international market is greatly hampered if historical records are available of the presence of this pathogen in the grain producing areas from where exports are to go. Here the SPS issues play deciding role. The major implications of the SPS Agreement are in the areas of developing pest risk analysis (PRA), fixing appropriate level of protection (ALP) and identifying nematode free areas. On contrary, generally more damage is experienced due to poor agricultural practices, monoculture and use of poor quality seed in third world countries. In some cases, combined infection of *Anguina tritici* and *Neovossia indica* in wheat seed galls were observed. Galls with both the fungus and nematode contained fewer nematodes than galls with only nematodes (Paruthi and Bhatti, 1980).

a) *Global presence of ECN*: The countries namely, Australia, Afghanistan, Brazil, Bulgaria, China, Egypt, Ethiopia, Hungary, India, Iran, Iraq, Israel, Lithuania, New Zealand, Pakistan, Poland, Romania, Russia, Syria, Switzerland, Turkey, and Yugoslavia suffer from this wheat galls. It has been observed that nematode damage (wheat galls) is negligible in the countries adopting modern mechanical and cleaning procedures to separate the nematode galls from visible wheat seeds. The use of high quality seeds has nearly eradicated this nematode from developed countries. However, nematodes cause severe crop losses to rye (25-65%) and wheat (20-50%) in 3rd world countries where poor agricultural practices monoculture and the use of poor quality seed are widespread. The nematode is damaging pest in 3rd world countries.It is a pest of regulatory significance in developed countries.Taking in to consideration the above version, large priority rating for a complete risk assessment and declaring gall free areas as per ISPM norms are required.

b) *Prevalence*: The occurrence of this disease is an indication of poor and primitive agriculture where seed is not replaced for years together. The disease is presently prevalent in Bihar, UP, Rajasthan, MP and Chattisgarh.

c) *Example of sporadic incidences*: In 1992- Gaya, Patna, Nevada and Munger districts of Bihar got seriously affected by this disease. Darbhanga and Madhubani districts in Bihar experienced heavy damage owing to severe intensity of this disease in 1997. In the year 1999, village Paloi of Powai Tehsil in Panna district of MP suffered huge losses due to ear cockle disease.Chattarpur, Satna and Tikamgarh also got seriously affected (Singh et. al., 2000).

d) *Present status in India*: ECN is not being found in Punjab (Kaur and Sharma, 2003) and Haryana but prevalent in other states. Role of new varieties especially PBW series in tackling incidence of ECN in the states of Punjab and Haryana can not be ruled out as these tolerant varieties have also played vital role in wiping out ECN from both the states. It has been experienced that farmers replace seed every 4-5 years with new variety in both the states and cultivate PBW series and other latest varieties, the inoculum of CCN, which is otherwise present in produce/seed, goes out of cultivation, thereby separating the galls from seed lot. Though ECN is still being reported from states of UP, Bihar, MP, Chattisgarh, Rajasthan etc but their prevalence and incidence are going down year after year with the seed replacement, use of new varieties which are tolerant and awareness created by the extension functionaries.

e) *Key Host*: Wheat including bread and durum, Emer, Rye, Barley and Triticale are hosts but Oat is not a host (Paruthi and Gupta, 1987; Dhawan and Pankaj 1998).

f) *Life Cycle*: Juveniles (J2) emerge from seed galls in soil and crawls on to the newly germinated seedlings.They(J2)establish infection sites between young leaves and feed as ectoparasites causing leaf crinkling and distortion to the plants.Later, they penetrate the flower buds at the time of flower bud initiation.J2 stimulates the formation of galls in floral tissues in place of seed development.Juvenile development is completely inside the galls.Newly formed females deposit eggs, which hatch producing J2, which remains encased in the galls in anhydrobiotic condition and perpetuate plant infection in following years.Dried cockles get harvested with fully developed healthy seeds.Each gall may contain up to 12, 000 J2 in it.

g) Syndrome of this disease

Initially diseased plants grow along soil surface (prostrate) and after few days, they take upright growth. Basal swelling appears in 20-25 days old seedlings. Crinkling, curling and twisting of leaves are invariably seen at seedling stage. Diseased plants stay dwarf and conspicuous patches can be noticed in affected fields. Diseased plants produce profuse non-productive tillers as well as small, broad, less awned or ownless ear heads. These ear heads remain green for long and produce nematode galls in place of grain. On harvesting and threshing, these galls fall in to the soil or get mixed up with healthy grains as harvest. If these grains are used as seed next year, the disease perpetuation and appearance take place. The ear cockle and tundu diseases depend on the presence of both the organisms and on temperature, humidity, age of seedlings and depth of placement of galls (Midha and Swarup, 1972). Karnal bunt has also been found to occur along with seed galls. Galls with both the fungus and nematode contained fewer nematodes than galls with only nematodes (Paruthi and Bhatti, 1980).

h) Tran boundary experience with seed galls in wheat export

Wheat is a commercial commodity from trade point of view.There are so many non tariff barriers which come in the way of international trade.To cite an incidence which occurred in South America? Brazil is one of the world's largest wheat importers.Brazil relies on imports for the majority of its wheat consumption with Argentina that supplies about 90% of Brazil's import needs. Brazil once banned(blocked) US wheat on basis of presence of TCK smut, Cereal stripe and flag smut as well as seed galls.Importation of US wheat from the states of Washington, Oregon, Idaho, California, Nevada, and Arizona remained prohibited due to phytosanitary concerns.On March 15, 2001 Ministry of Agriculture, Brazil lifted ban on US soft wheat on the condition that exports of these wheat varieties must now come with additional declaration in the phytosanitary certificate that the wheat comes from an area free of *Anguina tritici*, and can not be shipped out of west coast port.On contrary, Argentina enjoyed many advantages in the Brazilian market, such as proximity, lower transportation cost, shorter delivery times and protection from the 10% mercosul duty and 25% merchant marine tax.So there were tug of wars among Brazil, Argentina and USA. Moreover, US wheat had an opportunity from May to September in Brazil as local harvest came after that.

The *Anguina tritici* do occur in Australia.Hundred years back economic problems due to this nematode were recorded in southern states like South Australia, Victoria, Tasmania, and Western Australia.Now it can be found only in small pockets of the western Australian belt particularly with short growing seasons, very dry summer and a tendency towards continuous wheat production.Crop rotation and seed cleaning has either eliminated the ECN incidence in other parts of Australia.Some markets in Middle East still require a shipment from Australia to be free of *Anguina tritici* because of perception of the effect on quality, and not for phytosanitary reasons. Though *A tritici* is found in most of the Middle East countries, so without area freedom, this nematode could not be considered quarantine organism.*A tritici* is easily controlled in modern agriculture and trade concerns are not phytosanitary in nature, whether it is appropriate that it be given a major quarantine pest.In US, 30 years ago this nematode was recorded.In Feb 2003, though 18 countries passed regulations for this nematode that are phytosanitary in nature but still a question mark stands as to put this nematode under phytosanitary.

i) Pest free areas certification as an alternative to seed certification

Seed certification may serve as an alternative to pest free areas certification programme which would be easier to implement in this case compared to a pest free area certification programme in the cropping zone. TBT (Technical Barriers to Trade) is an agreement whereby certain conditions permit the denial of entry of imports based on non phytosanitary issues such as quality of produce and packaging also.Though phytosanitary issues are related to the denial of entry based on presence of organisms/prohibited pests in the product of the exporting countries but TBT may also raise its head.

j) PRA and sanitary documents needed

Supporting documents, other than phytosanitary certificates, are required in present day global trade. Some importing countries may require few supporting documents other than phytosanitary certificate, for example a CITES certificate or a treatment certificate. The facilitators advise that the relevant standards, where necessary, should be followed. PRA is need of the situation and it has to be done. This abbreviation (EESEY) satisfies the requirement of PRA. The determination of the status of a quarantine pest may be summarized by the above alphabets-Entry, Establishment, Spread, Economics and if pest is meeting all these requirements (Yes), then it can be categorized as quarantine pest.

k) IPM approach for preventing/managing disease

Use galls free seed or clean the gall infested seed lot. Certified seed obtained from reliable sources/agencies which are expected to be gall free, be employed. If farmer's want to use their own seed (previous years harvest)they must ensure that there is no gall in the seed lot. Galls are smaller than the healthy grains and by sieving, they can be separated. In the light infestation(less than 2%), water floatation technique can be easily employed to remove galls from seed.Being lighter in weight, these galls float on the surface of water and can be easily separated. In case of severe infestation in seed lot, 2-5% brine solution treatment can be used for gall separation. For this purpose, in 10 lit of water in a Bucket, apply 200-500g of common salt (sodium chloride) and dissolve it. On stirring, the seed in brine solution with wooden or iron stick, galls will float on the surface, collect them and finally burn them. Seed, thus obtained, should be washed 2-3 times in plain water and shade dried before sowing. The farmers may replace their seeds with certified seeds of improved varieties of the area after every four to five years.

Conclusion

CCN has been an important nematode and various approaches as mentioned above have been effective in containing this nematode.Inspite of best efforts in containing it, it has been observed that CCN is reappearing in erstwhile known areas and spreading in newer areas especially in the states of Rajasthan, Haryana and Punjab. The reason could be salinity with aridity, lowering of water table, changes in cropping pattern and pushing wheat in sandy soil in non traditional areas of semi arid regions of Rajasthan and Haryana.In our country *Anguina* infested wheat galls are present but it is required that pest free areas be demarcated as per norms fixed by ISPM and other requisites so that whenever seed gall comes in the way of export, at least documents will be in position to support our version. In present day global trade, many petty issues become the deciding factors and influence the trade, hence preparedness to face that situation is highly appreciable, required and desirable. The climate change which is occurring at snail's pace is bound to ultimately change the disease and pest patterns in coming years and preparedness to face unforeseen situation arising due to this is need of hour.

References

Dhawan, S.C. and Nagesh, M. (1987). Tolerance in wheat against cereal cyst nematode, *Heterodera avenae. Indian J. Nematol.*17: 336-338.

Dhawan, S.C. and Pankaj. (1998). Nematode pests of wheat and barley. pp. 12-25 in nematode diseases in plants (Trivedi, PC, ed) New Delhi, India: CBS Publishers and Distributors.

Handa, D.K. ; Mathur, B.N.; Sharma, G.L.; and Yadav, B.D. (1980) *Indian Journal of Mycology and Plant Pathology*.40 (2):VII.

Kaur, Damanjeet and Sharma, R.C. (2003). Effect of seed replacement on the infestation of ear cockle nematode of wheat in Punjab. *Seed Research.* 31(2) 228-233.

Mathur, B.N. and Handa, D.K. (1984). First Intl Congress of Nematology, Guelph, Ontario, Canada, Aug 5-10, pp. 62 (abstr).

Midha, S.K. and Swarup, G. (1972): Factors affecting development of ear cockle and tundu diseases of wheat. 2: 97-104.

Nath, R.P. and Pathak, K. N. (1993). Wide spread occurrence of ear cockle disease of wheat in Bihar. *Indian Journal of Nematology*, 23: 124-130.

Paruthi, I. J. and Bhatti, D.S. (1980). Concomitant occurrence of *Anguina tritici* and KB fungus (*Neovossia indica*) in wheat seed galls. *Indian Journal of Nematology*, 10: 2: 256-257.

Paruthi, I.J.and Bhatti, D.S.(1985)Estimation of loss in yield and incidence of *Anguina tritici*. *International Nematology Network Newsletter* 2:13-16.

Paruthi, I. J. and Gupta, D.C. (1987). Incidence of tundu in barley and kanki in wheat field infested with *Anguina tritici*. *Haryana Agricultural University Journal of Research*. 17: 78-79.

Reports of the Coordinated Experiments, (1997-1998; 1998-99; 1999-2000; 2000-2001; 2001-2002; 2002-2003; 2003-2004; 2004-2005). Crop Protection Programme, AICW&BIP, Directorate of Wheat Research, Karnal.

Singh, A.K.; Sharma, A.K. and Nagarajan, S (2000). Ear cockle disease of wheat-Management at ease. Wheat Extension Bulletin-10, Directorate of Wheat Research, Karnal- 132 001.

Twenty Years of Coordinated Wheat Research.1961-1986.Editors:J.P.Tandon and A.P Sethi, Wheat Project Directorate, AICWIP, IARI, New Delhi 12, PP. 189-206.

Vasudeva, R.S.and Hingorani, M.K.(1952).Bacterial disease of wheat caused by *Corynebacterium tritici* (Hutchinson) Bergey *et al.Phytopathology*.42: 291-293.

Phyto-nematodes in Crops: Their Identification, Treatment and Management, 2011
G.L. Sharma (editor), pp. 25-54
New India Publishing Agency, New Delhi (India)
E-mail : info@nipabooks.com; Website : www.nipabooks.com

Chapter 3

Phytonematode: A Thrust in Pulses Crops

RASHID PERVEZ AND S.S. ALI

Introduction

India has the distinction of being the largest producer of pulses in the world. Among the important pulse crops cultivated in the country are: chickpea, pigeonpea, lentil, mungbean, urdbean, fieldpea, rajmash and lathyrus. The pulse crops occupied an area of 22.85 million ha with a total production of 13.07 million tones and average productivity of 572 kg/ha. Being a high protein grain, pulses have special status in the diet, particularly for vegetarian population. Apart from being a good source of protein, they are also good source of energy, essential amino acids and vitamins in the daily diets of the people, especially of low-income group. Pulses have importance in sustainable agriculture by way of fixing atmospheric nitrogen in the soil.But with intensification of agriculture and introduction of irrigation and other facilities the area and production under pulses is declining so much so that per caput availability of pulses has declined from 69 g day^{-1} to about 36 g day^{-1}.

Among the various pest and diseases which are the limiting factor in the cultivation of pulse crops, plant parasitic nematodes are one of the constraint in reducing both the quality and quantity of the crops. They are considered to be a hidden enemy of crops due to their microscopic size, hidden habitats and lacking in manifestation of clear cut symptoms on the aerial parts of the plants.Among nematodes diseases of pulse crops, root-knot disease is predominant and widespread throughout the pulse growing regions of the country.Out of 13 species of root-knot nematode reported in India, only *Meloidogyne incognita and M.javanica* are the causal agent of root-knot disease in pigeonpea, mungbean, urdbean, cowpea, chickpea, lentil, common bean, fieldpea and some of the minor pulses like lathyrus, horse gram and rice bean.

Target Nematodes Pest

The key target nematode pests of pulse crops in India are:root-knot nematodes, *Meloidogyne incognita and M.javanica,* reniform nematodes, *R. reniformis,* cyst nematode, *Heterodera cajani* and root-lesion nematodes, *Pratylenchus thornei* (Table 1). These include one or more species each of *Tylenchorhynchus, Helicotylenchus, Hoplolaimus, Telotylenchus,* (all ecto-parasites); Rotylenchulus reniformis (semiendoparasite); Pratylenchus (migratory endoparasite); Meloidogyne, and Heterodera (sedentary endoparasites). High population densities of these nematodes in the root or rhizosphere are often associated with patchy, stunted, chlorotic and sickly growing crops (Table 2).

Symptoms

The symptoms of nematode damage are generally confused with those of nutritional deficiencies, drought or salinity. Examination of both root and soil is necessary to detect nematodes. Again, some nematodes are always found associated with roots and soil but serious consideration of the taxonomic identity, pathogenic potential and population density of the different species of nematodes is essential before deciding whether and which nematode species may be responsible for the crop damage.

Root-knot disease of pulses are characterised by the presence of simple multiple knots on the root system especially on root-tips and can be easily seen with naked eyes. Size of knots is variable depending on genotype reaction and nematode species. Generally *M. javanica*

Table 1: Major nematode pests of pulse crops in different zones of the country (After Ali and Askary, 2005)

Zone	Pigeonpea	Chickpea	Other Pulses
1. North East Plain Zone (Bihar, West Bengal, Eastern U.P., Orissa)	1. Cyst nematode (*Heterodera cajani*) 2. Root-knot nematode (*Meloidogyne* spp.) 3. Reniform nematode (*Rotylenchulus reniformis*)	1. Root-knot nematode (*M.incognita, M javanica*) 2. Reniform nematode 3. Stunt nematode (*Tylenchorhynchus* spp.)	1. Root-knot nematode 2. Cyst nematode
2. North West Plain Zone (Haryana, Punjab, Rajasthan, Western U.P.)	1. Root-knot nematode 2. Cyst nematode	1. Root-knot nematode 2. Stunt nematode 3. Lesion nematode (*Pratylenchus* spp.) 4. Cyst nematode (*H. swarupi*) only Rajasthan	1. Root-knot nematode 2. Cyst nematode
3. Central Zone (Gujarat, Maharashtra, M.P., Parts of Rajasthan)	1. Cyst nematode 2. Root-knot nematode 3. Reniform nematode	1. Root-knot nematode 2. Lesion nematode 3. Cyst nematode (*H. swarupi*) only Rajasthan 4. Cyst nematode (*H. swarupi*) only Rajasthan	1. Root-knot nematode 2. Cyst nematode
4. South Zone (Andhra Pradesh, Tamil Nadu and Karnataka)	1. Cyst nematode 2. Root-knot nematode 3. Reniform nematode	1. Root-knot nematode 2. Cyst nematode	1. Root-knot nematode 2. Cyst nematode

Table 2: Plant parasitic nematodes associated with different pulse crops in India

1. *Aphelenchoides aligarhiensis*
2. *Aphelenchoides* sp.
3. *Aphelenchus radicicolus*
4. *Aphelenchus radicicola*
5. *A. avenae*
6. *Aphelenchus graminis*
7. *Aphelenchus* spp.
8. *Basiriolaimus seinhorstii*
9. *B. indicus*
10. *B. Columbus*
11. *Basiria* spp.
12. *Bitylenchus brevilineatus*
13. *B. vulgaris*
14. *Boleodorus* spp.
15. *Carcholaimus bedizens*
16. *C. symmetricus*
17. *Criconema* sp.
18. *Criconemella ornate*
19. *Criconemoides* sp.
20. *Ditylenchus* sp.
21. *Ditylenchus archilisposomus*
22. *Dolichorhynchus phaseoli*
23. *Geocenamus brevidens*
24. *Helicotylenchus abunaamai*
25. *H. ciceri*
26. *H. dihystera*
27. *H. elegans*
28. *H. indicus*
29. *H. retusus*
30. *H. maronatus*
31. *H. sharafati*
32. *Helicotylenchus* spp.
33. *Hemicriconemoides cocophillus*
34. *Hemicriconemoides* spp.
35. *Heterodera avenae*
36. *H. bihari*
37. *H. cajani*
38. *H. swarupi*
39. *H. mothi*
40. *H. trifolii*
41. *Heterodera* spp.
42. *Hirshmanniella mucronata*
43. *H. oryzae*
44. *Hirshmanniella* spp.
45. *Hoplolaimus dimorphicus*
46. *H. indicus*
47. *H. seinhorsti*
48. *Hoplolaimus* spp.
49. *Longidorus pisi*
50. *Longidorus* spp.
51. *Macroposthonia ornate*
52. *Macroposthonia* spp.
53. *Melenchus* spp.
54. *Meloidogyne incognita*
55. *M. javainca*
56. *Meloidogyne* spp.
57. *M. arenaria*
58. *Merlinius brevidens*
59. *Merlinius* spp.
60. *Neoditylenchus* spp.
61. *Nothocriconemella* sp.
62. *Nothotylenchus* sp.
63. *Nygolaimus harishi*
64. *Ottolenchus facultativus*
65. *Paraphlenchus amblyurus*
66. *Paratrichodorus christiei*
67. *Paratylenchus mirzai*
68. *P. rotundicephalus*
69. *Paratylenchus* spp.
70. *Pratylenchus coffeae*
71. *Pratylenchus indicus*
72. *P. mulchandi*
73. *P. thornei*
74. *P. zeae*
75. *Pratylenchus* spp.
76. *Pseudhalenchus* spp.
77. *Psilenchus* spp.
78. *Rotylenchulus reniformis*
79. *Rotylenchulus secundus*
80. *Rotylenchus* spp.
81. *Scutellonema erectum*
82. *S. brachyurum*
83. *Scutellonema* spp.
84. *Siddiqia citri*
85. *Telotylenchus crenatus*
86. *Telotylenchus* spp.
87. *Tylenchorhynchus* sp.
88. *T. elegans*
89. *T. mashhoodi*
90. *T. brassicae*
91. *T. ciceri*
92. *T. nudus*
93. *T. sabourensis*
94. *Tylechorhynchus* spp.
95. *Tylenchus* spp.
96. *Xiphinema lambertii*
97. *X. basiri*
98. *Xiphinema* spp.

exhibits smaller knots than *M. incognita* whereas later produce profuse knotting all along the entire roots and on laterals. Uneven growth in patches, poor stand, wilting, poor pod formation, lesser yield and lesser rhizobium nodules are commonly associated with root-knot disease of pulses. Heavily infested plants show stunted growth, lesser branches and pale green leaves in patches. Heavily galled roots show wilting symptoms in the field and seed obtained from such plants are mostly deformed, less developed, smaller in size and lesser in weight. Symptoms advance with crop age and more severe in low fertility soil. Fields infested with root-knot disease in preceding kharif exhibit poor germination of pulses grown in rabi season and succumb to root-knot and wilt disease. Populations of *M. javanica* modify Fusarium wilt resistance in chickpea, pigeonpea and lentil

Plants infected with cyst nematodes Heterodera cajani show delay in flowering and pod formation for more than a week as compared to healthy ones in pigeonpea. The roots of such plants bear pearly white cyst. The plants exhibit yellowing of foliage, extensive necrosis and early senescence. Patches of stunting, yellowing of plants with few flowers are the visual symptoms when chickpea crops are infested with *H. ciceri* and pigeonpea with *H. cajani.*

The roots exhibit necrotic lesions when infested by Pratylenchus spp. The necrotic spots coalesce and extend to the entire root in case of heavy attack. At seedling stage, infestation is manifested by patchy appearance, stunted growth and pale green colour of the foliage. Stunting becomes more pronounced as growth advances.

The attack of Pratylenchus spp. on chickpea is not evident on above ground and is seldom severe.As compared to root-knot nematodes, the damage caused by lesion nematodes is less impressive. The plant show stunting, chlorosis and lack of vigor. On roots, infection is characterised by dark brown to back lesion. The above ground symptoms resemble those due to malnutrition.

Infection caused by reniform nematode (*Rotylenchulus reniformis*) is also called dirty root disease in pigeonpea. It is due to the fact that when the roots of infected plants are washed with water, the gelatinous matrix with soil is not completely washed out. Leaves of such plants become light green or slight yellow and are generally visualised in the field as patches of stunted plants.

In early stages of crop in chickpea, mungbean, urdbean and lentil plants exhibit uneven growth but at maturity it is difficult to see any visual symptoms on the plant. Early senescence, sometimes, even at flowering stage appears in patches in fieldpea and common bean. The presences of female nematodes on rootlets are pronounced after a gentle wash on entire root system. This is seen at flowering and at maturity stage and is the only remarkable symptom of reniform nematode infection.

Economic Loses

A large number of pathogenicity experiments have shown that population densities of root-knot, reniform, pigeonpea-cyst or root-lesion nematodes, individually, above one infective nematode per cm^3 soil at sowing time can cause significant reduction in growth and yield. Many field trials on crop loss assessment, in different states of India under All India Coordinated Research Programme (Nematodes) and All India Coordinated Project on Improvement of Pulses (AICPIP) Pigeonpea, Chickpea, Mullarp, in which yields in untreated plots were compared with those in identical plots treated with standard nematicides, have shown yield losses ranging from 5-60% depending upon nematode species and their population densities. As an estimate the annual yield loss in India would be 2.4 lakh tonnes, worth approx. Rs.360 crores, considering total production of about 12 million tones and average price of Rs. 15 per kg. According to Rayen (1994) the potential return for the monetary gains due to crop improvement in case of nematode diseases is estimated worth 131.7 m US $ on global basis, and an appreciable amount is in south Asia, about 12.8 m US $, followed by WANA countries (about 4.6 m US$).

In pigeonpea cv. ICP 6 the reduction in plant growth and grain yield were 27.6% and 30.1% respectively due to cyst nematode Heterodera cajani (Saxena and Reddy, 1987). Under All India Coordinated Research Project on Improvement of Pulses a number of field trials were conducted to know the yield losses of pigeonpea due to nematode pest in different agro-ecological situations in the country. The yield loss of pigeonpea depends upon genotype, soil texture and type of nematode involved.

Mechanism of Diseases

Second stage juveniles of root-knot nematodes, *M. incognita* and *M. javanica* are the infective stage. These second stage juveniles move towards the root-tip and invade roots, penetrate the epidermal cells and cortex, become sedentary, moult three times and develop to sacciform females or worm like free living males, after establishing in the vascular cylinder where they induce formation of giant cells on which they feed. Galls are initiated within 48 hours of infection but are not essential for nematode growth and development. Fertilised females extrude a gelatinous matrix in which many eggs (500 to 1500) are laid down during the life span, forming an egg mass. The nematode complete its life cycle in a month at optimum temperature, many generations can be completed in a growing season leading to a very high post harvest soil population. Knots produced at 25-30^{0}C are larger than those produced at 15-20^{0}C. In case of *M. incognita* and *M. javanica* eggs soon hatch and initiate another generation.

Root diffuses are not necessary for hatching *M. incognita* and *M. javanica* have life cycle of apparently one month at optimum temperature of 25-30^{0}C and many generations are completed in one cropping season. An over production of small cells also occurs adjacent to giant cells, leading to the formation of gall in which the nematode is embedded.

Knots join together and more than one nematode may be found in the same knot. By pressing against surrounding cells the parasites interrupt the function of xylem and phloem and eventually the cells may become necrotic.

Primary galls show a decrease in insoluble polysaccharides including starch. The large galls show maximum accumulation of PAS positive granules in cytoplasm of mature giant cells associated with adult ovipositing females. The nematode and egg matrix are PAS positive. The total protein localise richly at the infection site.

The giant cells show a gradual increase of protein in their cytoplasm as the nematode reaches maturity but the protein decreases during degeneration of the giant cells. The nematode also gradually accumulates more of proteins, becoming maximum in ovipositing females. Nucleic acids in galls increase many times, DNA concentrated in nuclei of hyperplastic parenchyma, adjacent a giant cell and nematode, RNA was mainly localised in giant cell and the nematode. With the advancing age of roots, increase amount of ascorbic acid are seen in galls. The ascorbic granules are mainly localised in giant cells and nematodes. Peripheral tissue in larger galls shows positive reaction for ascorbic acid. More lipids have been observed in galled roots and as soon as galls enlarge, their lipid content also increase.

H. cajani juveniles were observed in the cortex of pigeonpea seedlings 48 days after inoculation. Movement was predominantly intracellular; syncytia formed in the staler region cells near the feeding site became angular with thickened cell walls. Giant cells had dense granular cytoplasm with 4 to 5 nuclei. There was extensive disruption of the xylem vessels. Juveniles, which were established in the cortex, developed mostly into males and those in the stellar region into females. (Koshy and Swarup, 1979).

Second stage juveniles penetrate the roots and reaching the central cylinder of roots within 72 hours and begin to feed and swell. The nematode gradually passes through its third and fourth stage and becomes an adult female. Due to enlarging of the body of female, cortex damages and erupts from epidermis and rupturing the discontinuity of the xylem vessels. The swollen, lemon shaped females are formed after 12 days while adult males mature in 10 days. Fifteen days after penetration, infective juveniles can be seen in the soil. Eggs are deposited in egg sacs, also within the female body, which gradually transforms from white to a brown coloured protective cyst. A female produces 100 to 300 eggs depending on the length of host plant. (Reddy *et al.*, 1990).

Pathogenicity

(1) *Root-knot nematode*: *M. javanica* J_2 100/plant and above was determinate to the growth and development of pigeonpea. Roots inoculated with 5000 and 10, 000 *M. javanica* had 12.02 and 12.73; 41.81 and 56.97 and 13.36 and 23.28% higher peroxidase, polyphenol oxidase and total phenol contents over control (Patel *et al.* 1992). There was greater suppression in plant growth, pollen fertility and water absorption capability in pigeonpea cv. UPAS 120, than in cv. ICPL 4 following inoculation with *M. incognita* showing greater effect than *R. reniformis* (Tiyagi *et al.*, 1989).

(2) *Cyst Nematode*: Pigeonpea plants inoculated with *H. cajani* juveniles at 100 and 5000 levels were found stunted with reduced internode length and smaller leaves. Population levels of 1000 (2 juveniles per cm^3 soil) can cause significant reduction in shoot length and weight, root weight, number of leaves and leaf area (Sharma and Nene, 1988). The adverse effects of the nematode on plant growth were visible within 40 to 45 days after inoculation. Nematode infected plants produced smaller roots less than those with control pots. Root and shoot weights were

found to be more sensitive to nematode stress than shoot length. A 25% reduction in overall biomass of pigeonpea cv. ICP 2376 was recorded at 1500 *H. cajani* larvae. Final cyst, total eggs and larval numbers at 1000 and 5000 did not differ significantly. Pigeonpea cv. T-21 when inoculated @ 10, 100 and 1000 freshly hatched second stage larvae of *H. cajani* reduced, root length as well as fresh and dry shoot and root weight. Initial inoculum level of 10 juveniles per kg of soil resulted in significant reduction in fresh root weight (Zaki and Bhatti, 1986a). However, studies on the effect of cyst nematode, *Heterodera cajani* on the growth of pigeonpea indicated that an inoculum level of 100 larvae per 500 g of soil significantly reduced plant growth with an increase in the initial inoculum level. There was corresponding increase reduction in plant growth (Upadhyay, 1986). Damage threshold levels for *H. cajani* on pigeonpea were considered 0.1 to 1.0 nematode per cm^3 of soil. Concentration of N and P increased in roots but decreased in tips in pigeon pea plants inoculated with *H. cajani*. Accumulation of K decreased in roots and increased in tips of infected plants. Change in the concentration of micronutrients in both crops did not show any definite trend except in case of Fe. These are present in micro quantities (1% of total dry weight) and metabolic tones are less mobile in the plant and are less prone to change due to infection. Decrease of Na, S, Zn, Mn and Fe from infested pigeonpea roots can be attributed to increased leaching of the elements from roots to soil (Zaki and Bhatti, 1986c). Eight days old seedlings of pigeonpea cv. 21 when inoculated with 100 juveniles of *H. cajani* per 1000 g of soil altered the amino acid amount in infected plants. The alteration in the relative amount of individual amino acids in infected plant may be attributed to highly specialized host parasite relationship. Specific metabolic requirements to meet the high multiplication rate and formation of metabolic sink (Syncytia) in the conducting vessels leading to an imbalance in the normal amino acid composition (Zaki and Bhatti, 1987).

(3) *Reniform nematode*: Pigeonpea cv. UPAS 120 seedlings were inoculated with 10 juveniles of R. reniformis in logarithmic series upto 10, 000 under pot culture condition. The nematode produced significant pathogenic effects at a minimum inoculum level of 500 nematodes/pot by reducing root length, dry weight of shoot and number of rhizobium root nodules. Yield/dry weight of grain

was reduced at the highest inoculum density (Mohanty & Padhi, 1987). In pigeonpea T15-15 inoculum level of 100 reniform nematodes per plant was detrimental to development of the crop. 10, 000 nematodes/plant/800 g of soil significantly decreased polyphenol oxidase (81.36%) and β-glycosidase (59.39%) activities as well as total phenol contents (47.07%) while peroxidase activity did not alter much due to nematode infection (Patel, R.G. and Patel, D.J. 1990). *R. reniformis* (5, 50, 500, 1000 or 5000 nematodes per pot) or *H. seinshorsti* (5, 50, 1000 or 2000 nematodes/plant/pot) significantly reduced plant growth. *H. seinshorsti* severely reduced root biomass, while *R. reniformis* reduced shoot weight (Sharma, S.B. and Nene, Y.L. 1988). Pigeonpea cv. 2830-2 and susceptible cv. 1924 were inoculated with 10, 100, 1000, 10, 000 *R. reniformis*/plant. Damaging levels for the susceptible and resistant varieties were 1000 and 10, 000/ 700 g soil respectively.

Interaction with Other Microorganisms

Nematodes not only damage the plants directly but also interact with other soil borne microorganisms and cause more damage to pulse crops (Table 3). Salam and Khan (1986) reported increase in susceptibility of six cultivars of pigeonpea to *F. udum* in presence of *M. javanica* and suppression of bacterial nodulation. Inoculation of *M. incognita* in combination with *F. udum* also enhanced the suppression of plant growth including bacterial nodules (Dwivedi *et al.* 1992). In sequential inoculation of *M. incognita* and *F. udum* on pigeonpea revealed maximum effect on plant growth where nematode and fungus were inoculated simultaneously (Perveen *et. al.*, 1999). Edward and Singh (1979) reported that *H. cajani* caused less damage to pigeonpea variety T-21 than in association with *F. udum*, whereas, Hasan (1984) reported the synergism between *H. cajani* and *F. udum* in pigeonpea. Sharma and Nene (1989) reported the enhanced pathogenicity in wilt susceptible genotype ICP 2376 by *H. cajani* while wilt resistant and tolerant genotypes ICP 8863 and BDN 1 were unaffected. *H. cajani* reproduction was reduced in presence of *F. udum*. The greatest damage to plants occurred when *M. incognita, H. cajani* and *F. udum* were present altogether than individual pathogen (Siddiqui and Mahmood, 1996; Siddiqui *et al.*, 1999). Jain and Sharma (1996) reported loss of wilt resistance in a pigeonpea line ICPL 270 in reniform nematode infested soil at ICRISAT farm. Anver and Alam (1997) studied combined effect of *M. incognita* and *Rotylenchulus*

Table 3: Nematode interaction with fungal diseases (After Ali and Askary, 2005)

Crop	Nematode	Fungus	Disease	Reference(s)
Chickpea	Root-knot nematode	(a) *Fusarium oxysporum* f.sp. *ciceri*	Wilting	Goel and Gupta (1986); Patel *et al.* (1992) Ali and Gurha (1994)
		(b) *F. solani*	Black root-rot	Dalal and Bhattti (1985)
		(c) *Rhizoctonia solani*	Wet root-rot	Ramnath and Dwivedi (1981)
		(d) *R. bataticola*	Dry root-rot	Goel and Gupta (1985)
		(e) *Sclerotium rolfsii*	Collar rot	Mukhtar and Khan (1989)
	Lesion nematode	*F. oxysporum* f.sp. *ciceri*	Wilting	Devi (1993)
Pigeonpea	Root-knot nematode	*F. udum*	Wilting	Salam and Khan (1986); Sharma and Nene (1990); Perveen *et al.* (1999)
Lentil	Root-knot nematode	*F. oxysporum* f.sp*lentis*	Wilting	De *et al.* (1999)
Fieldpea	Root-knot nematode	*F. oxysporum* f.sp. *pisi*	Wilting	Siddiqui *et al.* (1999)

reniformis on the pigeonpea plants. Although, there was more damage to plant when both the nematodes were present at the same level of inoculum, however, synergistic effect was not observed. There are two reports in the literature where nematodes have been suspected to be the transmitters of the sterility mosaic virus of pigeonpea. Narayanswamy *et al.* (1963) reported sterility mosaic disease of pigeonpea in the field where crop was grown successively and suspected that the virus could be transmitted by the nematode species. Narayanswamy and Ramakrishnan (1965) reported that the sterility mosaic virus was not transmitted by sap and insects but by *R. reniformis or Tylenchorhynchus* spp. However, after that there is no report to prove this transmission by nematodes.

Host Plant Resistance

Research in the field of nematology and plant breeding has enabled to introduce varieties of pulse crops, which has proved resistance to nematodes. In fact researchers while releasing a new variety keep in mind that it should be resistant to both fungus and nematode because if the variety is only resistant to fungus, the nematodes can break the resistance by changing the host physiology. Therefore, it becomes necessary for the breeders to keep both the objectives in their mind prior to the release of a new variety.Resistant plants acts as a barrier for nematodes in a number of ways such as it retards the rate of reproduction, delays the maturity of the parasite, production of more males and making the nematodes unable to complete its development. A list of pulse cultivars resistant to certain nematodes which have been tested in All Indian Coordinated Research Project on Pulses at various locations have been given (Table 4).

Management of Nematodes

A variety of approaches based on physical, chemical, biological, and cultural methods have been found effective, to different degrees, in reducing nematode population densities and improving crop performance. None of the methods, except on very small areas or quantities of plant material, have been found sufficient or economic to completely control, or to maintain the population densities at levels below the economic tolerance level. Therefore, reliance has to be placed on a judicious combination of different practices suitable for location specific conditions.

Table 4 : Resistance against key plant parasitic nematodes infesting pulse crops.

Crop	Nematode	Sources of resistance	References
Chickpea (*Cicer arietinum*)	*M. incognita*	BG-390, 1000, 1005, 1044, 1051; KW-154; 157, H-88-50, 92-68; AKG-52; GL-88341; BGM-49; RSG-506-1, 669; GLK-88016; IET GL-93011, JAKI-9303, 9324; JG-935; Giet (B); Phule G-5; ICP-94; GG-1111; RSGM-796; L-55; KW-101; GB-90236; FG-444; BGD-17, 117; BGP-118	Anonymous, 2001
	M. javanica	ICC 16614, ICC 6444, IPC 44, IPC 105, IPC 96-49, IPC 86-98, IPC 96-69, IPC 96-70, FG-22, 444; BG-1032; RPG-142	Ali & Ahmad., 2000
	R. reniformis	BG-87, GC-1111, WCG-11, PMC-16, IPC 96-69, ICC 6928	Ali & Ahmad., 2000
	Pratylenchus thorn ei	H-91-80, BG-1035, GNG-1078, Saki-7313, KW-80-2, ICC 16614, ICC 6428, ICC 5824, ICC 6825, ICC 6910, ICC 6918, ICC 6928, ICC 6938, ICC 6956, ICC 6950, ICC 6953, ICC 6962, ICC 7962, ICC 6983, ICC 6990, PDE 2, IPC 96-67	Ali & Ahmad., 2000
Pigeonpea (*Cajanus cajan*)	*M. incognita*	PPH-4, T-92, AF-293, MRG-66, Pant-A-104, ICPH-8F2, ICPL-151(B) – All HR	Anonymous, 2001
	M. javanica	Co RG-9060; AKPH-6190, 9186; AL-600; METH-121, MTH-9607, TAT-97-16, TT-97-48, VRG-7	Anonymous, 2001
	R. reniformis	BWR-376	Anonymous, 2001
	H. cajani	BDN-699, TAT-9402, AKT-9221, URG-13, GAUT-97-2E, H-90-14	Anonymous, 2001
		ML-56, Pusa-105	Chavada *et al.*, 1988

	H. cajani	ML 131, ML 337, PDM 84-139, PDM 145, PDM 146	Anonymous, 2004b
		FM-1, MC-131	Devi & Gupta, 1987
	M. javanica	DDU-5, UN-81-7; PDU-10, VGG-2	Devi & Gupta, 1987
	R. reniformis	Ratanapur-1, UG-201, 135	Routaray *et al.*, 1986
Mothbean (*Vigna aconitifolium*)	*R. reniformis*	RMO-430	Anonymous, 2001
Pea (*Pisum sativum*)	*M. incognita*	DDR-16	Anonymous, 2001
	M. javanica	DDR-48	Anonymous, 2001
	R. reniformis	DDR-36, 37, 39, 42, 45, 46; HUP-2; HUDP-19, LEP-224; HFP-942; DPFPD-2, 20; AZAD/P	Anonymous, 2001
Rajmah (*Phaseolus vulgaris*)	*M. incognita*	ACPR 94040, AVT I & II, HUR 137, IPR 964, RSJ 176	Haider *et al.*, 2002
	R. reniformis	RSJ-178, HUR-150	Anonymous, 2001

Physical Control

Solarization

Solarization by covering the soil with transparent polyethylene sheets during the summer month (April, May, June)significantly reduced the population densities of *Pratylenchus* spp. and *Tylenchorhynchus* spp. parasitic to pigeonpea.Population density reduction of 93% of *H, cajani* of eggs and juveniles, 99% of *H.retnsus,* 98% of *Pratylenchus sp.* and 100% of *R. reniformis* were achieved. Irrigation before covering soil with polyethylene sheet improved the effects of Solarization in reducing population densities of *H. cajani.* Solarization for two seasons reduced the population densities each year as about the same as single season Solarization and residual effects of Solarization on nematode populations did not last for more than a crop season (Sharma and Nene, 1990). Summer fallow (February-

June)reduced densities of *H.cajani* juveniles by 45% at 0-15 cm but densities at 15-30 cm and 30-45 cm were not affected. The eggs and juveniles population in the cysts was reduced by 18% at 0-15 cm and 11% at 15-30 cm. However, *H.retusus, H.seinhorsti, H.cajani & R.reniformis* survived in soil inside buried plots with no plants for 305 days (Sharma and Nene, 1992).

Seed Treatment

Chemical : The nematicidal seed treatments also reduced the population of nematodes in the soil by 45.2 to 70.6% (Anonymous 1981, 1987). Nematicidal effect on mungbean has also been observed in lowering down the population of root-knot nematodes (Sultan *et al.*1985).Carbofuran 3G @ 2 kg a.i. ha^{-1} as seed and soil application was highly effective in managing *Meloidogyne incognita* population and increasing yield in chickpea (Devi 1993). Pigeonpea seeds when treated with carbofuran (2%)eliminated root galling due to *Meloidogyne incognita* (Mishra 1986).Seed treatment and soil application of chemicals have been proved useful in lowering down the nematode population and increasing the yield of pigeonpea, mungbean, chickpea, fieldpea and urdbean (Table 5).

Table 5 : Management of nematodes through seed treated with chemicals

Crop	Method	Dosage	Increase in yield (%)	Reference
Pigeonpea	Monocrotophos	0.1%	18.6	Anonymous (2001)
	Carbofuran	0.1%	38.9	-do-
	Latex of Calotropis	1%	20	Ali , 1997
Chickpea	Carbosulfan 25 ST	3%	35.10	Anonymous (2002a)
	Carbofuran	3%	39.7	Anonymous (1999)
	Latex of Calotropis	1%	18	Ali , 1997
Mungbean	Monocrotophos	0.1%	36.34	Anonymous (2000)
Urdbean	Monocrotophos	0.1%	17.64-26.14	Anonymous (2000)
	Carbofuran	0.1%	17.64-26.14	-do-
Fieldpea	Carbofuran	0.1%	32.89	Anonymous (2002b)

Botanicals : Seed soaking on aqueous extracts of oil seed cakes (neem, mustard and karanj) and leaf extracts of *Canabis sativa, Eclipta alba, Datura metel, Argemone mexicana and Azadirachta indica* reduce the penetration of *Meloidogyne incognita* juveniles in chickpea (Mojumder and Mishra, 1991). Mojumder and Mishra (1992) tested neem seed coat both as soil and seed treatment against *M.incognita* and found significant reduction in root-knot galling and significant improvement in plant growth.The reduction in root- knot galls was more when soil treatments were combined with seed treatments.It also provided better plant growth.Soil treatment with neem seed coat at 2% w/w of soil resulted in maximum reduction in number of rootknot galls followed by seed kernel and seed coat in soil infested with *M. incognita* where chickpea was grown.All the treatment significantly reduced the number of root-knot galls and were significantly superior to check.Neem seed kernel was found most effective followed by neem cake. This finding indicated that chickpea is a good host of *M. incognita,* the infestation of which can be managed by applying aqueous extracts of neem seed kernel or neem cake @ 100 ml of standard extract per kg soil (Mojumder and Mishra 1993a). Mojumder and Mishra (1993b) reported that seed soaking of chickpea seed in aqueous extracts of neem cake, neem seed kernel & seed coat also reduced penetration and developing of *M. incognita.* At 90 days there was more than 50% reduction in number of galls in almost all the treatments. The population of plant parasitic nematodes viz., *Meloidogyne incognita, Rotylenchulus reniformis, Tylenchorhynchus brassicae and Helico tylenchus indicus* and the frequently of pathogenic fungi viz., *Macrophomina phaseolina, Rhizoctonia solani, Phyllostica phaseolina & Fusarium oxysporum f. sp. ciceri* were significantly reduced by oil seed cake of neem, castor, mustard and duan (Eruca sativa) (Tiyagi and Alam 1995).

Seed treatment with neem based laboratory formulations of neem seedkernel, neem seed cake @ 20% w/w reduced root-knot and reniform *(Meloidogyne incognita and Rotylenchulus reniformis*) nematode multiplication in two important pulse crops, mung bean and chickpea in greenhouse trials. All these treatments reduced major phytonematode species in the field, viz., *M.incognita, R.reniformis, Tylenchorhynchus mashhoodi, Helicotylenchus indicus and Hoplolaimus indicus* in both the crops in field trials in 4sq.m.microplots.In either case, neem seed kernel was the most effective followed by neem seed cake and neem seed coat (seed shell). There was 78, 67 and 52% reduction in total plant parasitic nematode population in the mung bean crop and the increase in the grain yield was 74, 58 and 28% neem seed kernel, neem seed cake and neem seed coat (seed shell) respectively. In the case of chickpea crop, the percentage reduction in the total plant parasitic nematode was 87, 76 and 25 and the increase in the grain yield was 122, 78 and 48 respectively in neem seed kernel, neem seed cake and neem seed coat (Mojumder and Raman, 1999).

Growing neem seedling along with chickpea cultivars Pusa 209 and Pusa 267 in $4m^2$ microplots (one seedling/plot)significantly reduced the plant parasitic nematode population such as *M.incognita, R.reniformis, T.mashhoodi, H.indicus and H.indicus (Basirolaimus indicus)*.There was also a considerable increase in grain yield in both cultivars.However, there was not much difference between the cultivars in terms of treatments. (Mojumder and Mojumder, 2002).

Cultural Control

The most important of the cultural practices, besides summer plugging are crop rotation and manipulation or augmentation of soil organic matter.

Crop rotation : Many of the economically important nematodes, like root-knot, cyst, reniform, lesion nematodes etc. have wide but somewhat delineated host ranges. It is often possible to find commercially acceptable and agronomically suitable alternative crops which may be non-hosts, poor hosts, tolerant or antagonistic crops to the target nematode species. Pulse crops can be grown in rotation with cereals and millets if the field is infested with root-knot, reniform or cyst nematodes. The period of rotation, i.e., avoidance of the host crop depends upon the rate of decline of the nematode population in the absence of the host. The nematodes that do not depend upon the host root diffusates for their hatching decline faster than those which do (Perry, 1987). Usually, a two year rotation against cyst nematodes and a one year rotation against root-knot and reniform nematodes is sufficient to bring down the population densities to levels below the economic injury thresholds; although much longer rotations may be necessary to achieve almost total control.

Cultivation of non-host legume and other crops for two to four or more years have been suggested for the management of cyst nematodes causing severe losses to the majority of pulse crops (Brown, 1958; Hooper, 1983; Schmitt and Noel, 1984). Sharma and Swarup (1984) reported that barnyard millet, *Echinocloa colona*; Kodo millet, *Paspalum scorbiculatum*; Italian millet, *Setaria italica*; *Chionachne* spp.; and teosinte, *Zea mexicana* were non-hosts of *Heterodera cajani*. Having narrow host range, *Heterodera* ciceri can easily be managed by crop rotation. Da Ponte (1972) recommended rotations with graminaceous crops or Crotalaria for the management of root-knot nematodes. Rhoades (1964) found weed-free cultivation of *Crotalaria spectabilis* more suppressive to root-knot nematode populations than fallow for the same period. Castillo *et al.* (1976) reported that one crop of paddy was sufficient to effectively reduce root-knot nematode infestations in succeeding susceptible legumes. Haider *et al.* (2001) also made similar observations in Samastipur, India. Since root-knot nematode has a wide host range (Saka and Carter, 1987), its management through crop rotation has been felt difficult.

Intercropping : Certain nematode antagonistic crops such as *Tagetes* spp., sesame, mustard etc. grown as intercrops have been reported to reduce root-knot and reniform nematodes. Intercropping of mustard with chickpea has been found to reduce infestation of root-knot nematodes in chickpea and pea (Haque & Gaur, 1987; AICRP (Nematodes), 2001). However, appropriate crop geometry is essential to avoid shading effects of mustard, which may adversely affect the growth of pulses. Experience has shown that crop rotation is more effective than intercropping. Upadhyay and Dwivedi (2001) have reported that inter cropping of sorghum with pigeonpea was highly effective in reducing the population of *Heterodera cajani* and in increasing the yield (40.4%). However, our presently running experiments indicate that it may only be possible in long duration pigeonpea.Egunjobi *et al.* (1986) showed that *M.javanica* populations were lower when cowpea and maize were grown under mixed rather than sole cropping.

Cropping System : Cropping systems have a regulatory effect on the nematode populations and it could be an effective nematode management tactic. Host crops for cyst, root-knot, reniform, and lesion nematodes grown prior to pulse crops result in high build up of the nematode populations. A rapid build up of root-knot and reniform nematodes occurs when pulses are cultivated after vegetables. Non-host crops, when included in the system, reduce the nematode populations. Cereals such as pearl millet, sorghum, and wheat are generally not good hosts of the root-knot nematodes. Keeping the fields fallow and weed-free cause marked reduction in nematode populations. Intercropping of sorghum with cyst nematode-tolerant pigeonpea could be effective in increasing the productivity of traditional production systems in the nematode infested regions.

Densities of plant parasitic nematodes in traditional low input cropping system were three times lower than in high input cropping system. Sorghum/pigeonpea (intercropped) or soybean/pigeonpea (intercropped) were identified as system very conducive for the build up of *H. cajani*. The densities of plant parasitic nematode did not increase significantly in plots that were either fallowed or sown with pearl millet in the rainy season. A summer fallow period between February to June reduced the soil densities of all plant parasitic nematodes by 42%. *R.reniformis* population was reduced by 80%.*H.cajani, Tylenchorhynchus spp.* and all total plant parasitic nematodes densities were significantly higher in plots treated with inorganic fertilizers than in plots treated with farm yard manure. *R. reniformis* densities were higher in irrigated than in rainfed fields (Singh *et al.,* 1994).

A significant reduction in nematode population was observed when the chickpea was sown late. This indicated that the late-sown chickpea may escape the nematode infestation and provide good yield. More than 50% reduction was noticed in the plant parasitic nematode population when the field was kept fallow (Ali, 1995).

Chickpea cv. R 56 was grown in beds where egg plant, okra, mung bean and cluster bean were established along with fallow plots.Population growth pattern in *M.incognita* was greatly influenced by preceding crops. Chickpea favored free multiplication of *M.incognita*.The fallowing in the preceding season did not check the population development of the test nematode on chickpea (Alam and Saxena 1986).The effect of four kharif crops (groundnut, sesame, soybean and tomato) and four rabi crops (wheat, mustard, chickpea and tomato) in different combinations giving 16 different rotations on the population of *M.incognita* was studied in microplots. The rotation followed two years. There was manifold increase in level of population in tomato monoculture, while a substantial increase in population was observed in sequences having tomato in combination with chickpea and soybean. The maximum build up of *M.incognita* was on tomato followed by chickpea. The rotations having tomato in monoculture or in combination with soybean, chickpea and groundnut resulted in significantly low yield (Sharma *et al.,* 1992).

Root-knot nematodes on chickpea have been effectively controlled by inclusion of cereals or grasses in cropping system. Sesame, mustard and winter cereals are poor hosts of *M. javanica and M. incognita,* and

2-3 year rotations may be useful for disease management (Sharma *et al.*, 1992).

M. javanica population was suppressed drastically under mustard-chickpea inter cropping system as compared to adjacent plots of sole chickpea (Anonymous, 1990). Effect of intercropping on pigeonpea infested with *M. incognita* revealed that maximum yield was obtained in plots where pigeonpea was intercultured with groundnut followed by sesame. The least number of nematode galls per plant were recorded in the sesame intercropped plants. (Upadhyay and Dwivedi, 1997).

Organic amendments : Incorporation of organic materials or manures into soil is known to reduce plant parasitic nematode populations.Amendment of soil with non-edible deoiled seed cakes of neem *(Azadirachta indica), Melia azadirach,* karanj *(Pongamia glabra),* Mahua *(Madhuca indica),* castor *(Ricinus communis)* etc. improve soil fertility and organic matter status besides reducing nematode population density through their constituent or decomposition products which are toxic to nematodes and promotion of the antagonistic microbial activity. The high doses (10 - 30 q ha^{-1}) required are a constraint, but again lower doses in combination with other practices can be advantageous.

Biological Control

Recently, there has been much emphasis on the exploitation of biocontrol agent especially certain bacteria and fungi against nematodes.A few microorganisms like *Pasteuria penetrans, Paecilomysis lilacinus, Verticilium chaladosporium, Trichoderma viridi, T. harzianum, Aspergellus niger* etc., have shown great promise under limited experimental conditions. As such the mass culturing and release of biocontrol agents at regular intervals may be quite expensive. However, proper establishment and colonization of indigenous or introduced microorganisms is possible by simultaneous modification of the soil environment. Prior testing of such organisms to human and cattle health must of course from a part of the protocol before recommending them. Some bacteria and fungi have been reported as potential biocontrol agents for *H. cajani and M. incognita* (Sharma and Swarup, 1988). *P. lilacinus* reduces the damage caused by *R. reniformis or M. incognita* (Khan and Hussain, 1988). Sharma and Swarup have been recorded a number of fungi and other organisms (*P. penetrans, Glomus fasciculatum, Catenaria auxiliaries, C. vermicola, Allomyces anonalus*, etc.) parasitising the cyst of Heterodera spp. Studies conducted under controlled as well as under field conditions indicated that *T. harzianum, A.niger and P.lilacinus* act effectively against *M. incognita and H.cajani* infesting chickpea and pigeonpea (Haseeb and Shukla, 2002, 2003).

Chemical Control

Chemicals have been found to give quick and demonstrable control of nematodes, several chemicals belonging to diverse groups like halogenated hydrocarbons (DD, EDB, DBCP and MBr.), organophosphates (Phorate, Phenamiphos, fensulfothion and thionazin) and oxime carbamates (aldicarb, carbofuran and oxamyl) have served as nematicides. Among the concentrations of 0.5, 1.0 and 2.0% (w/v weight of Carbofuran, aldicarb, phorate and fensulphothion treated with pigeonpea cv. T 21 seeds for 30 and 60 minutes in each concentration, aldicarb exhibited best response followed by phorate in reducing cyst population significantly (Zaki & Bhatt, 1986b). The application of fensulphothion, phorate, aldicarb, vydate, carbofuran and DBCP either in the form of seed treatment or granules or drench application resulted in better plant growth than with uninoculated check (Panda and Seshadri, 1979). Application of phenomiphos, oxamyl, phosalone and Danusban @ 1000 ppm/plant (soil, spray and soil + spray) resulted in marked reduction in *H. cajani* population infesting pigeonpea (Mishra, and Prasad, 1979). Fensulphothion significantly reduced the *M. incognita* population and resulted in better plant growth in pigeonpea cv. Bahar (Jaiswal *et al.* 1987).

New Methodologies

Three commercially adopted chickpea cvs.C 235, RSG 2 and RS 11 were subjected to physical and chemical mutagens either singly or in various combinations.RS 11 did not produce a single resistant mutant, whereas C 235 was more responsive than RSG 2 and RS 11. Those isolated mutants were stable in morphological traits as well as in nematode reaction and could be utilized as parents for incorporating resistance (Bhatnagar *et al.*, 1988).Baroda dhaki local and L 550 were irradiated with gamma rays (10, 20 and 30 KR)and fast neutrons and subjected to ethylmethane sulfonate (EMS, 0.05, 0.1 & 0.2%) and methylmethane sulfonate (0.001 and 0.01%) treatments, either singly or in combination.Plants were screened for resistance to *M. javanica*. L 550 was more responsive to mutagens.The mutant MV/7-1 produced the highest yield 918.2g) under infested field conditions, it was obtained from the 30 KR + 0.1% EMS treatment (Bhatnagar *et al.,* 1988).

Peroxidase plays an important role in the resistance mechanism of plants. It is a key enzyme required for lignin synthesis as well as other trapezoids involved in phytoalexin production.Peroxidase catalyzes several reaction including those involved in the metabolism of phenols and indoles.On the other hand, protein content of galled roots has been used as an index of root-knot nematode infestation in okra plants.IC 4928 to IC 4848 (25 genotypes) were screened on the basis of increase in peroxidase activity. IC 4941, IC 4942, IC 4944 were reported tolerant against *M. incognita* (Siddiqui and Husain, 1992). Chickpea cv. K 850 was inoculated with 1000-2500 J2, *M. javanica;* after 60 days, biochemoical analysis revealed that there was an increase of 10-18% and 26-54% in total protein and amino acid contents, respectively, which was found greater in the stem and at the higher levels of infection. Increase in protein contents of chickpea was dependent upon the level of infection by root-knot nematodes (Upadhyay and Banerjee, 1986).

Future Thrusts

- Evaluation of improved cropping systems for the management of root-knot, lesion and cyst nematodes problem in pulse crops at regional basis.
- Developing integrated management for pigeonpea, chickpea and lentil in wilt and root-rot disease complex involving both nematode and fungus pathogens.
- Search of high level resistance in pulses genotypes against root-knot nematode and *H.cajani.*
- Enhancing 15-20% yield through low input and practical nematode management.
- Identify hot spots of nematode in major pulses growing areas through surveys and to disseminate nematode management technology to the farmers.
- Development of integrated nematode management technology against polyspecific nematode communities.
- To investigate and monitor the spread of major nematode pests.

References

Alam, M. M. and Saxena, S. K. 1986. Effect of preceding crop on the population growth pattern of *Meloidogyne incognita and Tylenchorhynchus brassicae* on chickpea cv.R.56.*International Chickpea Newsletter* 14: 33-34.

Ali, S.S. 1995. Nematodes associated with mung bean in Gwalior district. *Indian Journal of Pulses Research* 8 (1) : 100-101.

Ali, S.S. 1995. Nematode problems in chickpea. Kanpur : Indian Institute of Pulses Research. Kanpur, India. 184p.

Ali, S. S. 1997. Status of nematode problem and research in India. In Diagnosis of key nematode pests of chickpea and pigeonpea and their management (Ed. S. B. Sharma). ICRISAT, Patancheru, India. pp 74-82.

Ali, S. S. and Ahmad, R. 2000. Screening of chickpea germplasm against nematode. *International Chickpea & Pigeonpea Newsletter* 7: 8-9.

Ali, S.S. and Gurha, S.N.1994.Interaction between *Fusarium oxysporum f.sp. ciceri and Meloidogyne incognita* in chickpea wilt. International Symposium on Pulses Research. 2-6 April, New Delhi (Abstract) p178.

Anonymous 1981. All India Coordinated Research Project on Nematode Pests of Crops and their Control. *Indian Council of Agricultural Research. Proceedings IIIrd Workshop held at Tamil Nadu Agricultural University, Coimbatore,* 10-13 February, 1981, p 65.

Anonymous 1987. All India Coordinated Project on Plant Parasitic Nematodes with Integrated Approach for their control. *Proceedings of the Workshop. College of Agriculture, Pune, Mahatma Phule Agricultural University, Rahuri (Maharashtra), 9-12 June,* p., 128.

Anonymous 1990. Annual Report 1990. Directorate of Pulses Research. Indian Council of Agricultural Research. pp., 1-2.

Anonymous 1999. All India Coordinated Research Project on Pigeonpea. Indian Council of Agricultural Research. pp., 126-130.

Anonymous 2000.All India Coordinated Research Project on Mullarp Crop. Indian Council of Agricultural Research. *Proceedings of the Workshop held at Tamil Nadu Agricultural University,* Coimbatore.27-30 April, 2001, pp. 153-159.

Anonymous 2001a.All India Coordinated Research Project on Mullarp Crops. Indian Council of Agricultural Research. *Proceedings of the Workshop held at Rajasthan Agricultural University,* ARS, Durgapura, Jaipur3-6 September, pp. 226-228.

Anonymous 2001b. Annual Report 2001. All India Coordinated Research Project on Pigeonpea Crop.Indian Council of Agricultural Research. p.130.

Anonymous 2002a. Annual Report 2002. All India Coordinated Research Project on Chickpea Crop.Indian Council of Agricultural Research.pp249-250.

Anonymous 2002b.Annual Report 2002.All India Coordinated Research Project on Mullarp Crop. Indian Council of Agricultural Research. *Proceedings of the Workshop held at C.C.S.Haryana Agricultural University*, Hissar. 10-12 September, 2002, p 238.

Anonymous 2004a.Annual Report 2004.All India Coordinated Research Project on Mullarp Crop. Indian Council of Agricultural Research. *Proceedings of the Workshop held at Agriculture Training Center*, Narendrapur, Kolkata. 9-11 September, 2005, p 238.

Anonymous 2004b. Annual Report 2004. All India Coordinated Research Project on Pigeonpea Crop. Indian Council of Agricultural Research. *Proceedings of the Workshop held at ANG Ranga Agricultural University, Hyderabad.* 4-6 May, 2005, p 231.

Anver, S. and Alam, M.M. 1997. Effect of root-knot and reniform nematodes alone and in combination on the growth of pigeonpea. *Nematologia Mediterranea.* 25 (1) : pp. 13-15.

Bhatnagar, C.P., Handa, D.K.and Mishra, A.1988.Kabuli chickpea mutants resistant to root-knot nematode, *Meloidogyne javanica. International Chickpea Newsletter.* 19 : 16-17.

Brown, E.B. 1958. Pea root eelworm in the eastern counties of England. *Nematologica* 3: 257-268.

Castillo, M.B.; Arceo, M.B. and Litsinger, J.A. 1976. Nematodes in cropping patterns. IV.Populations of plant parasitic nematodes associated with cropping patterns under different rice-growing environments in Manaog, Pangasinan. *Philippine Phytopath.*, 12: 24-29.

Chakrabarti, U. and Mishra, S.D. 2001. Seed treatment with neem products for integrated management of *Meloidogyne incognita* infecting chickpea. *Current Nematology* 12: 15-19.

Chavda, J.C.; Patel, B.A. and Patel, D.J. 1988. Screening of pigeonpea lines to *Rotylenchulus reniformis. International Nematology Network Newsletter.* 5: 28-29.

Dalal, M.R.and Bhatti, D.S.1985.Effect of interaction between *Meloidogyne javanica and Fusarium solani* on chickpea.4th Nematology Symposium, M.L. Sukhadia University, Udaipur, 17-18 May. *Indian Journal of Nematology,* 5 : 287.

De, R.K., Ali, S.S.and Dwivedi, R.P.1999.Interaction between *Fusarium oxysporium f.sp.*lentil and *Meloidogyne javainca* in lentil.Indian Phytopathological society, 25th Annual meeting and national symposium on role of resistance in intensive agriculture. Feb.15-17, 2000 at Directorate of Wheat Research, Karnal in collaboration with ICAR, NASS, Karnal, (Abstract), pp 42-43.

Devi, S.1993.Evaluation of carbofuran 3G as a seed and soil treatment against *Meloidogyne incognita* on gram.In:National Symposium on Recent Approaches in Integrated Nematode Management of Agricultural Crops. 6-7 August, 1993.Chowdhry Charan Singh Haryana Agricultural University, Hissar. 33pp.

Devi, L.S.and Gupta, P.1987.Evaluation of green gram and black gram varieties against cyst nematode, *Heterodera cajani. Indian Journal of Nematology* 17: 341.

Dwivedi, R.K., Upadhyay, K.D. and Dwivedi, K. 1992. Interaction of root knot nematodes, *Meloidogyne incognita* and fungus, *Fusarium oxysporum f.sp. udum* on pigeonpea. *Indian Journal of Nematology* 22 (2): 96-100.

Edward, J.C. and Singh, K.P. 1979. Interaction between *Heterodera cajani and Fusarium udum* on pigeonpea. *Allahabad Farmer* 50 : 23-24.

Egunjobi, O.A., Akonde, P.T. and Caveness, F.E. 1986. Interaction between *Pratylenchus sefaensis, Meloidogyne javanica and Rotylenchulus reniformis* in sole and mixed crops of maize and cowpea. *Review de Nematology*.9 : 61-70.

Goel, S.R.and Gupta, D.C.1986.Interaction of *Meloidogyne javanica and Fusarium oxysporum f.sp.ciceri* on chickpea. *Indian Phytopathology*, 39:112-114

Goel, S.R.and Gupta, D.C.1985.Interaction of *Meloidogyne javanica and Rhizobium bataticola* on chickpea *(Cicer arietinum)*. *Indian Journal of Nematology* 16 (1) : 133-134.

Haider, M.G., Askary, T.H.and Nath, R.P.2001.Nematode population as influenced by paddy based cropping sequences. *Indian Journal of Nematology* 31: 68-71.

Hasan, A. 1984. Synergism between *Heterodera cajani and Fusarium udum* attacking *Cajanus cajan*. *Nematologia Mediterranea* 12 : 159-162.

Haseeb, A. and Shukla, P.K. 2002. Management of wilt disease of chickpea by the application of some chemicals, bio-pesticides and bio-agents under field conditions. *Current Nematology* 13: 61-63.

Haseeb, A. and Shukla, P.K. 2003. Nematode-fungus wilt disease complex of pigeonpea and its management in western Uttar Pradesh. *Proc. Zonal Conf. ISMPP&Sem.'Diseases of national importance with special reference to guava wilt and mango malformation'held at CISH, Lucknow* during 4th-5th April 2003.

Jain, K.C. and Sharma, S.B. 1996. Loss of Fusarium wilt resistance in a pigeonpea line ICPL 270 in reniform nematode infested soil at ICRISAT Asia Center. *International Chickpea and Pigeonpea Newsletter* No. 3 : 90.

Jaiswal, B.K., Nath, R.P., Haider, M.G. and Pathak, K.N. 1987. Efficacy of some pesticides on *Meloidogyne incognita and Rotylenchulus reniformis* infesting pigeonpea. *Indian Journal of Nematology* 17 (1) : 60-61.

Khan, T.A. and Husain, S.I. 1988. Studies on the efficacy of seed treatments with pesticides, oil cakes, neem leaf and culture filtrate of *Paecilomysis lilacinus* for the control of diseases caused by the presence of *Rotylenchulus reniformis, Meloidogyne incognita and Rhizoctonia solani* either individually or concomitantly on cowpea. *Indian Journal of Nematology* 18: 192-198.

Koshy, P.K.and Swarup, G.1979.Factors affecting emergence of larva from cyst of *Heterodera cajani* Koshy. *Indian Journal of Nematology*. 1 : 209-219.

Kundu, G.G. and Mishra, S.D. 1985. Field evaluation of some soybean germplasm for resistance to stem fly, Melanagromyza sojae (Zehnt.), root-knot nematode, *Meloidogyne incognita* and yellow masaic virus. *Indian Journal of Entomology* 47: 230-232.

Mishra, S.D. 1986. Control of root-knot nematode, Meloidogyne incognita infesting pigeonpea by nematicidal seed treatment. In: *Proceedings of National Symposium on New Dimensions in Parasitology, University* of Allahabad, 18-19 January, 1986 (Editors Kapoor, V.N. and Mishra, S.L., Malhotra, S.K.) Allahabad, Uttar Pradesh, 34-36.

Mohanty, C.B. and Padhi, N.N. 1987. Pathogenic effect of reniform nematode at varying levels of inoculation on pigeonpea. *International Nematology Network Newsletter* 4 (3) : 15-16.

Mojumder, V. and Mishra, S.D. 1991. Nematicidal efficacy of some plant products and management of *Meloidogyne incognita* in pulse crops by soaking seed in their aqueous extracts. *Current Nematology* 2: 27-32.

Mojumder, V. and Mishra, S.D. 1992. Management of root-knot nematode, *Meloidogyne incognita* infecting chickpea with neem seed coat. *Annals of Agriculture Research* 23 (4): 388-390.

Mojumder, V. and Mishra, S.D. 1993a. Soil application of aqueous extracts of neem products for the management of *Meloidogyne incognita* in chickpea. *Current Nematology* 4: 105 -108.

Mojumder, V. and Mishra, S.D. 1993b. Management of *Meloidogyne incognita* infesting pulse crops through seed treatments with neem product. *World Neem Conference India* 2: 699-703.

Mojumder, V. and Mishra, S.D. 2001. Seed coating of pigeonpea with neem formulations for the management of *Heterodera cajani* and other phytoparasitic nematodes. *Current Nematology* 12: 7-10.

Mojumder, V. and Mojumder, V. 2002. Effect of neem seedlings on major phytoparasitic nematodes associated with chickpea and on crop yield. *Pesticide Research Journal* 14: 150-152.

Mojumder, V.and Raman, R.1999.Comparative efficacy of neem based nematicidal products for management of nematode pasts of pulse crops. In: *Azadirachta indica A.* Juss. P.P. Singh (Ed.), I.A.R.I., New Delhi, India, 223-234 pp.

Mukhtar, J.and Khan, A.A.1989.Disease complex in chickpea involving *Meloidogyne javanica and Sclerotium rolfsii. International Nematology Network Newsletter* 6 (4) : 31-32.

Narayanswamy, P. and Ramakrishnan, K. 1965. Studies on the sterility mosaic disease of pigeonpea. I. Transmission of the disease. *Proceedings of Indian Academy of Sciences* (Section B) 62 : pp. 73-86.

Narayanswamy, P., Seshadri, A. R. and Ramakrishnan, K. 1963. Preliminary notes on suspected nematode transmission of redgram sterility mosaic virus. *Madras Agricultural Journal* 50 (2): 109-110.

Nayak, D. K.; Routaray, B. N. and Das, S. N. 1987. Relative susceptibility of some horse gram cultivars against reniform nematode, *Rotylenchulus reniformis. International Nematology Network Newsletter*, 4: 19-20.

Panda, M. and Seshadri, A. R. 1979. Efficacy of few nematicides in controlling the infestations of reniform nematodes in some pulse crops. Nematology symposium, Bhubhaneshwar, Orissa. March 1979. *Indian Journal of Nematology* 9 (1) : 84-85.

Patel, B.A.; Chavda, J.C. and Patel, D.J. 1989. Screening of certain green gram lines against reniform nematode, *Rotylenchulus reniformis. International Nematology Network Newsletter* 6: 25-26.

Patel, G.A., Patel, D.J. and Patel, B.A. 1992. Pathogenicity and biochemical changes induced by infection of *Meloidogyne javanica* on pigeonpea. First Afro-Asian Nematology Symposium, Afro-Asian Society of Nematologists. Department of Botany, Aligarh Muslim University, Aligarh, India, November 29-December 3, (Abstract). pp. 39.

Patel, R.G. and Patel, D.J. 1990. Pathogenicity and biochemical changes induced by *reniform nematode, Rotylenchulus reniformis* Linford and Oliveira, 1940 in pigeonpea. *Current Nematology* 1 (1) : 11-14.

Perveen, K., Haseeb, A. and Shukla, P.K. 1999. Effect of *Meloidogyne incognita and Fusarium udum* on the disease development and growth of pigeonpea. *Current Nematology* 10 (1, 2) :33-40.

Ramnath and Dwivedi, R.P.1981.Effect of root–knot nematode on development of gram caused by *Fusarium oxysporum f.sp. ciceri* and root-rot by *Rhizoctonia sp. Indian Journal of Mycology and Plant Pathology*. 11: 46-47.

Rayen, J. G. 1994. A global prospective on pigeonpea and chickpea sustainable production systems: Present status and future potential. Int. Symp. on Pulses Research, April, 2-6, New Delhi (Abstr.) 1-2 pp.

Reddy, M.V., Sharma, S.B. & Nene, Y.L.1990. Pigeonpea:Disease Management. In:The pigeonpea (Editors:Nene, Y.L., Susan, D., Hall, Sheila, U.K.), CAB International, ICRISAT, pp. 303-347.

Rhoades, H.L. 1964. Effect *Crotalaria spectabilis and Sesbania exaltata* on plant nematode populations and subsequent yield of snap beans and cabbage. Proc. Fl. State Hor. Soc., U.S.A. 77: 233-237.

Routaray, B.N.; Sahoo, H. and Das, S.N. 1986. Evaluation of green gram and black gram varieties against reniform nematode, *Rotylenchulus reniformis. Indian Journal of Nematology* 16: 27-29.

Saka, V.E. and Carter, C.C. 1987. Hosts and non-hosts of root-knot nematode *(Meloidogyne incognita)* Coo. Publ. Dept. Plant Pathol., North Carolina State Univ. and U.S. Agency Int. Dev., Raleigh, 62 p.

Salam, M. A. and Khan, M.W. 1986. Reaction of some cultivars of pigeonpea against *Fusarium udum and Meloidogyne javanica. International Nematology Network Newsletter* 3 (4) : 16-17.

Saxena, R. and Reddy, D.D.R. 1987. Crop losses in pigeonpea and mungbean by pigeonpea cyst nematode, *Heterodera cajani. Indian Journal of Nematology* 17 (1) : 91-94.

Schmitt, D.P.and Noel, G.R.1984.Nematode parasites of soybeans.In: *Plant and Insect Nematodes Nickel W.R. (Ed.).* Marcel Decker, New York, pp. 13-59.

Sharma, R. and Swarup, G. 1984. Cyst Forming Nematodes of India. Cosmo Publ., New Delhi, 149 p.

Sharma, R. and Swarup, G. 1988. Pathology of cyst nematodes. Malhotra Publishing House, New Delhi, 88 pp.

Sharma, S.B. and Nene, Y.L. 1988. Effect of *Heterodera cajani, Rotylenchulus reniformis and Hoplolaimus seinhorsti* on pigeonpea biomass. *Indian Journal of Nematology* 18 (2) : 273-278.

Sharma, S.B. and Nene, Y.L. 1989. Interrelationship between *Heterodera cajani and Fusarium udum* in pigeonpea. *Nematropica* 19 (1) : 21-28.

Sharma, S.B. and Nene, Y.L. 1990. Effect of soil solarization on nematodes parasitic to chickpea and pigeonpea. *Journal of Nematology,* 22 (45): 644-658.

Sharma, S.B. and Nene, Y.L. 1992. Spatial and temporal distribution of plant parasitic nematodes on pigeonpea in alfisols and vertisols. *Nematropica* 22: 13-20.

Sharma, S.B., Reddy, B.M.R. and Krishnappa, K.1992.Incidence nod importance of plant parasitic nematodes on pigeonpea and groundnut in Karnataka state in southern India. *Afro-Asian Journal of Nematology* 2: (1-2), 21-26.

Siddiqui, Z.A. and Husain, S.I. 1992. Response of 20 chickpea cultivars to *Meloidogyne incognita* race 3. *Nematologie Meditereanea* 20 (1): 33-36.

Siddiqui, Z.A. and Mahmood, I. 1996. Effect of Heterodera cajani, *Meloidogyne incognita and Fusarium udum* on the wilt disease complex of pigeonpea. *Indian Journal of Nematology* 26 (1) : 102-104.

Siddiqui, Z.A., Mahmood, I.and Mahmood.I. 1999. Effect of *Heterodera cajani, Meloidogyne incognita with Fusarium udum and Bradyrhizobium japonicum* on the wilt disease complex of pigeonpea. *Indian Phytopathology* 52 (1) : 66-70.

Singh, M., Sharma, S.B. and Anders, M.M. 1994. Plant - parasitic nematode densities in cereal and legume based cropping systems on vertisols, *Afro-Asian Journal of Nematology* 4 (1) : 44-50.

Sultan, M.S., Singh, I. and Sharma, S.K.1985.Efficacy of some nematicides for The control of root-knot nematode infesting mung. *Nematologia Mediterranea* 13: 2, 257-259.

Thakar, N.A. and Patel, C.C. 1985. Reaction of some varieties of mung bean, black gram and okra to root-knot nematodes, *Meloidogyne incognita and M. javanica. Indian Journal of Nematology* 15: 102-103.

Tiyagi, S.A., Anver, S. and Alam, M.M. 1989. Studies on the pathogenicity of root-knot and reniform nematodes on pigeonpea. *International Nematology Network Newsletter*. 6 (4) : 12-15.

Tiyagi, S.A., and Alam, M.M. 1990. effect of root-knot, reniform and stunt nematodes on plant growth, water absorption capability and chlorophyll content of chickpea. *International Chickpea Newsletter*. 22: 40-42.

Upadhyay, K.D. and Banerjee, H. 1986. Some chemical changes in chickpea plant infected with root-knot nematode, *Meloidogyne javanica. Indian Journal Nematology*. 16 (2) : 286-288.

Upadhyay, K.D.and Dwivedi, K.1987.Root-knot nematode, *Meloidogyne javanica* breaks wilt resistance in chickpea variety Avrodhi.*Current Science* 56 (17) : 915-916.

Upadhyay, K.D. 1986. C.S. Azad University of Agriculture and Technology, Kanpur. In : Nematology in India. Indian Agricultural Research Institute, New Delhi, pp. 93-96.

Upadhyay, K.D. and Dwivedi, K. 2001. Population dynamic of *Heterodera cajani* associated with pigeonpea in various cropping systems of Central Uttar Pradesh. *Indian Journal of Nematology* 2: 120-121.

Zaki, F.A. and Bhatti, D.S. 1986a. Pathogenicity of pigeonpea cyst nematode, *Heterodera cajani*, Koshy 1967 on some pulse crops. *Indian Journal of Nematology* 16 (1) : 30-35.

Zaki, F.A. and Bhatti, D.S. 1986b. Control of pigeonpea cyst nematode, *Heterodera cajani* Koshy 1967, by chemical seed treatment. *Indian Journal of Nematology* 16 (1) : 106-108.

Zaki, F.A. and Bhatti, D.S. 1986c. Effect of pigeonpea cyst nematode, *Heterodera cajani* Koshy 1967 on macro and micro nutrients of pigeonpea and moth. *Indian Journal of Nematology* 16 (1) : 103-105.

Zaki, F.A. and Bhatti, D.S.1987.Amino acid composition of *Heterodera cajani* infected pigeonpea plants. *International Pigeonpea Newsletter* 6 : 51-52.

Phyto-nematodes in Crops: Their Identification, Treatment and Management, 2011
G.L. Sharma (editor), pp. 55-114
New India Publishing Agency, New Delhi (India)
E-mail : info@nipabooks.com; Website : www.nipabooks.com

Chapter 4

The Nematode Pests of Horticultural Crops with Special Reference to the North-Eastern Region

A.K. SINHA, B. MAHANTA AND P.P. NEOG

Introduction

Plant paprasitic nematodes are one of the limiting factors in the production of horticultural crops in the country. The visible symptoms of nematode attack are: Reduced individual plant growth, varying degrees of chlorosis, wilting of the foliage and sometimes death of the plants.

The roots, damaged by nematodes are not efficient in the utilization of available moisture and nutrients in soil resulting in reduced functional metabolism. Nematode infested roots are easily infected by various pathogenic fungi and bacteria and aggrevate root damage. These deleterious effects on plant growth result in reduced yield and poor quality of crops as in citrus, banana, grapevine, pineapple, papaya and vegetable crops. Decline or death of highly prised ornamentals and turf is often the result of nematode infestation.

The study of plant parasitic nematodes on horticultural crops in India dates back to 1933 when Ayyr reported the root-knot nematode on vegetable from the southern part of the country.Jones (1961) detected the presence of golden nematode on potato, *Globodera rostochiensis* in Nilgiris, a pest of great economic importance.This discovery triggered nematological research in India. Siddiqi (1961) reported the citrus nematode, *Tylenchulus semipenetrans* for the first time from India. Subsequently this nematode has been reported from different parts of the country. Sinha and Chawla (1986) also reported this nematode from Assam.Another important plant parasitic nematode (burrowing nematode) *Radopholus similis* was described on banana from Kerala state, Nair *et al.* (1966); fortunately this nematode has not yet been recorded from Assam.

Major Nematode Pests

Several plant parasitic nematodes (PPN) have been found to be associated with horticultural crops in India, not much is known about their relative importance as potential pathogens even though high populations of same have been consistently reported around plant roots, there is no information on economic loss is caused by them individually and in concomitant association with each other.A good number of Plant Parasitic Nematodes are also recorded in Assam on different horticultural crops *viz., Aglenchusas samensis, Helicotylenchus dihystera, H.paracrenacauda, Hoplolaimus indicus, Macroposthonia onostris, M.onoensis, Meloidogyne incognita, Pratylenchus indicus, Tylenchorhynchus annulatus, Xiphinema insigne, X.paradicicola, X.americananun* on vegetable crops. *Cephelenchus leptuys, Hemicriconemoides mangiferae, Helicotylenchus dihystera, Hoplolaimus indicus, H.columbus, Macroposthonia onostris, M. medani, Meloidogyne incognita, Paratylenchus nannus, Pratylenchus indicus, P.thornei, P.musii, P.neopratensis n.s Rotylenchulus reniformis, Scutellonema unum, S.labiatum, Tylenchorhynchusannulatus, T.leviterminalis, Xiphinema insigne, X.mammillocaudatum* on banana.*Aulosphorada homensis, Cephalenchus leptus, Helicotylenchus, dihystera, H.paracrenacauda, Hemicriconemoides citro-philus, H.mangiferae, H.cocophilus, Hoplolaimus indicus, Meloidogyne incognita, Macroposthonia onostris, M.onoensis, M.medani, Pratylenchus indicus, P.penetrans, P. zeae, P. neopratensis, Rotylenchulus reniformis, Scutellonema labiatum, Tylenchulus semipenetrans, Tylenchorhynchus annulatus, T.leviterminalis, Xiphinema insigne, X. americanum, X. brasiliense* on citrus. *Caloosia paxi, Cephalenchus leptus, Helicotylenchus dihystera, Hemicriconemoides mangiferae, Hoplolaimus indicus, Macroposthonia onoensis, Meloidogyne incognita, Pratylenchus indicus, P. brachyurus, Rotylenchulus reniformis, Scutellonema unum, S labiatum,*

Tylenchorhynchus leviterminalis, Xiphinema insigne on pineapple *Aglenchus muktii, Helicotylenchus dihystera, Hemicriconemoides mangiferae, Hemicycliophora tarjani, Hoplolaimus indicus, H. columbus, Heterodera* sps *Meloidogyne hapla, M. incognita, Macroposthonia onostris, M. onoensis, Meloidodera floridensis, Paratylenchus nannus, P. curvitatus, P. pseuduncianatus, Paratylenchus curvitus, Pratylenchus brachyurus, Paratrichodorus pososus, Rotylenchulus* sp., *Scutellonema brachyurum, Tylenchorhynchus leviterminalis, T.mushoodi, Xiphinema insigne, X. elongatum, X. mammilocaudatum, X. brasilense* on tea. *Helicotylenchus dihystera, Hemicycliophora tarjani, Hoplolaimus seinhorsti, Macroposthonia onostris, Paratylenchus curvitatus, Pratylenchus indicus, Rotylenchulus reniformis, Trichodorus borai, Xiphinema insigne* on coffeae. *Helicotylenchus dihystera, Hemicriconemoides mangiferae, Hoplolaimus columbus, H. indicus, Meloidogyne incognita, Macroposthonia onoensis, Tylenchorhynchus leviterminalis, Tgoldeni, Xiphinema brasiliense, X.insigne* on betelvine.*Helicotylenchus dihystera, Hoplolaimus indicus, Macroposthonia onoensis, Scutellonema labiatum, Tylenchorhyn-chusgoldeni, Tylenchorhynchus astriatus, Xiphinema elongatum, X. radicicola* on blackpiper are also reported from different Agroclimatic Zones of Assam.

Pathogenic levels of Plant Parasitic Nematodes on economically important horticultural crops in Assam.

Vegetables

Sinha (1974) observed that plant height, number of buds and leaves, fresh weight of shoots and roots, dry weight of shoots and root length of brinjal were significantly reduced at higher inoculum level of *M. incognita*. Hazarika (1990) recorded the initial level of 100 J_2/plant of brinjal in 0.5 kg of soil significantly reduced the number of leaves, shoot length, fresh and dry weight of shoot and dry weight of root. He furthet observed that at the highest inoculum level (10, 000 J_2), the plant showed chlorosis, rugose in leave and shedding of leaves of brinjal plant. Pathak (1990) observed that an initial inoculum of 1000 J_2 of *M.incognita*/1 kg of soil was highly pathogenic to'okra'.Like Hazarika (1990), Pathak (1990) also observed reduction of number and size of 'okra' leaves accompanied by chlorosis and defoliation at the highest inoculum level of *M.incognita*.Maximum reproductive rate of *M. incognita* was observed at the lowest level of inoculum (10 J_2/plant) in brinjal (Hazarika, 1990) and on okra (Pathak, 1990). Sarma (1993) observed that an initial inoculum level of 1000 J_2 of *M.incognita* per 0.5 kg of soil was pathogenic to chilli, causing significant reduction in plant height, fresh and dry weight of shoot and root. Devi (1993) also found that an initial inoculum level of 100 $J_2$of *M.incognita* per kg of soil was proved to be pathogenic to carrot causing significant reduction in plant growth parameters (plant height, tap root length and weight of both shoot and tap roots). Pathogenic symptoms like stunting of plant, forking of tap roots (Fig. 1&2) and formation of irregular knots and circular constriction in the tap roots were observed. Hussain (1995) and Hussain & Bora (1995) found that 10 J_2 of *M.incognita* per 500 g of soil caused significant reduction of plant growth of french bean.

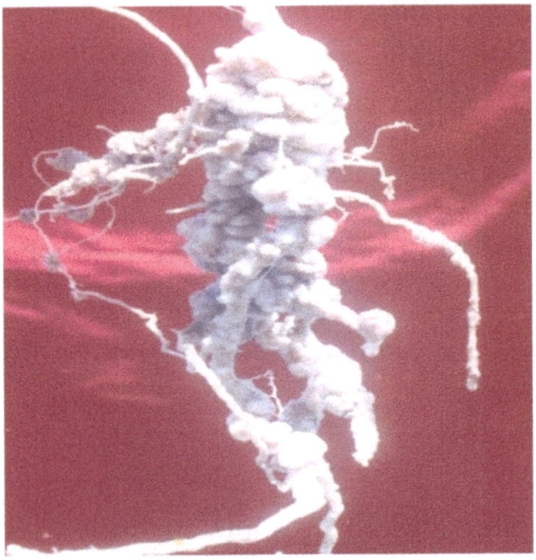

Fig. 1 : *Meloidogyne incognita* infested Ridge gourd-root.

Fig. 2 : Root-knot infested carrot forking

Bharali (1996) and Bharali & Phukan, (1996) found that an initial inoculum level of 100 J_2 of *M. incognita* caused significant reduction in growth characters and proved to be pathogenic to the cucumber plants.

Borah, Mahanta & Phukan (2002) found that the age of seedlings of tomato and brinjal had significant effect on the development of *M.incognita*. The damage caused by *M.incognita* was more to plants of tender age (3-4 weeks old) than that of older age.Association of pathogenic and above pathogenic inoculum levels of *M.incognita* and *R.solanacearum* increased severity of wilt on brinjal compared to associations of below pathogenic levels of these pathogens.The bacteria alone and below pathogenic level of the nematode could not reduce the plant growth parameters of brinjal significantly (Hussain, 2004). In presence of bacteria, nematode activities were inhibited compared to absence of the bacteria.

Plantation Crops and Fruit Crops

The life history of *Meloidogyne incognita* and *Pratylenchus brachyurus* in tea seedlings had been studied in the Tocklai Experimental Station (Anonymous, 1970, 1973). Pathogenicity of *M. incognita* on tea seedlings revealed that there was a near linearity between gall intensity and nematode population (Anonymous, 1974'a'). *Helicotylenchus, Rotylenchulus* and *Paratylenchus* had no pathogenic effect on tea seedlings (Anonymous, 1977 and 1978). Basu (1974, 1975 and 1978'b') recorded a numbers of weeds grown in tea estates as hosts of *Meloidogyne* spp. Basu and Banerjee (1978) recorded nine species of shade trees of rehabilitation crops as host of *M. incognita.* They also reported that as a result of this infestation growth of some of these plants were significantly reduced. Basu and Gope (1979) also recorded some weeds from tea gardens as the host of *Meloidogyne* spp. It was observed that in presence of root-knot nematode, the species of *Helicotylenchus* had pathogenic effect on tea seedlings (Anonymous, 1979). In a pathogenicity trial, Basu, Gope and Banerjee (1981) observed that plant height, stem girth and canopy expanse of *Indigofera teysmanii* progressively decreased with the increasing level of inoculum of *M. incognita.*In a study on the reaction of three different hosts to three different populations of *T.semipenetrans* (Assam, Delhi & Punjab) Sinha (1985) observed that the roots of *Ponicirus trifoliata* remaind practically free from nematode penetration irrespective of the populations used.The number of larvae penetrating the root system of the other two hosts *(Citrus jambhiri* and *Vitis vinifera*) ranged from 3 to 102. More penetration by Delhi and Punjab populations had been observed in *C.jambhiri* than Assam isolate while penetration of all the three populations was almost similar in *V.vinifera.*In an separate investigation on the effect of different hosts on the morphological changes of *T.semipenetrans,* Sinha and Chawla (1988) did not find any morphological difference among these three populations.Saikia (1992) observed that the length and weight of both shoot and root of betelvine were significantly reduced at or above the inoculum level of 100 J2 of *M.incognita* per plant.He further observed yellowing, pre-mature shedding of leaves and reduction of leaf number and size at the highest inoculum level. Dutta (1995) observed a progressive decrease in growth of tea seedlings with the increase of inoculum level from 10 to

10, 000 J_2 of *M. incognita* per 500 g of soil.Bora (1995) and Mili (1996) observed that an initial inoculum level of 1000 J_2 of citrus nematode, *T.semipenetrans* per kg of soil caused significant reduction in all growth parameters and proved to be pathogenic to khasi mandarin.Borah, Mahanta& Phukan, (1998) observed that an initial level of 3000 juveniles per kg of soil was found to be the pathogenic threshold level of *M. incognita* on papaya.

The pathogenicity of *T.semipenetrans* on *Citrus jambhiri* indicated that the growth parameters of the host crop decreased gradually with the increase in inoculam level of test nematode viz; 10, 100, 1000 and 10, 000 J2 per kg of soil (Deka, 2000, Deka, Sinha and Neog, 2003b) .An inoculum level of 1000J_2/kg of soil and above was found to cause significant reduction in all growth parameters of *C.jambhiri* and proved to pathogenic.However, maximum reduction were recorded at inoculum level of 10, 000 J_2/kg of soil.

Bora and Neog (2006) observed that with the increasing inoculum level of *M. incognita* there was a gradual reduction in all plant growth parameters of tea seedlings. Maximum reduction was recorded at the inoculum level of 10, 000J_2 followed by 1000 J_2/seedlings. The number of galls, eggmasses/root system and final nematode population in soil increased with the increased level of inoculum from 10-1000 J_2/seedling & then declined at 10, 000J_2/seedling grown per kg of soil (Table 1).

Crop Loss

Vegetable Crops

Das and Phukan (1987) estimated the losses in shoot length, shoot weight, root length and yield of 'okra' to be 16.56, 25.36, 14.70 and 19.16 per cent respectively in a field with initial population of 325 J_2 of *M. incognita*/200 ml of soil. A loss of 39.17 per cent in yield of brinjal was recorded in untreated plots over carbofuran treated plots (@ 3 kg a.i./ha) in a field with initial population of 378 J_2 of *M. incognita*/200 ml of soil (Hazarika, 1990) . Loss in other plant growth characters had been estimated to be 7.31 per cent in plant height, 21.34 per cent dry weight of shoot and 25.8 per cent in dry weight of root. The losses in 'okra' due to infestation of *M. incognita* had been estimated to be 12.08 and 17.87 per cent in 1990 and 1992 at initial population of 335 J2 of *M. incognita*/200 ml of soil and 295 J2/250 ml of soil respectively (Anonymous, 1991a, 1993a). Devi (1992) estimated 10.53 per cent loss in plant height, 27.52 per cent in tap root length, 40.8 per cent in

Table 1 : Pathogenic levels of plant parasitic nematodes in horticultural crops.

S. No.	Crops	Initial inoculum level of 2nd stage juvenile/plant/ kg of soil	Nematode species	References
Vegetables				
1.	'Okra'	1000/1	*M. incognita*	Pathak, 1990
2.	Brinjal	3000/3.5 1000/1	*M. incognita* *M. incognita*	Sinha, 1974 Hazarika, 1990
3.	Chilli	1000/.5	*M. incognita*	Sarmah, 1993
4.	Carrot	100/1	*M. incognita*	Dev, 1993
5.	Frenchbean	10/.5	*M. incognita*	Hussain, 1995
6.	Cucumber	100/1	*M. incognita*	Bharali. 1996, Bharali & Phukan, 1996
7.	Pea	1000/.5	*M. incognita*	Bhagawati & Phukan, 1991
8.	Cowpea	1000.5	*M. incognita*	Sarma & Sinha, 1995
	Plantation crops			
9.	Tea	1000/.5	*M. incognita*	Dutta, 1995
10.	Betelvine	100/2	*M. incognita*	
	Fruit crops			
11.	Citrus	1000/1	*T. semipenetrans*	Milli & Sinha, 1997, Deka *et al.*, 2003
12.	Papaya	3000/3 kg	*M. incognita*	Bora *et al., 1998*

production of tap roots and 56.64 per cent loss of marketable yield of carrot in untreated plots over treated plots (carbofuran 3 kg a.i./h) in a field with an initial population of 282 J_2 of *M. incognita*/200 ml of soil. Das (1994) estimated 17.13 and 19.38 per cent losses in yield of frenchbean in 1992 and 1993 at the initial population of 420 and 378 J_2 of *M. incognita*/200 ml of soil respectively. The losses in yield of okra were 15.82 and 18.13 per cent in 1993 and 1994 and of greengram 18.08 per cent in 1993 in field with initial population of 280, 295 and 280 J_2 of *M. incognita*/250 ml of soil respectively (Anonymous, 1995). The avoidablle yield losses in okra due to *M. incognita* were assessed to be 15.05 and 21.58 per cent in the field with an initial population of

300 and 340 J_2 /250 ml of soil in the year 1995 and 1996 respectively (Deka, 1997, Deka & Rahman, 1997). Dutta (1999) estimated the losses in plant height, tap root length and yield of carrot as.18.06, 27.48 and 44.3 per cent respectively in a field with an initial population of 548 J_2 of *M. incognita*/250 ml of soil. It was also observed that increased number of forked tap root was recorded in untreated plots comparing the treated plots. Some informations on crop losses due to plant parasitic nematodes in economically important vegetable crops in Assam are shown in (Table 2).

Table 2 : Yield losses due to Plant parasitic nematode in economically important crops in Assam

Sl. No.	Crop	Nematode species	Per cent yield loss	Reference
Vegetables				
1. Cowpea	*Meloidogyne incognita*	20.2 - 30.30	Anonymous, 1987	
2. Pea	-do-	15.05-29.60	Bhagawati, 1989	
3. Frenchbean	-do-	17.14-26.43	Das, 1994	
4. Okra	-do-	15.82 – 18.13	Das & Phukan, 1987	
5. Brinjal	-do-	39.17	Hazarika, 1990	
6. Carrot	-do-	56.64	Devi, 1993; Devi & Das, 1995.	

Plantation Crops

Losses in yield of betelvine leaves were 23.84 in 1990-91 and 28.30 per cent in 1991-92 at initial populations of 258 and 231 J_2 of *M. incognita*/200 ml of soil respectively (Anonymous, 1991 c, 1992). A loss of 21.59 per cent in number of leaves and 21.10 per cent in weight of betelvine leaves were recorded in a field having initial population of 205 J_2 of *M. incognita*/200 ml of soil (Saikia, 1992). A significant reduction of 17.95 per cent in number of leaf yield per vine and 29.06 per cent weight of fresh leaves were assessed in untreated plots compared to treated plots at a initial nematode population of 259.99 J_2 of *M. incognita*/200 ml of soil (Hazarika, Dutta & Saikia, 1999c).

Nematode Pests of Economic Importance

Fruit Crops

1. Root-knot nematode, *Meloidogyne* spp.: The root-knot nematode, *M. incognita* is a serious problem on grapevine, papaya, pine-apple and passion fruit. Plant parasitic nematode is responsible for 25 to 50 per cent yield loss in established grape vine-yards of 'Muscat Haneburg' in Tamil Nadu (Rajagupalan and Nagahathan, 1977). *M. indica* causes considerable damage to citrus in Andrapradesh and Delhi.

2. The reniform nematode *(Rotylenchulus reniformis)* : *R. reniformis* is a major problem on grapevine (Cv. Anag-e-shahi), papaya, pine apple, and passion fruit. It is responsible for decline in papaya and pine apple.

   R. similis causes rhizome rot of banana and is becoming a serious problem in Tamil Nadu, Kerala, Karnataka and Andra Pradesh (Parvatha Reddy and Sing, 1980). However, this nematode is not yet reported from Assam. It is also spreading slowly to Maharastra, Gujarat and Madhya Pradesh (Tewari and Dave, 1985). It causes retarted growth and extensive, root and rhizome necrosis resulting in serious economic consequences viz., fertilizers are not effectively utilized, the period from planting to harvesting is lengehened, maximum bunch weights are not attained, the quality of fruit is impoverished and fields have to be replanted every two to three years because of drastic reduction in plant numbers. Yield increase of 30 to 60 per cent have been obtained following the control of this nematode over a range of growing conditions. Root population of *R. similis* is indirectly correlated with the yield. (Charles *et al.*, 1985).

3. The citrus nematode *Tylenchulus semipenetrans* : *T. semipenetrans* causes "slow decline" disease of citrus and is considered to be one of the factors responsible for "die back" disease. In India, the citrus nematode was first recorded from UP, MP and Maharastra Siddiqui (1961). Subsequently it has been reported from Delhi, Haryana, Himachal Pradesh, Kerala, Orissa, Punjab, Rajasthan, West Bengal and Aassam (Chone *et al.*, 1966; Prasad and Chawla, 1965; Nair *et al.*, 1969; Swarup, 1969; Khuntia and Das, 1969; Das and Mukhapadhyaya, 1981, Sinha, 1985; Sinha and Chawla, 1986; Sinha, 1987; Sinha, 1988. The total annual reduction in citrus crop due to this nematode is estimated to be

15 per cent. The citrus nematode is responsible for "slow decline" in citrus. The symptoms include reduced terminal growth, yellowing of the leaves, general symptoms of mul nutrition; smalling of leaves, small and ununiform fruit, defoliate branch ends twig die back.

Heavily infested feeder roots of citrus may harbour hundreds of citrus nematodes per centimeter of root. Numerous lessons are produced on the root surface and develop a dark colour in the infected roots; soil particles cling tightly, even after thorough washing, to the gelatinous egg masses which cover the protruding part of the nematode body. In severely infected roots, the entire cortex may get separated from the vascular stele (Fig. 3 & 4).

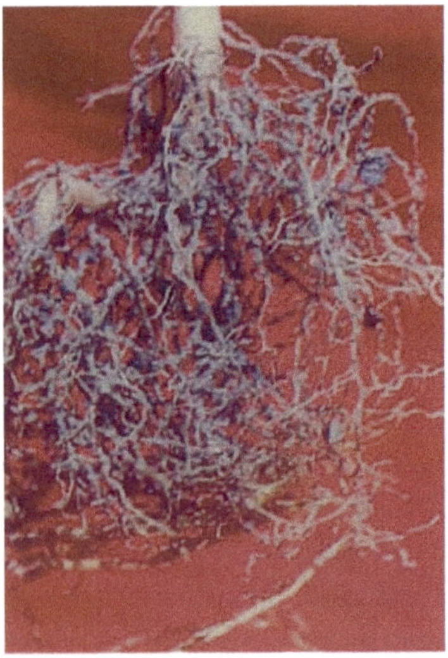

Fig. 3

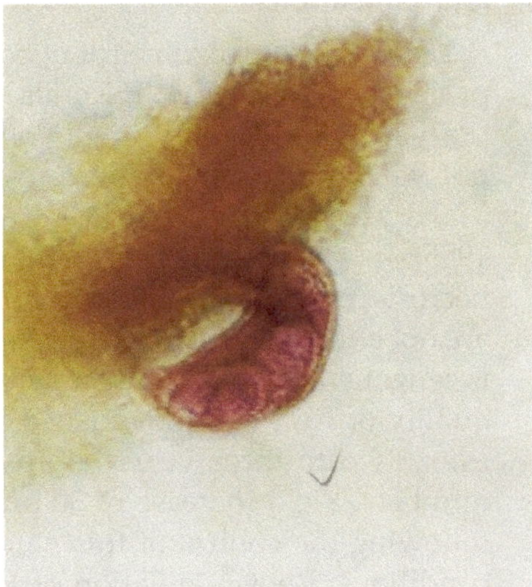

Fig. 4

Citrus Nematode Infested Citrus Root

1. The lesion nematode *(Pratylenchus coffeae)* : *P. coffeae* is responsible for root-rot of citrus in Uttar Pradesh (Sidiqui, 1984) and root necrosis in banana.

2. The spiral nematode, *Helicotylenchus multicinctus* is an important pest of banana throughout India and is responsible for 20 per cent loss in fruit yield also recorded in Assam. It causes extensive root necrosis, dieback and disfunction on leading eventually to debility of entire plant.

3. The lance nematode *(Hoplolaimus indicus)* : *H. indicus* causes damage to citrus in Punjab. The nematode at 1, 000 per 500 g of soil cause some growth reduction of citrus, but marked reduction occurred at or above an initial population level of 2, 000 (Gupta and Atwal, 1972) presence of *Hoplolaimus* spp. recorded from Assam also (Sinha, 1985; Sinha & Chawla, 1986).

Vegetable Crops

Vegetables in common usage edibl herbaceous plant parts. Plant parasitic nematodes take a heavy toll of this crop. The most important nematode pests of vegetable crops are root-knot, *Meloidogyne* spp.; *Rotylenchulus reniformis* and cyst nematodes, *Globodera* spp. The stunt nematode, *Tylenchorhynchus brassicae* has been found to reduce the yield of cabbage and cauliflower (Reddy, 1988).

1. Root-knot nematode *Meloidogyne* spp. : The root nematode specieses are the most important pests of vegetable crops. Four species viz., *Meloidogyne incognita, M. javanica, M. hapla* and *M. arenaria* have been reported and all form conspicuopus root galls. *M. incognita* and *M. javanica* are most wide spread in distribution and have a wide host range among vegetables, where as *M. hapla* attacks potato and *M. arenaria* infects chillies. Three races of *M. incognita* have been reported from India (Krishnappa and Setty, 1983). The root-knot nematode, *M. incognita* is also found widely distributed in Assam infesting various economically important vegetable crops.

   The avoidable yield losses due to *M. incognita* were estimated to be 40-46, 28-91, 34-45, 43, 29 and 20 per cent in tomato, okra, brinjal frenchbean, cowpea and peas, respectively (Parvatha Reddy, 1985, Parvatha Reddy and Singh, 1981, Bhatti and Jain, 1977, Krshnappa *et al.*, 1981; Reddy, 1985).

2. The Reniform nematode, *Rotylenchulus reniformis*: The reniform nematode is an important pest of vegetable crops in India. It infects tomato, brinjal, okra, cowpea, dolichos, French beans and other vegetables. This nematode has also been recorded from Assam. It causes 13 per cent loss in yield of cowpea (Palanisamy and Sivakumar, 1981). Two races of this nematodes have been reported from India (Dasgupta and Seshadri, 1971).

   The emergence of cowpea seedlings was delayed by 7 to 9 days and seedling stand was reduced to the tune of 6 to 11 per cent due to *R. reniformis* at one nematode per 1 g of soil (Nanjappa *et al.,* 1978).

The stunt nematode, *Tylenchorhynchus brassicae*: The stunt nematode has been found to reduce the yields of cabbage and cauliflower in Uttar Pradesh.

3. *Other nematodes*: The lesion nematode, *Pratylenchus delatteri* reduces the vigour of chillies and produces down-ward cupping and crinkling of leaves.

   The spiral nematode *Helicotylenchus dihystera* causes perceptible reduction in root growth of chillies.Okra, tomato, brinjal and onion were also found to be good host for this nematode (Muthukrishanan *et al.,* 1975).

   The lance nematode, *Hoplolaimus indicus* infects onion.

   The stem and bulb nematode, *Ditylenchus dipsaci* infects onion and garlic, while *D. destructor* causes dry rot of potato tubers in Jammu.

Ornamental Medicinal and Aromatic Crops

The acreages used for production of ornamental and aromatic crops are low in comparision with the total land area under cultivation but the cost returns are high because of the intensive cropping. Monoculture is a common feature of these crops and consequently they suffer from a wide range of nematode pests, but because of this high value the annual use of extensive phytosaanitary measures are essential.

1. Root-knot nematodes: *Meloidogyne* spp., Root-knot nematodes are a major problem on crossandra, tuberose, gladiolus, croton aster, champak, balsam, zinnia, calendula, celosia, jasmine, patchouli, *Solanum khasianum* etc.

   Meloidogyne incognitqa infected crossandra plants exabited retardation in shoot growth with smaller chlorotic leaves almost turning to white in later stage (Rajendra *et al.,* 1979).

   Root-knot nematode infected plants of tuberose showed stunting, yellowing and drying of leaves and rotting of bulbs (Jayaraman *et al.,* 1975). In severe cases, the emergence of penicles was suppressed resulting in an absolute loss of the flower yield. Further the emergence of side shoots from the bulbs was also affected.

M. incognita invades gladiolus roots, daughter corms and cormels which develop after flowering and nematodes will surving in corm tissue in the soil as a source of inoculum for next season (Parvatha Reddy *et al.,* 1979).

In patchouli, the root-knot nematode cause chlorotic leaves; premature drying and shedding of leaves. It was responsible for 47.0 and 36.7 per cent loss of shoot weight and shade dried leafs yield, respectively (Prasad and Reddy, 1984). The root-knot nematode also causes heavy galling to the root system. (Fig. 5&6), Sinha and Gogoi, (2007).

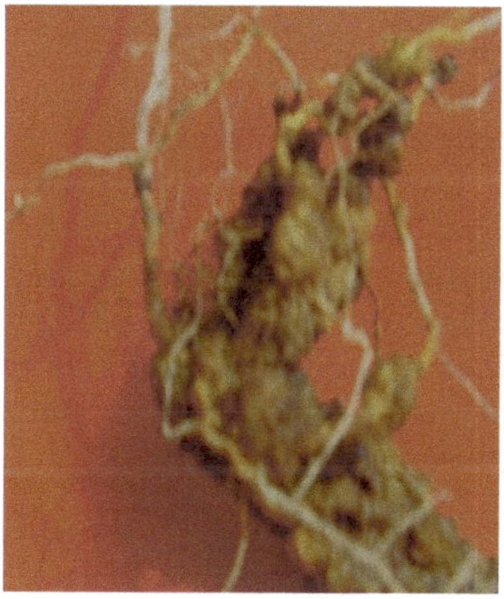

Fig. 5

Fig. 6

Root-knot nematode, *Meloidogyne incognita* on Pachouli

1. The lesion nematode, *Pratylenchus delatteri* was frequently associated with crossandra crop exhibiting stunting, chlorosis of leaves and wilting. The leaves showed mottled appearance which turned to brown and becoming pinkish eventually. Later the leaf tips dried and entire leaves turned yellow and shed. The plants were devoid of side shoots and remained completely defoliated. The roots exhibited brown to black lesions. Heavily infested plants did not produce tertiary spike and there by flower yield is reduced Srinivasan and Muthukrishnan, 1975).
2. The spiral nematode, *Helicotylenchus dihystera* caused heavy damage to crossandra and patchouli roots and it responsible for decline.

3. The bud and leaf nematode, *Aphelenchoides ritzema bosi*: *A. ritzemabosi* causes considerable damage to the foliage chrysanthemum, zinnia, salvia, aster and dahlia (Gill and Sharma, 1976). The veins hinder nematode movement within leaves and this the characteristic symptoms take the form of interveinal discolouration. They are pale green at first, gradually becoming yellow and then dark brown or black and the leaf dies.
4. The needle nematode, *Longidorus africanus*: In crossandra, *L. africanus* causes stunting of plants, branching, swelling and darkening of root tips (Muthu Krishnan *et al.*, 1977).

Control Method

The objective of control is to reduce the nematode population, by one or more methods, to below the economic threshes level. To be of economic value, any control must be profitable. The effects of nematode control are increased yields and quality for the particular crop. In addition, nematode control may result in control of other soil-borne pathogens such as fungi, bacteria and viruses which express themselves only in the presence of plant parasitic nematodes. Healthy root systems resulting from nematode control make it possible for maximum utilization of moisture and mineral elements in the soil.

Once nematodes are established in a field, the possibilities of complete eradication are remote and impractical on a field scale. Although the eradication of nematodes is not practical, several means of economic methods of nematode control are available to farmers to protect their crops.

Nematode control methods can be broadly divided into regulatory, physical, cultural, chemical, biological and host resistance. The integrated crop protection systems will utilize the best combinations of resistant cultivars, crop rotation, nematicides, biological control agents and cultural practices. In this way, a broader spectrum of nematodes can be managed simultaneously.

6.1. Regulatory Control

i) Quarantine

Regulatory control of plant parasitic nematodes is simply the legal enforcement of measures planned to prevent them from spreading or having spread, from multiplying sufficiently to become intolerably troublesome.The principle involved in enacting quarantines is exclusion of the nematode from entering into an area which is not infested.

The interchange of planting stocks with roots and also as bulbs, corms, tubers, seeds, etc., is one of the main reasons for the spread of nematodes. As an evidence to the spread of plant parasitic nematodes through exchange of plant material, plant quarantine inspections practiced in India have shown interception of many plant nematodes. From the imported germplasm material, economically important nematodes like *Heterodera gottingiana, Meloidogyne incognita, Ditylenchus dipsaci, Radopholus similis* (Sethi *et al.*, 1972) and *Globodera rostochiensis* (Renjhen, 1973) have been intercepted. Several plant nematodes like *Aphelenchoides ritzemabosi, Criconemella onensis, Heterodera* sp., *Merlinus quardrifer, Paratrichodorus porosus, Pratylenchus coffeae, P. neglectus, Tylenchorhynchus annulatus, Xiphinema krugi* and others have been intercepted from imported ornamental and other plants (Ramchandani *et al.*, 1971; Muthaiyan and Sridevi, 1982; Muthaiyan *et al.*, 1983, 1984) .

In India, there is a quarantine act against the cyst nematode of potato *(Globodera rostochiensis*) in Nilgiris. Infected seed potatoes are not allowed to be transported from Nilgiris to other parts of India for seed purposes.

ii) Seed certification

The production of vegetative plant-propagation material that is certified to be nematode free is accomplished by growing clean plants in clean soil. Strict sanitation must be practiced, and both the plant material and rooting medium must be periodically checked for the presence of pathogenic nematodes. Vegetative seed of banana free of the burrowing nematode and potato seed pieces free of the cyst nematode can be produced commercially.

6.2. Physical Control

Controlling nematodes by killing them with heat is one of the oldest method and is probably the most successful physical control measure developed so far. It is used for killing the nematodes in soil and within the plant tissues.

i) Heat treatment of soil

Sterilization of soil by steam is a long-established practice in greenhouse beds, seed beds and other small areas. Insects, weed seeds, nematodes, bacteria and fungi are killed by steam sterilization. The most important nematode pests controlled by greenhouse steaming are cyst nematode of potato attacking tomato and root-knot nematodes of tomato, cucumbers, lettuces and some ornamentals.

ii) Hot-water treatment of planting material

It is widely used for the killing of nematodes within plant tissues before planting and has proved useful with nematode-infested bulbs, tubers, seeds and roots of plants such as chrysanthemum, banana and citrus (Table 3). First, the time-temperature-mortality curves of the nematode have to be determined.This entails immersing the nematodes in to thermostatically controlled water bath at different temperatures of different length of time.Information must also be obtained on the plants reaction to temperature because if the treatment is to be successful the plants must be able to tolerate temperature lethal to nematodes.

Incubation of potato tubers at 45^{0}C for 48 hrs. has been shown to kill about 98.9 per cent *M.incognita* without affecting tuber viability (Nirula and Bassi, 1965) .Since most of animal enzymes are inactivated after short exposure to temperature near 50^{0}C, death can probably be explained on this basis.

Table 3 : Time and temperature recommendations for hot-water treatment for denematizing planting stock.

Nematode	Planting stock	Time (min)	Temp (^{0}C)
Tylenchulus semipenetrans	Citrus rooted cuttings	10	46.7
		25	55.0
Radopholus similis	Banana suckers	25	45.0
Aphelenchoides ritzemabosi	Chrysanthemum stools	15	47.8
		30	43.0
Meloidogyne spp.	Grape rooted cuttings	10	50.0
		30	47.8
	Potato tubers	120	46.0-47.5
	Sweet potatoes	65	46.7
	Yam tubers	30	51.0
	Begonia tubers	30	48.8
		50	45.0
	Tuberose tubers	60	49.0
	Caladium tubers	30	50.0
	Gladiolus corms	30	57.8

6.3. Cultural Control

The direct control of nematodes by chemical or physical means is not often economic or practicable, especially on low acre value crops and indirect or cultural methods have to be employed. Cultural methods are attempts to adapt crop husbandry practices so as to minimize the effects of nematodes. Usually the operation will be a normal one carried out at little or no extra expense.

i) Crop rotation

Crop rotation attempts to keep nematode populations to a low level at which crop damage is reduced to a minimum. By growing non-host crops, the numbers of plant nematodes are reduced, the number of years for this to occur depending on the initial population and the rate of population decrease. Rotation must also provide economically useful crops. In choosing rotations one must be careful not to promote a new set of pathogens in place of the ones to be controlled.

Out of five different crop-sequences tried, the sequence–French bean–ahu rice-sesamum-french bean was found to be effective in reducing the soil population, root galls and egg masses of *M.incognita* and increased yield of french bean (Ahmed 2000, Ahmed and Choudhury, 2004). Increased population of *M.incognita* was recorded in monoculture of french bean or french bean in rotation with cowpea and green gram.Management of root- knot nematode in vegetable based cropping system it was observed that the first crop okra, increased the nematode population in the field.In second crop with rice, cowpea, cucumber and brinjal; soil nematode population of *M.incognita* was reduced. The rotation with the crops, cowpea, cucumber and brinjal also increased the *M.incognita* population. Finally at the end of third crop it was observed that the best sequence of crop rotation was okra–rice-fallow followed by okra–cucumber–mustard and okra–cowpea -cabbage (Anonymous, 2003, 2005).

Crops which have been shown to reduce root-knot populations in the soil include varieties of cereals (sorghum, millet, maize, wheat, rice), cruciferous crops (cabbage, cauliflower, kohl rabi, mustard), onion, garlic, groundnut, cotton, roselle *(Hibiscus sabdariffa)*, pigeon pea and mustard.

Continuous cropping with potatoes increases the cyst population of *Globodera rostochiensis*. More than three year rotations with wheat, strawberry, cabbage, cauliflower, peas, maize and beans reduces the nematode population to a safe level. Likewise, decrease in root-knot nematode population on tomato, brinjal, okra, chillies and sponge gourd occurs following marigold, spinach and bottle gourd (Khan *et al.,* 1975).

There was sudden decline in population of *Tylenchorhynchus brassicae* when cabbage and cauliflower were rotated with wheat (Siddiqi *et al.,* 1973). The mustard (rabi), radish (summer), sesamum (kharif) sequence considerably reduced the population of *Tylenchorhynchus* spp. (Haque & Gaur, 1985).

Planting of sugarcane or pangola grass following the destruction of banana infected with *R. similis* eliminates the nematode after 10 weeks. Rice or green gram grown after banana suppressed the population of *R. similis, Pratylenchus coffeae* and *Helicotylenchus multicinctus* (Rajendran *et. al.,* 1979). Considerable reduction in population of *T. brassicae* was noted in potato and tomato, or *Hoplolaimus indicus* in cabbage, chilli and brinjal (Alam *et al.,* 1980).

A number of crops and other plants are reported resistant to the reniform nematode, *Rotylenchulus reniformis* which include sugarcane, *Coffea robusta,* Jack bean *(Canavalia ensiformis*), velvet bean *(Stizolobium deeringianum*), chillies, maize, sorghum, *Lecaena glauca,* finger millet *(Eleusine coracana*), *Crotolaria, cyanodon dactylon,* onion, garlic, *Brassica compestris, B.rapa, Capsicum frutescens,* carrot, *Anethum graveoleus,* coriander, *Foenicutum vulgare, Spinacea oleracea, Beta vulgaris* and *Raphanus sativus* (Khan and Khan, 1973). Two successive crops of maize or sorghum preceding susceptible crops are effective in controlling *R. reniformis.* Rotations with sugarcane and pangola gross are recommended for control of the reniform nematode.

ii) Fallow

Many nematode pests can be controlled by depriving them of plants on which to feed. From practical standpoint, however, keeping land fallow for the necessary length of time is not economically feasible or even dangerous because of soil erosion by wind or water. Nevertheless, crops that have been damaged by root-infecting nematodes of any kind should not be allowed to stand in the field for a long time after harvest but should be destroyed immediately, preferable by ploughing to expose the roots to the drying action of wind and sun. Weeds which are susceptible to nematodes should not be allowed to grow during fallow.

Best nematode control is obtained when fallow is maintained during hot, dry weather. The cycles of drying and wetting are more effective than continuous summer fallow in reducing nematodes. Turning the soil during the fallow will give additional kill of the nematodes.

Fallowing for a period of three months after banana effectively suppressed the burrowing nematode population (Rajendran *et al.*, 1979). Decrease in *M.incognita* population on tomato, brinjal, okra, chillies and sponge gourd occurs when the field was left fallow (Khan *et al.*, 1975).

iii) Selection of healthy propagating material

Propagation stocks, including rhizomes, roots, seeds, transplants and tubers, are of special importance to the horticultural industry because they may harbour nematode pests.All such stocks should be obtained from reliable sources which practice rigid nematode control. The potato cyst nematode *(Globodera rostochiensis)* and the burrowing nematode of banana *(Radopholus similis)* can be eliminated by selecting nematode-free planting material.

Soil amendments

The efficacies of different oil cakes as organic amendments in decreasing the pathogenic effect of *M. incognita* on tea seedlings had been determined (Anonymous, 1974 b). Sinha (1974) recorded reduced gall formation on brinjal roots by *M. incognita* due to the application of mustard oil cake, farm yard manure and cowdung. The effect of three localy available organic amendments, DCTW, mustard oil cake and sawdust @ 210 g/furrow (3 m length) and 25 g/spot and nitrogen equivalent to cake, added in the form of urea had been determined against *M. incognita* on brinjal (Anonymous, 1987). Spot application of mustard oil cake and furrow application of sawdust were found most effective among all the treatments in respect of reduction of nematode infestation and increase yield. Sarmah (1989) found that sawdust, mustard oil cake, poultry manure and press mud each at 0.5, 1.2 and 2.4 per cent (w/w) could reduce the incidence of *M.incognita* in cowpea effectively. Significant increase in plant growth (height, shoot and root weight) of cowpea was recorded in sawdust followed by poultry manure at 1.2 and 2.4 per cent (w/w). In a field experiment, Hazarika (1990) found that castor cake, mustard cake and neem cakes at spot (15 g/spot) and furrow (100 g/furrow of

3 metre) applications reduced the infestation of *M.incognita* and increased the yield of brinjal. The yield was maximum in spot application of neem cakes with maximum reduction of *M.incognita* population in soil. Mustard oil cake, neem cake, sawdust and choped plant parts of water hyacinth @ 100 g/furrow (2 m length) and 15 g/ spot were found to be effective in reducing the infestation of *M.incognita* and increasing the plant height, fresh & dry weights of shoot and root in brinjal (Sarmah, 1992) Das (1992) tried mustard cake, neem cake and DCTW at the rate of 750 and 1500 kg/ha and urea at 45 kg/ha for the management of *M.incognita* on soybean.Neem cake at 1500 kg/ha followed by mustard cake at 1500 kg/ha were most effective in reducing the gall and egg masses of *M.incognita* and increasing the yield of soybean under field condition. He further observed that urea and DCTW application had no effect on the formation of gall and egg masses on roots.

Application of CaCN2, NaCN and urea cyanamide is effective in controlling root-knot nematodes. This may be due to the release of HCN or NH3 which are toxic to the nematodes. Increased levels of potash have significantly reduced the number of galls by *M.javanica* in tomato (Gupta and Mukhopadhaya, 1971). Application of potash in combination with phosphorus or nitrogen or potash alone checks the reniform nematode multiplication on okra to a great extent (Sivakumar and Meerzainuddin, 1974).

Talukdar (1993) tested castor cake, mustard cake, and neem cake each at 100 g/2 m furrow and 15 g/spot application and urea (nitrogen equivalent to cakes) for the management of *M.incognita* on 'okra' and found that all the treatments of organic amendments were quite effective in reducing the nematode infestation and also increasing the growth parameters of the crop in comparison to urea treatment and control. The furrow application of each amendment was superior than that of spot application. Neem cake in furrow was the best followed by neem cake spot application followed by neem cake at spot, mustard cake and castor cake at both applications giving 58.88, 48.36, 35.86, 35.53, 34.87 and 31.91 per cent increase yield respectively over control. Significant reduction of infestation of *M.incognita* on chilli was obtained after application of mustard cake, neem cake, cattle manure and sawdust at 1.0 and 2 per cent (w/w). However, neem cake and mustard cake at 2.0 per cent were more effective than other treatments (Sarma, 1993). In a field trial, sawdust, neem cake and poultry manure applied @ 1 and 1.5 t/ha and waterhyacinth at 1 and

2 t/ha increased yield (28-66%) of carrot by reducing the infestation of *M.incognita.* Maximum reduction of *M.incognita* in soil was recorded in neem cake (38.5%) followed by sawdust (23%), poultry manure (22%) and waterhyacinth (14%) at their higher doses over control (Devi, 1993, Devi & Das, 1999). Barman (1993) found that neem cake, mustard cake and castor cake each in spot (15 g/spot) and furrow (150 g/3 m furrow) applications were effective in reducing the formation of galls (70-77%), egg masses (61-65%) in root system and juvenile population of *M. incognita* (31-38%) in soil over control and increasing the plant growth parameters (upto 34, 67 and 174% increase of plant height, dry weight of shoot and root respectively over control) of tomato under field condition. Spot applications of these organic amendments were more effective than that of furrow application. Neem cake spot application gave best result. These findings were further confirmed by Deka (1994) and Deka & Phukan (1997) where it was observed 70-76 per cent reduction of gall, 60-65 per cent of egg masses and 31-38 per cent reduction in population of *M.incognita* in soil and 27-70 per cent increase in yield of tomato over control after amending the soil with neem cake, mustard cake, castor cake each applied at 15 g/spot and 125 g/furrow (2.5 m size). Like the earlier works they also found that spot application was superior to furrow application and spot application of neem cake with best result. Significant reduction in the population of plant parasitic nematodes had been observed by the application of sawdust and neem cake @ 2 kg/plant at a distance of 0.45 cm from the base of the banana plant on the 6th month after planting and on the 4th month after first application (Anonymous, 1994 b). Hazarika (1995) also found effective result in reducing the infestation of *M.incognita* and increasing the yield of okra (38-80% over control) by applying castor cake, mustard oil cake and neem cake each at 100 g/2 m furrow and 15 g/spot. Mustard oil cake and neem cake at 10 g and 15 g per seedling were found to be effective in reducing galls, eggmasses, eggs per egg mass and final nematode population of *M.incognita* and in increasing the plant growth characters of tea seedlings (Dutta, 1995). Kalita (1995) also confirmed the findings of Barman (1994) & Deka (1995) on mustard oil cake, castor cake and neem cake applied at the spot and furrow application for reducing the infestation of *M.incognita* and increasing the yield of tomato.Neog (1997), Neog & Bora (1999) tried four organic amdments *viz.*neem cake, mustard oil cake, deaffeinated tea waste and ratna each at 1.0, 2.0 and 3.0 per cent (w/w) in comparison with carbofuran 3 G @ 3 g per kg of soil against *M. incognita* in tea nursery and observed that all

treatments of these organic amendments and carbofuran were effective in reducing the number of galls, eggmasses, egg per eggmass and final nematode population in soil. However, carbofuran was found to be most effective followed by neem cake and mustard cake at 3.0 per cent (w/w).

Organic amendments *viz.*mustard cake and neem cake tested each @ 10 g per seedling were found effective in reducing galls, eggmasses and eggs per eggmass of *M.incognita* along with increasing plant growth parameters of tea seedling. Out of three two organic amendments tested neem cake @ 10 g per seedling was found to be the most effective (Kalita, 2000). Sinha and Neog (2003) evaluated the effect of three organic amendments (neem cake, saw dust and mustard cake) with two doses (5 and 10 t/ha) alone and in combination with NPK against *T. semipenetrans* on Khasi mandarin.All the treatments were found effective in increasing the plant growth as compared to untreated control.Maximum fresh and dry weight of shoots as well as maximum reduction in nematode population with NPK + mustard cake 10t/ha followed by NPK+ sawdust 10 t/ha.Ahmed and Choudhury, (2005) evaluated the effect of four organic amendments *viz.*, saw dust, poultry manure, mustard cake and neem cake for the management of *M.incognita* in french bean (Ahmed 2000). Poultry manure applied as spot application (16.6 g/spot) was found to effective in increasing plant growth parameters whereas neem cakes as spot application (16g/spot) was found effective in reducing gall, eggmasses and soil population of *M.incognita*.Spot application of all the organic amendments was comparatively better than furrow application (100g/farrow) in reducing nematode population and increasing yield of french bean.

iv) Flooding

Keeping the land submerged in water decreases oxygen content of the soil and kill the nematodes by asphyxiation. Chemicals lethal to nematodes such as butyric and prophonic acids, hydrogen sulphide and NH3, often develop in flooded soils. Flooding for three months after banana effectively suppressed the burrowing nematode (Rajendran *et al.*, 1979). This method of control is also effective against root-knot and stunt nematodes on vegetables.

v) Trap cropping

Cowpeas causes the root-knot nematode eggs to hatch, the larvae enter the roots and develop to immobile stage. Then the crop is destroyed before the nematodes mature. *Crotalaria* is highly susceptible to invasion by the root-knot nematode but is resistant to the development of larvae into adults.

vi) Enemy plants

1. Mustard

Potatoes grown with white mustard in a pot of infested soil where less heavily attacked by nematodes than potatoes growing alone. Potato root diffusate was ineffective in the presence of teachings from the roots of mustard seedlings. Mustard oils increased the yield of potatoes by reducing the severity of nematode attack. The active principle involved in mustard is allyl isothiocyanate which is toxic to the nematodes.

2. Marigold

The root-knot development on tomato and okra was low when interplanted with *Tagetes erecta.*The population of *Tylenchorhynchus, Helicotylenchus, Hoplolaimus, Rotylenchulus* and *Pratylenchus* was also markedly reduced (Khan *et al.,* 1971a;Alam *et al.,* 1977). α-Terthienyl is the active principle in *Tagetes* spp. which is toxic to these nematodes.

3. Asparagus

Asparagus officinalis would not support populations of *Trichodorus christiei* for more than 40 to 50 days. Tomato, normally a good host of this nematode, supported only a low population when asparagus was growing in the same pots.A glycoside (asparagusic acid) is the active principle involved which is toxic to *T. christiei* and several other nematode species.

4. Crotalaria

Crotalaria root system exerts a toxic effect on *Meloidogyne* populations and that planted in advance of a crop can reduce subsequent damage. *Meloidogyne* larvae freely entered the roots of *Crotalariai* but failed to survive. Populations of *Xiphinema americanum* decreased under *Crotalaria spectabilis.*

5. Sesame

Atwal and Mangar (1969) showed that root exudates from sesame *(Sesamum orientale)* have nematicidal properties against *Meloidogyne incognita.* When okra was grown in *M. incognita* infested soil, it was only slightly attacked and there were fewer nematodes compared with when okra grown in the absence of sesame.

6. Cucumber

The cucurbitacins are triterpenoids that accumulate as aglycones or glycosides in cucumbers carrying a dominant gene for bitterness. Extracts from bitter cucumber attracted fewer *Meloidogyne incognita* larvae than did a comparable extract from nonbitter cucumbers which do not accumulate cucurbitacins. The cucurbitacins are repellent to *M. incognita*. Once larvae penetrate the roots it is possible that cucurbitacin (an insect antibiotic) acts as a toxin that either kills the invading larvae or repels them from the root.

7. Margosa (*Azadirachata indica*)

Margosa is reported to contain at least two nematicidal compounds-nimbidin and thionemone. In comparison test, margosa root exudates were found to be more toxic to *Hoplolaimus indicus, Helicotylenchus indicus, Rotylenchulus reniformis, Tylenchorhynchus brassicae, Tylenchus filiformis* and *M. incognita* than marigold root exudates.

8. Pangola grass

Another reported naturally occurring nematicide is pyrocatechol. This compound accumulates in high concentration in roots of pangola grass *(Eragrostis curvula)* and is a potent nematicide against *Meloidogyne javanica, M. hapla, M. arenaria thamesi* and *M. incognita acrita.*

vii) Time of planting

In some cases, crops may be planted in winter when soil temperatures are still below necessary for activation of nematodes. Early potatoes grow in soils too cold for cyst nematode activity. By the time the soil warms suffieiently for nematodes to become active, the plants are advanced enough to withstand a massive invasion of larvae. They continue to grow and eventually produce a crop.

Planting of potato during third or fourth week of March in Simla Hills would reduce the damage due to *M. incognita* (Prasad *et al.*, 1983). The yields were maximum which was concomitant with the lowest tuber infestation and lowest larval population in the soil at harvest.

6.4. Chemical Control

The primary advantage of chemical control is that the nematode population is reduced to a very low level within a matter of days after the chemical is applied, enabling the grower to plant a crop soon after treatment or in some cases at the time of treatment. Most crops are especially vulnerable to nematode attack during the seedling stage when the young root system is established. Crops planted in treated soil develop good root systems, so that usually, in case of annuals, the crop is mature before residual population of nematodes has increased to a level which would cause damage.

Mukherjee (1960) tried nemagon and nematox against root-knot species on tea but could not obtain any effective results from these chemicals. Mukherjee (1966) found that nemagon, nematox, basudin, nematox with combination with basudin and vapam in different doses significantly controlled *M. incognita* and *M. hapla* on tea nurseries. Mukherjee (1967) tried nemagon and E.D.B. granules against root-knot nematode in tea nurseries and found a significantly higher percentage of plants free from the nematode infestation in treated plots than control. Application of vapam and temik reduced attack of *P. brachyurus* on tea (Anonymous, 1973). Choudhury (1976) evaluated seven nematicides against *M. incognita* on brinjal and found them effective in reducing the infestation of the nematode significantly. He found maximum reduction in galls and egg masses with nemagon followed by metham sodium and hexanema. Dutta (1978) found that vapam improved the diseased condition of jute caused by the association of *M. incognita* and *Rhizocotnia solani*. Basu (1979) tried Furadon, Temik, Thimet and mocap against *M. incognita* infesting tea seedlings and found effetive result from Furadon only. In field trials aldicarb at 0.5 and 1 g and carbofuran at 1 g/tea seedling were found to be effective for the control of *M. incognita* with signifiant increased in shoot height (Basu and Gope, 1985) . Carbofuran, ekalux, diazinon and phorate each at 1, 2 and 3 per cent (w/w) tested as seed treatment were found to be effective in reducing the infestation of *M. incognita* on cowpea (Sarmah, 1989). Carbofuran 3 per cent w/w was found to be best among all the treatments for controlling *M. incognita* and increasing the yield of cowpea. Efficacy of carbofuran, diazinon, ekalux and phorate at 1, 2 and 3 per cent (a.i. w/w) as seed treatment for the control of *M. incognita* on pea in both pot and field conditions were determined by Bhagawati and Phukan (1990). They found that all these chemicals were effective even at low rate of application in reducing galls and egg masses in roots of pea and increasing yield. The best result was found with carbofuran at 3 per cent. In a field trial carbofuran, diazinon and phorate as root dip treatment @ 0.5 and 0.1 per cent concentration for half an hour and soil application of 1.5 kg and 3 kg a.i./ha and combination of root dip treatment (0.5%) + soil application @ 1.5 kg a.i./ha after 25 days of transplanting were found to be effective in reducing the infestation of *M. incognita* and increasing the yield of brinjal (Hazarika, 1990). Among different methods of applications, root dip treatment (0.5% for 1/2 hr) followed by soil application (1.5 kg a.i./ha) was found to be most effective. Carbofuran in root dip treatment followed by soil application was best of all giving better protection against *M. incognita* with increasing the yield. Pathak (1990) tried carbofuran, phorate and diazinon each at 1, 2 and 3 per cent a.i. (w/w) as seed treatment and found 3 per cent a.i. (w/w) of these chemicals were highly effective for the better control of *M. incognita* on okra with increasing yield. Among four different chemicals each at 1, 2 and 3 per cent a.i. (w/w) tested as seed treatment, carbofuran at 3 per cent was found to be most effective in reducing galls, egg masses and increasing yield of greengram followed by mocap, diazinon and phorate at the same dose (Bora and Phukan, 1990) .Four pesticides tried by soaking cowpea seed in 0.1 per cent solution for 12 hr were found to be effective in reduing the development and infestation of *M.inognita* on cowpea.Of these, triazophos was the best followed by phosphamidon, monocrotophos and dimethoate (Anonymous, 1991 a) .Gope and Borthakur (1991) reported that prolonged use of herbicide did not cheched the build-up of nematode popiulation in the rhizosphere of tea plants. On the contrary, with the progress of time the nematode population increased to higher levels in the herbicide treated plots than that in the control except diuron followed by paraquat on chieling and 2, 4-D or paraquat followed by chieling treatment. In a field trial Saikia (1991) as bare root dip treatment for 6 hr against *H.oryzae* and found effective. The best result was tested carbosulfan, chloropyriphos, quinalphos, dimethoate, phosphamidon at 0.2 and 0.3 per cent obtained from carbofuran 0.2 per cent followed by chloropyriphos 0.2 per cent with increase in grain yield by 37.7 and 34.4 per cent respectively.He further observed that although all these chemicals with 0.3 per ent could reduce the nematode population but could not produce higher yield probably due to phytotoxic effect of these chemicals at higher dose on rice plants. Of seven pesticides *viz.*, zolone, phosphomidon, monocrotophos, carbosulfone, triazophos, chloropyriphos and dimethoate each at 0.2 per cent tested as seed soaking treatment for 6 hr against *M. graminicola* on rice chloropyriphos was found to be best in both soil and roots followed by carbosulfan and monocrotophos (Kalita, 1991, Kalita & Phukan, 1996). In a separate trial, Kalita (1991) found that among 4 different pesticides tried at 0.2 per cent with single and double foliar sprays, carbosulfone at both the spray schedule was found to be best of all in reducing galls, egg masses on roots and nematode population in both soil and roots followed by monocrotophos.All the four pesticides *viz.*, hostathion, carbosulfone, monocrotophos and phosphomidon each 0.2 and 0.3 per cent as bare root dip treatment for half an hour/30 minuts were found effective in reducing the infestation of *M. incognita* and increasing plant growth of brinjal (Sarma, 1992) .In another field trial efficacy of carbofuran, phorate and diazinon each as bare root dip (0.1% solution for 1/2 hr), soil application 1.5 kg a.i./ha and combination of root dip (0.5%) and soil application (0.75 kg a.i./ha) at 30 days of planting was evaluated and found to be effective in reducing the infestation of *M.incognita* and increasing the plant growth parameters and yield of brinjal.Carbofuran root-dip+soil application was found to be most effective followed by carbofuran soil application and carbofuran root dip treatment (Sarma, 1992).

The efficacy of carbofuran, diazinon and phorate applied at the rate of 2 g and 3 g a.i./plant in four month interval in new and established plant of banana had been evaluated in two separate field experiments (Deka, 1992). All these chemicals with different doses could reduce the nematode population significantly. The yield in these treatments except diazinon 2 g a.i. increased significantly over control (Deka, 1992).

Talukdar (1993) tested monocrotophos, dimethoate, triazophos and carbosulfone each at 0.25 and 0.5 per cent concentration (v/v) as seed soaking treatment for 6 hr. under net house conditions and found them as effective in reduing the galls, egg masses and eggs per egg masses of *M. incognita* on okra and increasing the plant height. Monocrotophos at 0.5 per cent (v/v) was the most effective (40% for redution in number of galls, 16% egg masses, 30% eggs per egg mass). In a separate trial, Talukdar (1993) found phorate, carbofuran and Sebuphos each 2 per cent and 3 per ent a.i. (w/w) as seed treatment to be quite effective in reducing the galls, egg masses and eggs per egg mass of *M. incognita* and also increasing the plant height. Sebuphos at both the concentrations was the best causing 74.1 and 74.75 per cent reduction in gall and egg masses over control respectively. In a pot experiment the efficacy of carbosulfone, triazophos, monocrotophos and zolone at 0.05 and 0.1 per cent solution as seed soaking for 6 hr.was determined against *M. incognita* on cowpea (Anonymous, 1993 a). All these chemicals with both the doses were found effective in reducing the population of *M. incognita* on 15, 30 and 45 days after germination of seed. Barman (1993) found that carbofuran, phorate, diazinon and rugby as spot application @ 1.5 kg a.i./ha and carbosulfone, triazophos, monocrotophos and phosalone as root dip treatment @ 0.1 per cent concentration were quite effective in reducing the formation of gall (22-64%, egg masses (26-54%) and juvenile population of *M.incognita* in soil (15-41%) and increasing the plant growth characters (43-91% in dry weight of shoot and 20-112% in dry weight of root) and yield (13-33%) of tomato over control in a field with 456 J_2/200 ml of soil.In a pot experiment Barman (1993) found effective control of *M.incognita* on tomato with triazophos, carbosulfone, monocrotophos and phosalone each at .05 and 0.1 per cent of a.i.as bare root dip teatment of 3 weeks old tomato seedlings for 6 hr. giving 50-70 per cent, 42-75 per cent and 60-83 per cent reduction of gall, egg mass and final soil population of *M.incognita* respectively over control.Triazophos at 0.1 per cent was most effective for the control of *M.incognita* on tomato and increased 57 per cent and 25 per cent dry weight of shoot and root over control respectively. Das (1994) observed that carbosulfone, chloropyriphos, monocrotophos and triazophos each at 0.1 and 0.2 per cent concentration (v/v) tested as seed soaking for 6 hr were effective in reducing galls, egg masses, eggs per egg mass and penetration of juveniles into the root system of french bean. Triazophos at 0.2 per cent was the most effective. Deka (1994) tested triazophos, carbosulfone, methovip and quinolvip each at 0.1per cent (v/v) as root dip treatment for 1/2, 1, 2 and 3 hr in a pot experiment and found triazophos with all four dipping periods and carbosulfone and methovip with 2 and 3 hr diping periods to be very effective in reducing galls on roots of tomato including *M.incognita* in soil.Deka (1994) and Deka and Phukan, (1997) tested triazophos, carbosulfone, methovip and quinolvip each 0.05 per cent and 0.1 per cent concentrations as bare root dip treatment for 1/2 hr against *M. incognita* on tomato, effectively in reduced galls, egg masses including nematode population in soil.Triazophos, carbosulfone and methovip at 0.1 per cent caused significant increase in the growth of tomato plant. Dutta (1995) observed that carbofuran, phorate, diazinon each at 3 g and 5 g per tea seedling were found effective increasing the plant growth parameters as well as reducing the infestation of

M. incognita. In a field trial Kalita (1995) also found that triazophos and carbosulfone @ 0.1 per ent as root-dip treatment for 3 hr, carbofuran and phorate at 1.5 and 3 kg a.i./h as spot application and the combination of root dip treatment in 0.1 per ent carbosulfone and triazophos and carbofuran and phorate at 1.5 kg a.i./ha were quite effective in reducing the infestation of *M.incognita,* increasing the yield of tomato. Hussain (1995) also found effective result by (i) soaking the seed in 0.1 per ent solution of triazophos and monocrotophos for 6 hr and (ii) soil application with phorate and carbofuran at 3 kg a.i./ha and (iii) soil amended with neem cake at 1500 kg/ha for reducing the infestation of *M. incognita* on french bean.

Bora (1995) found that carbosulfone, triazophos and monocrotophos at 0.2 per cent were quite effective in respect of reduction in per cent penetration and development of citrus nematode, *T.semipenetrans.* Hazarika (1995) found that carbofuran and sebuphos as seed dressing treatment at 2 per cent (w/w), carbosulfone and monocrotophos as seed soaking in 0.5 per cent solution for 6 hr and carbofuran at 1 kg a.i./ha as spot application reduced the infestation of *M. incognita* and increased the yield of okra.Carbofuran 1 kg a.i./ ha as spot application was most effective.

Significant reduction in percent penetration and development of *T. semipenetrans* on khasimandarin upto 6 weeks were observed by Mili (1996) when the roots of seedlings were dipped in 0.2 per cent solution of carbosulfone, triazophos and monocrotophos for 45 minutes before transplanting.Seed soaking with carbosulfan, hostathion and quinalphos at 0.1 and 0.2 per cent a.i. (v/v) for 1 & 2 hr was found to be effective in reducing galls, eggmasses, eggs per eggmass of *M. incognita* and improved the plant growth characters of cucumber (Bharali, 1996, Bhorali&Phukan, 1997) .Hazarika (1996) observed that seed soaking treatment with triazophos, quinalphos and monocrotophos at 0.1 per cent for 6 hr was quite effective in reducing the penetration of *M.incognita* juveniles into soybean roots and also the development of the nematode as well as increasing the plant growth.Carbosulfan, triazophos and monocrotophos at 0.1 and 0.2 per cent (v/v) concentration as bare root dip treatment of citrus seedlings for 45 minutes reduced the per cent penetration and development of *T.semipenetrans* upto 6 weeks after planting. Carbosulfan at 0.2 per cent was found to be best throughout the period of observation followed by triazophos and monocrotophos at 0.2 per cent (Maznoor, 1991). In a field experiment Sarmah (1999) observed that treatment with carbofuran 3 G @ 3 kg a.i./ha as spot application and monorotophos, triazophos and multineem at 0.1 and 0.2 per cent as bare root dip treatment for 30 minutes were found to be effective in reducing host infestation of *M.inognita* and increasing the growth of brinjal. Among different treatments, carbofuran as spot application was found to be the best of all in reduing the formation of galls and egg masses in roots and final population of *M. incognita* in soil and increasing yield and growth of brinjal followed by triazophos as root dip treatment.In a pot experiment Das (1999) evaluated the efficacy of carbosulfan, triazophos and monocrotophos at 0.1 and 0.2 per cent concentration as bare root dip treatment for 45 minutes for the management of *T.semipenetrans* on Khasi mandarin and found that all chemicals with these concentrations except monocrotophos 0.1 per cent were quite effective in increasing the growth parameters and reduced the nematode population both in soil and root.However, the higher concentration of these chemicals were found to be more effective. Monocrotophos, triazophos, carbosulfan, chloropyriphos and multineem at 0.05 and 0.1 per cent as bare root dip treatment for 30 to 60 minutes were quite effective against *M.incognita* on brinjal of these, triazophos at 0.1 per cent with 60 minutes dipping period was found to be most effective (Sarmah 1999).

Investigations carried out in 1995, 1996, 1997 and 1998 for the management of *M. incognita* on brinjal through nursery treatment with carbofuran and dozomet @ 0.3 g and 0.6 g a.i./m^2 resulted that 11.1 to 79.07 per cent increased yield of brinjal, could be obtained (Anonymous, 1997, 1999).

Pesticides *viz.*, carbofuran 3 G and phorate 10 G each at 3g per seedling were found effective in increasing the growth parameters of tea seedling as well as reducing root galls, eggmasses, eggs per egg masses and final population of *M.incognita* (Kalita 2000).

Hussain and Bora (2007) stated that effect of four different nematicides on root galls, egg mass, eggs per egg mass, nodule numbers, rate of penetration and multiplication of *M.incognita* on french bean. They observed that the number of root galls, egg masses were minimum in the treatment with phorate 10 G @3 kg a.i./ha.The results further indicated that the nodules were observed 21 days after germination and recorded maximum in control and minimum in the treatment with phorate 10 G a.i./ha. In case of number of eggs per egg mass minimum was recorded in the treatment with phorate 10 G@ 3 kg a.i./ha and maximum in control.

Avermectins

Excitement is being generated by the results of a new class of insecticide-nematicide which have demonstrated a broad spectrum of antihelminthic activity.Promising results for the control of root-knot nematodes have been reported.Avermectins, produced by the soil microbe *Streptomyces avermitilis,* block the transmission of electrical potentials between interneurons and excitatory motoneurons in the ventral nerve cord.They inhibit transmission between inhibitory motoneurons and muscle by opening the chloride channels on the membranes but have little effect on excitatory neuromuscular junctions. The block of interneuron to excitatory motoneuron transmission can be reversed by picrotoxin, a known gamma-aminobutyric acid (GABA) antagonist.

i) Nursery bed treatment

Considerable loss occurs from nematode infestations in plant nurseries. An effective and economical control can often be achieved by the elimination of nematodes from nursery beds by soil treatments.

Nursery bed treatment with aldicarb or carbofuran both at 4 kg a.i. per ha was effective for the control of *T. semipenetrans* on citrus.

Nursery bed treatment with aldicarb or carbofuran both at 4 kg.a.i. per m^2 were effective in increasing seedling growth and reducing root-knot nematode population on tomato, brinjal and chillies (Jain and Bhatti, 1983; Ramakrishnan & Balasubramanian, 1981; Anon, 1983). In the main field, the above treatments were also effective and increased the fruit yield.

DBCP at 50 I per ha and metham sodium at 250 I per ha were effective in reducing the reniform nematode population in tomato nursery and in increasing the growth of seedlings (Sivakumar *et al.,* 1977). Nursery bed treatment with phorate, aldicarb, fensulfothion and carbofuran each at 5 g per m^2 gave good and healthy crossandra seedlings free from *Pratylenchus delatteri* (Vadivelu and Muthukrishnan, 1979).

ii) Seed treatment

Seed treatment with nematicides is being used increasingly to cut down the cost of chemical without any adverse effect on the plant growth. Seed treatment gives adequate initial protection from the nematode larvae to seedlings leading to better plant growth and increased yields.

Seed treatment with nematicides can be effectively employed to reduce the severity of root-knot nematode infestation in leguminous vegetables (Parvatha Reddy, 1984). Fenamiphos, aldicarb and carbofuran all at 1 per cent concentration were effective in controlling the root-knot nematode *(Meloidogyne incognita)* infecting cowpea, french bean and peas and in increasing their pod yields. Seed treatment with aldicarb and carbofuran both at 6 an 12 per cent was effective against the root-knot and reniform nematodes infecting okra (Sivakumar *et al.,* 1976). Aldicarb and carbofuran at 3 per cent were also effective against root-knot nematodes infecting chilli and bottle gourd (Kandasamy and Sivakumar, 1981; Anon, 1983). Jain and Bhatti (1981) reported effective control of *M. javanica* on okra by seed treatment with fenamiphos at 2, 4 and 6 percent. Seed treatment with carbofuran and aldicarb sulfone was effective against root-knot nematode control in parvel and bitter gourd respectively (Anon, 1985a).

Seed soaking treatment for 12 hr. in aqueous solutions of dimethoate and UC-54229 was effective against root-knot nematodes infecting okra and cowpea (Anon, 1985a).

Application of aldicarb, carbfuran and phorate as seed treatment resulted in an increase in yield of cauliflower and significant reduction in the population of *Tylenchorhynchus brassicae* (Singh *et al.,* 1984).

The removal of the burrowing nematode infested tissue along with some of the surrounding healthy tissue would normally disinfect banana sets. The pared set can be dipped in mud slurry and sprinkled with 1. 2 g a. i. carbofuran or 1 g a. i. aldicarb per set. The sets can be dried in shade and used for planting.

Dipping of gladiolus corms in thionazin or fensulfothion solution (0. 5 g a. i. per I) have reduced root-knot nematode infestations.

iii) Seedling bare-root dip treatment

Nematode damage is more harmful to seedlings than to older plants. For example, the yield loss was greater when young tomato plants were infected with root-knot nematodes. Hence treatment with nematicides applied only to seedling roots may, therefore, lesser nematode damage and more economical.

Dipping of banana suckers in 0. 1 per cent aldicarb solution increased plant height, bunch weight, fingers per bunch, induced early flowering and effectively reduced the burrowing nematode population (Venkitesan and Charles, 1983).

Carbofuran and oxamyl at 1, 000 ppm for 15 to 30 min (Parvatha Reddy and Singh, 1970) ; fenamiphos at 250 to 750 ppm for 30 min (Thakar and Patel, 1985); dimethoate at 500 ppm for 6 hr., phosphamidon and dichlofenthion both at 1000 ppm for 8 hr. (Jain and Bhatti, 1978) and thionazin at 500 ppm for 15 min (Reddy and Seshadri, 1975) was effective control of root-knot nematodes on tomato when used as bare-root dips. Alam *et. Al.*, (1973) reported that oxamyl was effective for the control of root-knot nematodes infecting brinjal and okra. VC-13 was found effective against *M. incognita* on chillies, brinjal and tomato (Saxena *et al.* , 1974;Alam *et al.* , 1973).

Bare-root dip treatment of brinjal seedlings with aldicarb, carbofuran and turbofos at 500 to 100 ppm was effective in reducing the reniform nematode population (Prasad and Krishnappa, 1981).

Shetty and Reddy (1985) reported that root dip treatment with fenamiphos and aldicarb sulfone both at 2000 ppm for 2 hr gave effective control of *M. incognita* infecting *Solanaum khasianum.*

Root dip treatment for 8 hr. in 0. 05 and 0. 1 per cent concentration of aldicarb, fesulfothion, fenamiphos and phorate protected crossandra seedlings from *Pratylenchus delatteri* for 30 days after transplantation (Vadivelu and Muthukrishanan, 1979).

iv) Soil treatment in the main field

Most of our important plant-plant parasitic nematodes are inhabitants of the soil. A chemical can be brought into contact with soil-inhabiting nematodes by mechanical mixing. Experiments have demonstrated that nematodes can be controlled by mixing granular chemicals with the soil mechanically. This method might enable the grower to prepare the land to a good tilth and incorporate both fertilizer and nematicide in one operation. Thorough mixing can be achieved in field scale either by disking or rotary tillage.

Significant reduction in the citrus nematode, *Tylenchulus semipenetrans* population (76 to 84 per cent) in sweet orange and the root-knot nematode, *M. incognita* population (81 per cent) in grapevine was observed with soil application of aldicarb and carbofuran both at 6 kg. a. i. per ha. The significant increase in yield was also observed in sweet orange (161 to 224 per cent) and grapevine (37 to 212 per cent) from the above treatments (Baghel and Bhatti, 1983a, b). Rajendran and Naganathan (1981) have reported that carbofuran at 2 kg a. i. per ha gave significant reduction in soil population of the reniform nematode, *Rotylenchulus reniformis* and increased the papaya fruit yield by 38. 4 per cent over control. Pre-plant soil application of metham sodium at 40 I per ha was effective in controlling *R. reniformis* infecting grapevine and increased the bunch weight by 63 per cent over control. The burrowing nematode on banana, *Radopholus similis* can be controlled by soil application of aldicarb or carbofuran at 2 to 4 kg a.i. per ha.

Several chemicals have been tested for their efficacy against root-knot nematodes infecting vegetable crops. Eldicarb, carbofuran, ethoprophos and fenamiphos each at 1 to 2 kg a. i. per ha was found effective in reducing the root-knot nematodes of tomato, cauliflower (Ahuja, 1983), brinjal (119 to 143 per cent) , okra (56 to 87 per cent), French bean (100 to 112 per cent), cowpea (40 to 50 per cent), peas (58 to 85 per cent) (Singh *et al.*, 1978; Rao and Singh, 1978; Singh and Parvatha Reddy, 1981b, 1982b;Parvatha Reddy, 1985) and Chillies (Handa and Mathur, 1981).

Aldicarb at 0. 5 kg a. i. per ha and carbofuran at 1. 5 kg a. i. per ha were effective in reducing *Tylenchorhynchus brassicae* and *T. dubius* population and in increasing yields of cabbage (Verma *et al.* , 1978).

Reddy and Seshadri (1972) reported that thionazin at 4 kg a. i. per ha proved effective against *R. reniformis* infecting tomato.

Aldicarb at 1 g per pot gave best control of *M. incognita* and increased shoot height, number of branches and leaf yield in geranium (Doraiswamy *et al.*, 1979).

Aldicarb, carbofuran and phorate each at 2 to 3 kg a. i. per ha significantly reduced *Helicotylenchus dihystera* and increased leaf yield of crossandra (Kumar and Nanjan, 1983). Phorate, carbofuran, fenamiphos and disulfotan at 1 g per plant effectively controlled *Pratylenchus delatteri* and increased crossandra flower yield (Vadivelu and Muthukrishnan, 1979).

Aldicarb and phorate each at 1. 5 kg a. i. per ha were effective in controlling *Aphelenchoides ritzemabosi* on chrysanthemum (Gill, 1981).

Gunasekaran (1979) reported the effectiveness of aldicarb at 1. 5 kg. *a.i.* per ha in controlling phytonematodes infecting patchouli.

For the control of *M. incognita* on tuberose, aldicarb, fensulfothion and phorate were effective and increased flower yield (Rao and Singh).

v) Foliar application

Foliar application of methyl parathion, chlorpyriphos and quinalphos each at 0.05 per cent were effective in the control of *Aphelenchoides ritzemabosi* infecting chrysanthemum and zinnia (Gill, 1981; Gill and Walia, 1980).

6.5. Host Resistance

Plant resistant is regarded as an extremely feasible method for controlling nematodes. It is cost-effective, economical and environmentally safe means of reducing losses from diseases caused by these pests. Use of resistant plants enables the grower to control the parasitic nematodes without increasing production costs associated with the purchase of expensive chemicals, applicators and in the numerous mechanical operations that go into the production of a crop. Resistance is especially valuable in controlling nematodes in low value crops and minor crops. Resistance in crops to nematodes is governed either by one or a few major genes, which generally convey high levels of resistance and are easy to incorporate into new cultivars. Genetic resistance has been exploited in developing nematode resistant crop cultivars and is important in crop improvement programme.

i) Screening of germplasm

A number of attempts have been made to identify the sources of resistance against parasitic nematodes among horticultural crops. Sources of resistance have been identified in citrus, grapevine, banana, potato, tomato (Fig. 7), brinjal, chilli, cowpea (Fig. 8), french bean, okra, cucurbits, carrot, muskmelon, water melon, ridgegourd, ash gourd and pumpkin (Table 4). Nematode-resistant varieties of a very few horticultural crops have been developed that too against one or two nematodes out of many that exists.

ii) Nature of resistance

Nature of resistance to *Meloidogyne incognita* in cowpea selections IC 9642-B and TVU 2430-P was investigated by Singh and Parvatha Reddy (1985a). The resistance was associated with reduced larval invasion, root galling, egg mass production and fecundity, delayed development of larvae to adult females coupled with high concentration of magnesium.

Table 4 : Nematode-resistant cultivars of horticultural crops

Crop	Nematode	Resistant cultivars	Reference
Banana	*Radopholus similis*	Kadali, Pedalimoongil, Kunnan, Ayiramkapoovan, Peykunnan, Pisang Seribu, Tongat, Vennettu Kunnan, Anaikomban	Charles *et al.*, 1983 Rajendran *et al.*, 1979
Citrus	*Tylenchulus semipenetrans*	Trifoliate orange, Swingle Citromelo	Parvatha Reddy, 1983
Grape-vine	*Meloidogyne incognita*	Black Champa, Dogridge, 1613, Saltcreek, Cardinal, Banquabad	Parvatha Reddy and Singh, 1984
	Rotylenchulus reniformis	Mukchilani, Dakshi, Joazbeli	Darekar & Patil, 1982
Tomato	*Meloidogyne* spp.	Nematoex, SL-120, NTR-1, SL-12, Patriot, VFN-8, VFN Bush, Piersol, Radiant, Nemared, Ronita, Anahu,	Rao & Tikoo, 1975 Patel *et al.*, 1979
		Bresch, Helani, Campbell-25, Punuui,	Anon, 1985
		Pelican, Hawaii-7746, Hawaii-7747	Jain *et al.*, 1983
	Rotylenchulus reniformis	Radiant, SL-120, Patriot, Bresh, Ronita, Piersol, F-38-E-2	Anon, 1983
Brinjal	*M. incognita*	Giant of Banaras, Black Beauty, Gola	Alam *et al.*, 1974
	M. javanica	Ghatikia White, Vijaya Hybrid, N(17)-4, Gacha Baigan, PBr-91-2, Dingra's Multiple Purple	Anon, 1983

Contd...

Chilli	*M. javanica*	579, CAP-63, Pusa Jwala	Jain *et al.*, 1983 Anon, 1983
	R. reniformis	Pusa Jwala	Anon, 1983
Potato	*M. incognita*	Kufri Dewa	Raj & Gill, 1983
	Globodera rostochiensis	Kufri Swarna	Anon, 1985b
Carrot	*M. incognita*	Black	Arya & Tiagi, 1982
	R. reniformis	Pusa Jwala	Anon, 1983
Cowpea	*M. incognita*	Barsati Mutant, Iron, New Selection, C-152, 82-1-B, IC 9642-B, TVU 24,O-P, Dixie Cream	Sharma & Sethi, 1976a; Darekar & Patil, 1981; Singh & Parvatha Reddy, 1982a
French Bean	*M. incognita*	Banat, Blue Lake Stringless, Bountiful Flat, Brown Beauty, Cambridge Countess, Gallaroy, Kenya-3, Pinto W5-114, Seafarer, Sutton's Masterpiece	Singh *et al.*, 1981
Okra	*Meloidogyne* spp.	Harichikni, Kanki Local Green, Vaishali Badher	Anon, 1985
Musk melon	*M. incognita*	Scarsol	Khan *et al.*, 1971
	M. javanica	S-445	Jain *et al.*, 1983
Water melon	*M. incognita*	Shahjanpuri	Khan *et al.*, 1971
Ridge gourd	*M. incognita*	Panipati, Meerut Special	Khan *et al.*, 1971
Ash gourd	*M. Incognita*	Jaipuri, Agra	Khan *et al.*, 1971
Pumpkin	*M. incognita*	Jaipuri, Dasna	Khan *et al.*, 1971

Some of the resistant vegetable crop varieties reported from Assam shown in Table 5

Table 5: Crop varieties identified as resistant to plant parasitic nematodes.

Crop	Nematode	Resistant varieties
Tomato	*Meloidogyne incognita*	PAU 1-6-1-4, PAU 7-3-1-4, PAU 8-2-1-25, NT-3, NT-8, NT-12, Punjab NR 7, PAU-6, PAU-8, PAU-9, PAU-10, PAU-12, PAU-13, PAU-14, PAU-15, Hisar N-1, Hisar N-2, HisarN-3, Kewala, Matabo, Marsol and Anahu
Pea	*Meloidogyne incognita*	HUDP 16 and LEP 227
Cowpea	*Meloidogyne incognita*	HFC 719, EC 43487, EC 4571 A, EC 955-B, EC 241049, NIC 15304, NIC 15305, NIC 15318, NIC 15321 , NIC 15322, V 38 (547-2), EC 390213, IC 249591, IC 253181, IC 253286, IC 253271, IC 257424, IC 259063, IC 259095, IC-259588 and L/B 46

Histological studies revealed that there was a direct correlation between the number of cork layers and resistance to *M. incognita* in cowpea. Cortical sclereids were noticed in resistant selections while they were absent in the susceptible cultivars (S-288 and S-488). There were more number of starch grains in the cortex of susceptible cultivars than in that of resistant selections (Singh *et al.,* 1984). Histopathological investigations revealed that there were less number of giant cells which were smaller in size with less number of nuclei in resistant selections as compared to susceptible cultivars of cowpea. Death of cells (hypersensitive reaction) around infecting *M. incognita* larvae occurred in the roots of resistant selections. Histochemical studies revealed that there were more of insoluble polysaceharides, proteins and nucleic acids in susceptible cultivars as compared to the resistant selections of cowpea.

Various compounds are responsible for imparting resistance to root-knot nematode infection. Higher concentrations of phenolic and orthodihydroxy phenolic compounds and more active IAA-oxidase and peroxidase systems have been observed in resistant tomato varieties Bajaj and Mahajan, 1977; Masood and Husain, 1976;Ganguly and Dasgupta, 1981).

ii) Inheritance of Resistance

Inheritance of root-knot nematode *(M. incognita)* resistance was investigated by Singh and Parvatha Reddy (1986) by using two resistant (IC 9642-B and TVU 2430-P) and two susceptible (S-88 and S-488) lines of cowpea. It was conditioned by dominant allele at a single locus. Raj and Patel (1978) investigated the nature of resistance in a range of cowpea lines with multiple disease resistance, including P426 and Iron. Resistance to *M. incognita* was governed by a single recessive (P 426, Iron) or dominant (Barsati Mutant) gene.

6.6. Biological Control

Biological control aims at increasing the parasites, predators and pathogens of nematodes in the soil, in order to increase the mortality of plant nematodes. This can be done by changing the environment by addition of organic amendments or by introducing other organisms directly. This should enhance the biotic component of environmental resistance. In practice, this has not yet been possible on a large scale.

Some outstanding work has been carried out on the role of organic amendments in controlling the parasitic nematodes on horticultural crops at Pantnagar (Singh and Sitaramaiah, 1973) and Aligarh (Khan *et al.,* 1973).

Incorporation of organic amandments into the soil (F.Y.M.) or compost at 20 to 25 tons per ha; oil cakes at 15 to 20 Qtls per ha) has reduced the citrus nematode, *Tylenchulus semipenetrans* population. Irrigation of citrus orchards with sewage water also reduced the citrus nematode population. The organic amendments reduce soil populations of *T. semipenetrans* by increasing the microbial activity which is unfavourable for survival of the nematode.

Singh and Sitaramaiah (1966) applied finely divided oil cakes to root-knot infested soil and noted a reduction in disease incidence in okra and tomato. Amendment of soil with oilcakes of neem, groundnut, mustard and castor was effective in reducing the population of *Tylenchorhynchus brassicae* around the roots of cabbage and cauliflower (Siddiqi *et al.,* 1976). The above cakes were also effective in reducing parasitic nematode population (*Hoplolaimus, Tylenchorhynchus, Meloidogyne* and *Helicotylenchus* species) and in increasing yields of tomato, potato, carrot and turnip. Mishra and Padhi (1985) reported that neem and karanj oil cakes at 2 ton per ha were most effective in reducing *Rotylenchulus reniformis* infecting french bean.

Other organic amendments such as saw dust (Sitaramaiah *et al.,* 1976) and rice-hull ash (Sen and Dasgupta, 1981) have given encouraging results. Best results in respect of root-knot reduction due to *M. javanica* and increase in yield of okra and tomato were obtained by amending the soil with saw dust at 25 Qtls. per ha 3 weeks before planting and then applying N through urea at 120 kg per ha (Singh and Sitaramaiah, 1971). Rice hull ash at 2. 5 ton per ha increased the yield of tomato by 133 to 317 per cent and reduced the root-knot incidence by 46 to 100 per cent (Sen and Dasgupta, 1981).

Paecilomyces lilacinus is an effective parasite of *Meloidogyne* eggs. Egg parasites are more dramatic in reducing the nematode population. Nematode eggs of the group Heteroderidae and those deposited in a gelatinous matrix are more vulnerable to attack by these organisms than are those of migratory parasites. Once in contact with the cysts or egg masses, the fungus grows rapidly and eventually parasitizes all the eggs that are in the early embryonic developmental stages.

M. incognita on Potato

P. lilacinus infected eggs of *Meloidogyne incognita* had destroyed the embryos within 5 days in laboratory tests. In green house tests, a reduction of up to 70 per cent root-galls and egg masses was observed in the plants from fungus-treated tubers. Since up to 60% of the eggs within egg masses were infected with the fungus, the size of subsequent nematode generations was reduced. In field tests, the fungus significantly reduced the tuber and root galling due to the root-knot infection. 86 per cent of the egg masses collected from the roots of fungus-treated plots were infected with *P. lilacinus* and over 54 per cent of the eggs within these egg masses were destroyed. Yield of marketable potatoes (non-infected tubers) was significantly higher in the fungus-treated plots. The fungus spread rapidly in the field which is a desirable characteristic of a successful and competitive biological control agent. The fungus treated plots had higher yield and there were significantly lower numbers of cysts. *Tylenchulus semipenetrans* on citrus:The fruit diameter of sweet orange cv. 'Valencia'from fungus treated trees was greater than that from trees treated with nematicides. The number of nematodes in roots of citrus and soil around citrus trees, inoculated with *P. lilacinus* was significantly lower than that in the non-treated or the nematicide treated treesThe fungus treated plots had higher yield of tomato also controlled *M. incognita* than did control plots and those treated with carbofuran.

Results on substrate study showed that the fungus grows abundantly in sterilized mashed potato, chopped water lily plants, ipil ipil *(Leucaena glauca)* leaves and corn grits. The fungus is by for the most promising and practicable biological control agent for management of root-knot and cyst nematodes.

Explanations for the possible mechanisms involved in the control of nematodes by the use of organic amendments are as follows:

1. The volatile fatty acids (formic, acetic, propionic and butyric acids), ammonia and hydrogen sulphide that are released during the microbial decomposition of organic amendments are directly toxic to the nematodes.
2. Microbivorous nematodes rapidly reproduce on the abundant bacteria and decomposition products.
3. Changes in the physical and chemical condition of the soil may alter the host-nematode relationship.
4. Organic amendments improve soil conditions for more rapid root growth This enhances the utilization of soil nutrients and marks the effects of nematode damage.
5. Amino acids released during the decomposition of organic amendments may be responsible for reducing the nematode population in soil (Parvatha Reddy *et al.,* 1975).

6.7. Integrated Management

To alleviate nematode problems, integrated nematode management (INM) procedure are implemented based on the principles of prevention, population reduction and tolerance. INM seeks to stabilize populations of target nematodes at acceptable levels resulting in favourable long term socioeconomic and environmental consequences. In practice, integration of different methods of nematode control is the rule than the exception. In most cases, practical control of a nematode disease involves integration of several diverse control measures.

i) Potato

The integrated control of the potato cyst nematode, *Globodera rostochiensis* can be achieved by a rational combination of crop rotation, resistant cultivars and nematicides (Table 6).

The control of potato rot nematode, *Ditylenchus destructor* was achieved by the combination of disease escape; hygiene, in the form of seed certification and crop rotation:

Table 6 : Integration of control methods for *Globodera rostochiensis*

S. No.	Control method (s)	Resulting population (% initial population)	Kill (%)	Population after growing and harvesting a susceptible cultivar, calculated at two assumed nematode multiplication rates* (% initial population)	
				30x	70x
1.	4 yrs. without potatoes	3	97	90	>100
2.	1 yr. with resistant potatoes	20	80	>100	>100
3.	Nematicide treatment (Fumigant)	25	75	>100	>100
4.	1 and 2	0.6	99.4	18	42
5.	1 and 3	0. 75	99.25	22.5	52.5
6.	2 and 3	5	95	>100	>100
7.	All three methods	0.15	99.85	4.5	10.5

* The observed maximum reproductive rate is thought to lie between 25 and 75 times but could be greater.

i. Healthy seed, planted late, harvested early and stored as cool and dry as possible.
ii. Proper rotation-potatoes not more frequently than once in 3 or 4 years and avoid other susceptible crops.
iii. Field hygiene-removal of old infested tubers and weed control.

i) Tomato

The application of aldicarb and carbofuran each at 1 kg a. i., per ha in combination with neem cake and urea each at 10 kg N per ha, at transplanting, produced maximum yield with lowest gall index (2.5) and nematode population, 90 days after planting (Routaray and Sahoo, 1985).

Combinations of cultural (deep ploughing to a depth of 20 cm) and chemical methods (nursery bed treatment with aldicarb at 0.4 g a. i. per m^2 and main field treatment with aldicarb at 1 kg a. i. per ha) gave effective control of root-knot nematodes in tomato which also registered maximum yield (Jain and Bhatti, 1985).

Red gram (*Cajanus cajan*) and *Crotalaria spectabilis* were used in conjunction with clean fallow and a nematicide (fensulfothion) for managing root-knot nematode populations in the production of tomato transplants.

ii) Brinjal

Deep ploughing (20 cm) together with spot application of aldicarb at 1.0 kg a. i. per ha increased growth of brinjal plants and significantly reduced the root-knot nematodes. Deep ploughing (20 cm) together with spot application of aldicarb at 1.0 kg a. i. per ha increased growth of brinjal plants and significantly reduced the root-knot nematodes.

Nursery treatment with metham sodium at 25 ml per sq. m. along with deep ploughing (20 cm) in the main field gave significant reduction in root-knot nematode population and better brinjal yields.

iii) Okra

Deep ploughing (20 cm) together with seed treatment (carbofuran at 3% W/W) was found to be effective for the control of root-knot nematodes in okra.

Deep ploughing (20 cm) followed by fallowing for one month, fallowing for one month after weeding, integration of aldicarb application at 0.8 kg a. i. per ha at sowing after eigher of the cultural practices or deep ploughing (20 cm) together with carbofuran or aldicarb seed treatment resulted in the control of the reniform nematode (*Rotylenchulus reniformis*) and better yield of okra (Lakshmanan and Sivakumar, 1981).

iv) Sweet potato

Control of root-knot damage to sweet potatoes involves integration of at least three methods:

i) The selection of seed roots that are either free of nematodes or freed of nematodes by hot water treatment.

ii) Planting of the seed roots into beds of sand or coarse textured soil that is either nematode-free or, if infested, is preplant fumigated.

iii) Transplanting of the clean 'slips' into nematode-free soil or pre-plant fumigated soil.

It is seen from the above discussion that the integration of the different methods offers effective control possibilities which are of great practical value in the suppression of harmful nematode population densities as well as to prevent the buildup of measurable populations of nematodes.

Integrated Nematode Management on different Horticultural Crops in Assam

In green house trials it was observed that inhibitory effects of marigold on the population of *Meloidogyne* species in tea soil was temporary (Anonymous, 1976). Fertilization of potted citrus plants with NPK fertilizers @90 g of each per 5.2 kg soil increased growth and reduced multipliation of *T. semipenetrans* (Sinha, 1984). Among different levels of NPK, the double doses of nitrogen (@ 300 g of nitrogen/plant) combined with normal doses of phosphorus and potash and double doses of NPK (240 g of P_2O_2 and 200 g K_2O/plant) were found to be most effective in increasing the growth of citrus plant and reduction in population of *T. semipenetrans* (Phukan, 1992, Anonymous, 1993b). Mustard oil cake, sawdust and cowdung alone and in combination of with NPK were found to be effective in reducing the citrus nematode population (Phukan, 1992, Anonymous, 1993b). Mustard oil cake @ 10 t/ha in combination with NPK was most effective in reducing the population of *T. semipenetrans* in both soil and roots of citrus plant. Three different plant extracts *viz., Datura* sp., *Tegetes patula,* and neem *(Azadiracta indica)* showed nematicidal activity at different concentrations. Datura 1 : 5 dilution gave best result in reducing the population of *T. semipenetrans* (Phukan, 1992).

Bio-control of *M. incognita* by using fungal parasite *Paecilomyces lilacinus* (Thom) Samson showed an effective suppressing nematode activities. Saikia & Roy (1995) found that the plant growth characters and yield of okra increased progressively with the increase of inoculum level of *P. lilacinus* alongwith *M. incognita* when the production of galls and egg masses were decreased significantly with the fungal doses of 4 and 7 g/kg soil. *P. lilacinus* inoculum @ 8 g/kg of soil was superior to improve the plant growth parameters of betelvine and reduced the *M. incognita* population. Inoculation of fungus prior to nematode inoculation proved to be more effective (Saikia, 1992, Anonymous, 1992). Borkakaty (1993) observed that inoculation of *P. lilacinus* @4 g/kg of soil alone or in combination with mustard oil cake @1 and 0. 5 t/ha increased plant growth with corresponding decrease in number of galls, egg masses and egg per egg mass of *M. incognita* on brinjal. The fungus when inoculated @ 2 g/kg soil in combination with mustard oil cake at the above dose was less effective. He further observed that mustard oil cake @ 2 tons/ha alone could not offset the attack of the nematode.

In a field experiment with integrated approaches, soil solarization with polythene covering for 2 to 3 weeks in the month of November was as effective as ploughing in respect of plant growth, yield of the tomato crop and population of *M. incognita* (Deka, 1994). Nursery treatment with carbofuran 3 G @ 3 g/m2, resistant variety alone or combination of nursery treatment and resistant variety gave significantly higher yield and reduced the infestation of *M. incognita* on tomato. Sarmah (1995) found that *Paecilomyes lilacinus* grown in rice grain when applied @ 4 g/kg soil alone or in combination with either carbofuran @ 1 kg a.i./ha or neem cake @ 0.5 t/ha in *Meloidogyne incognita* inoculated soil increased the growth of brinjal plants cv. JC-1 and yield significantly with corresponding decrease in production of galls, egg masses and also eggs per egg mass. He further observed the residual effect of *P. lilacinus* in reducing the activity of *M. incognita* even on the succeeding crop. Deka (1996) observed that leaf extract of neem, castor and keharaj each at N and N/2 concentrations and 0. 5 and 0.1% carbosulfan were effective as seed soaking treatment for reducing the penetration of *M. incognita* into pea roots, number of gall and egg masses and increasing the plant weight. In a separate trial incorporating leaves of 5 different plants in the soil Deka (1996) found that *Ricinus communis* was quite effective for the management of *M. incognita* on pea. Mili (1996) observed 33 -85. 50 per cent mortality of *T. semipenetrans* when exposed to 1:4 to 1:1 per cent concentrating of plant extracts of datura, tegetes and neem. Leaf extract of neem was found to be best followed by leaf extract of datura, stem and bud extracts of tegetes at 1 : 1 concentration in respect of mortality of the J_2 of *T. semipenetrans*.

A field experiment was conducted during 1995-96 and 1996-97 to study the efficacy of *P. lilacinus* in controlling the root-knot nematode on betelvine (Anonymous, 1996, 1997). Where lowest final nematode population was recorded in the treatment with three applications of the fungus. To develop an integrated pest management for root-knot nematode on betelvine an experiment was conducted for four consecutive years at Jorhat (Anonymous, 1996, 1997, 1998, 1999). All the treatments *(viz.*, oilcake + 3 doses of *P. lilacinus,* oil cake + carbofuran (0. 1% SD) +one dose of *P. lilacinus,* oilcake + carbofuran 0. 1 SD + 3 split dose of *P. lilacinus* and carbofuran @ 1.5 kg a.i./ha) were significantly superior to control. The best results were recorded in treatment with oilcake +carbofuran+ three application of *P. lilacinus.* Different neem products alone and their combinations as seed soaking and soil incorporation were tested against *M. incognita* on okra in field condition (Deka, 1997, Deka & Rahman, 1997a). The combination of two neem products were found superior the neem produts applied alone and the best result was obtained in the treatment with Achook @ 1 per cent+Neem cake @ 500 kg/h followed by Achook @ 1per cent +Nimin coated urea. It was further observed that Achook @ 1 per ent gave better result than other neem produts when applied alone. A microplot experiment was conducted to determine the efficacy of seed coating of okra seeds with Achook, neem cake, *Azadirachta indica* powdered leaves & *Melia azedarach* powdered leaves each at 25, 50 and 100 per cent (w/w) and also soil application of carbofuran 3G @ 3 kg a.i./ha as base for comparison the efficacy of these neem parts and neem products against *M. incognita* (Deka, 1997, Deka & Rahman, 1998). It was observed that among all the treatments carbofuran was found to be best followed by Achook 100 per cent in improving the plant growth with reduction in nematode population. Besides, neem cake @100 per cent, Achook @ 50 per cent and *A. indica* powdered leaf @100 per cent were also found effective. Plant extracts obtained from leaf stem and buds of both datura and tegetes and leaf extract of neem tested against *T. semipenetrans* showed that the leaf extract of neem at higher concentration (1:1) was found to be effective in respect to the mortality of nematode followed by leaf extract of datura, stem and bud extract of tagetes at same (1:1) concentration (Maznoor, 1997). *P. lilacinus* grown in mustard oil cake and rice bran when applied against *T. semipenetrans* on khasi mandarin at different levels (4&8 g/kg of soil) showed increased plant growth as well as reduced nematode population effectively with the increase in the level of *P. lilacinus* (Maznoor, 1997). However, the level of 4 g fungus culture grown in either media was found to be optimum in reducing the nematode population.

All the physical treatments *viz.*, sundrying of soil for 10 days, heat treatment of soil by hot pan for 3 hr at 60°C, stem sterilization of soil by 15 lb pressure per square ince for half an hour and the treatment with carbofuran 3 G @ 3 g per kg of soil tested for the management of *M. incognita* on tea nursery, were found effective in increasing the plant growth characters of tea nursery as well as reducing the number of galls, egg masses, eggs per eggmass and final nematode population as compared to that of untreated control (Neog, 1997, Neog & Bora, 2000b).

Split application of *P. lilacinus* mass cultured in mustard oil cake had been found to be effective against *M. incognita* resulting increased growth characters including yield of betelvine (Hazarika *et al.*, 1998). Increased frequency of split application of *P. lilacinus* caused increase in yield of betelvine leaves with corresponding decrease of soil nematode population and root-knot index. The experiments conducted in 1995, 1997 and 1998 on the integrated control of *M. incognita* on tomato treated nursery and neem cake gave a clear finding that treated nursery was significantly superior over untreated nursery (Anonymous, 1997, 1999). Treated nursery combined with neem cake @ 100 kg and 200 kg/ha and nitrogen 3.5 kg and 7.0 kg/ha significantly superior in increasing the yield and lowering the gall formation over untreated nursery.

A field trial was conducted to study the efficacy of *Trichoderma viride* in managing the population of *M. incognita* on betelvine (Anonymous, 1998, 1999). The results revealed that the treatment with *T. viride* could neither reduce the nematode population nor the nematode infestation. However, the build of nematode population in the treatments with three applications of *T. viride* was much lower in comparision to the growth of population in control. The yield of leaves recorded in this treatment was found to be at par with the yield recorded in the treatment with carbofuran. *P. lilacinus* and carbofuran individually and in combination were tested against *M. incognita* on betelvine under Baroj conditions (Hazarika , Dutta & Saikia, 1999b). *P. lilacinus* @ 1 ton (mass cultured on mustard oil cake) /h together with carbofuran 1 kg a.i./h was found to be the best in controlling nematode population, increasing the yield of betelvine.

The growth parameters of Khasi mandarin were found to be increased with the increase in the level of *P. lilacinus* applied against *T. semipenetrans* (Das, 1999). Besides, the nematode population in soil and roots were also found to be reduced with the increase in the level of the fungus culture. The level at 4 g fungus culture grown in either mustard oil cake or rice bean/kg soil was found to be optimum in reducing the nematode population. Amongst all the treatments *P. lilacinus* @ 8 g mass culture on mustard oil cake was found to be best. In a field trial Dutta (1999) found that seed treated with 1 per cent (w/w) carbosulfan 25 STD followed by poultry manure, neem cake and mustard oil cake @ 0. 5 and 1 ton/ha improved the plant growth and yield of carrot and reduced the number of galls and final nematode population of *M. incognita*. Dutta (1999) further observed that application of neem products as seed treatment alone or combination with urea significantly increased the plant growth parameters of carrot as well as reducing the formation of root galls on carrot. Urea + Achook 1 per cent was the effective treatment followed by Nimin coated urea + multineem 1 per ent and Nimin coated urea + Neemazol. Das (1999) tested the water extracts of six different plants *viz.*, biholongoni, mustard, ghandhali boan, citronella, pineapple, and posotia to determine their effect against citrus nematode *T. semipenetrans* for 24 hr of exposure of citrus seedlings and observed that citronella leaf extrat at 1:1 concentration was found to be the best in respect of mortality of *T. semipenetrans* followed by posotia at same concentration. In a field experiment the *P. lilacinus* was found to be effective against root-knot nematode in betelvine. The treatment with 4 g inoculum of *P. lilacinus* per kg soil was found to be best where increased plant growth and yield of betelvine and decreased root-knot infestations were recorded. The fungus and Furadon were found to be compatible (Anonymous, 2000).

P. lilacinus grown in rice bran applied @ 14 g per plant alone or in combination with either neem dry leaf @ 5 g per plant, mustard oil cake @ 5 g per plant and neem cake @ 5 g per plant in *T. semipenetrans* inoculated soil increased all growth parameters *Citrus jambhiri* with the corresponding decrease in nematode population (Deka, 2000, Deka, Sinha & Neog 2003a. Mustard oil cake @ 10 g and neem cake alone @ 10 g per plant increased growth parameters as well as reduced the nematode population. Neem dry leaf alone @ 10 g per plant did not have any significant effect on plant growth. However, neem dry leaf @ 5 g per plant in combination with *P. lilacinus* @ 14 g per plant significantly increased the plant growth characters and reduced the activity of the nematode. In a similar type of investigation Bora & Phukan (2000) also found that plants inoculated with VAM fungus,

G. fasciculatum prior to nematode inoculation had increased shoot length with reduction in number of galls and egg-masses in roots of brinjal.

All the four combination of nematodes and organic amendments viz. , mustard cake @ 5g + carbofuuran 3 @ 1. 5g/seedling; mustard cake @ 5g+phorate 10 G @ 1. 5g/seedling; neem cake @ 5g+carbofuran 3 G @ 1. 5g/seedling and neem cake @ 5 g +phorate 10 G @ 1. 5g/ seedling proved to be effective, increasing plant growth parameters of tea seedling and controlling the root-knot nematode. Neem cake @ 5g + carbofuran 3 G @ 1. 5 g per seedling was found to be the best among them (Kalita, 2000).

Deka (2000), Deka, Sinha&Neog (2003a) conducted an experiment to evaluate the efficacy of *P. lilacinus* alone and in combination with some botanicals (*P. lilacinus* alone @ 14 g/plant, mustard oil cake alone @ 10 g/plant, mustard oil cake along @ 5g +*P. lilacinus* @ 14 g/plant, neem dry leaf alone @ 10 g/plant, neem dry leaf @ 5g+*P. lilacinus* @ 14 g/plant; neem cake alone @ 10 g/plant, neem cake @ 5 g+*P. lilacinus* @ 14 g/plant) against *T. semipenetrans* on citrus *jambhiri*. Maximum plant growth and reduction in nematode population was recorded with neem cake 5 g + *P. lilacinus* 14 g/plant. All the treatments, except neem dry leaf @ 10 g/plant were significantly effective in increasing plant growth and reducing nematode population as compared to untreated control.

The interaction between *G. fasciculatum* with *M. incognita* on brinjal indicated that mycorrhizal plants had higher plant growth parameters than non-micorrhizal and nematode inoculated plants (Bora 2000 & Bora & Phukan, 2003). Plant with VAM at higher dose (600spores/ plant) 10days prior to nematode had significantly increased plant growth parameters of brinjal. The treatment with VAM @ 600 spores+ nematode after 10 days was found to be best in reducing galls, eggmasses and nematode population in soil.

Sinha & Neog (2003) studied the effect of different levels of NPK fertilizers against *T. semipenetrans* on the seedlings of Khasi mandarin. They observed that maximum fresh and dry weight of shoots and roots were recorded in the treatment with $N_2P_1K_1$ (300 g N, 120 g P_2O and 100g K_2O/plant) followed by the treatment with $N_1P_2K_1$ (150 g N, 240 g P_2O and 100 g K_2O/plant) which were significantly different from control. Maximum reduction of nematode population in soil and root was observed in the treatment with $N_2P_2K_2$ (300 g N, 240 g P_2O

and 200 g K_2O/plant) followed by $N_2P_1K_1$ which was significantly different from control. *P. lilanicus,* grown on mustard oilcake on rice bran and applied on Khasi mandarin, at different levels, against *T. semipenetrans* resulted in improved plant growth and reduction in nematode population (Maznoor, Sinha & Bora, 2003). Maximum reduction in nematode population was obtained at 8 g fungus culture per kg soil.

In an investigation on integrated management of *M. incognita* and *R. solanacearum* complex in brinjal, the treatment with summer ploughing + half dose of carbofuran + half dose of neemcake + half dose of streptocycline + full dose of *T. harzianum* (1st combination of treatment was found as the best treatment in respect of improvement of plant growth parameters of brinjal, followed by the treatment with summer ploughing + half dose of carbofuran + half dose of neemcake + half dose of streptocycline (2nd combination treatment treatment) (Hussain, 2004). The first combination of treatment was found most effective in reduction of galls, egg masses and final population of *M. incognita* in both soil and roots. The bacterial population and percent wilt incidence was found maximum in the treatment with the 2nd combination followed by first combination of treatment.

Studies on the efficacy of certain botanicals in combination with nematicides for the management of *M. incognita* on brinjal under field condition showed that all the treatments (neem cake@15 g/spot, vermicompost @ 1. 5t/ha, neem seed kernel 1. 5t/ha, sawdust @15 g/spot, carbofuran @ 2 kg a.i./ha as spot application) and in combination (neemcake @ 15 g/spot +half dose of carbofuran;vermicompost @ 1.5 t/ha + half dose of carbofuran; neem seed kernel @ 1. 5t/ha + half dose of carbofuran and sawdust@15g/spot+half dose of carbofuran) had significant effect on plant growth characters and yield of brinjal with corres-ponding decrease in nematode reproduction rate both in soil (17.15) and roots (38.93) over control (Saikia, 2005). Among all the treatments, the combined application of neem cake @ 15g/spot + half dose of carbofuran showed significant difference over control in respect of increasing plant growth parameters and yield of brinjal and reducing the nematode population. Saikia (2005) found that *G. fasciculatum* @ 600 spores/m_2, *P. penetrans* @ 1g/m_2, *Trichoderma harzianum* @ 2 g/kg of soil and carbofuran @ 2 kg a.i./ha alone and in combination increased the plant growth parameters and yield of brinjal with corresponding decrease in gall number and rate of nematode reproduction of *M. incognita* both in soil (12.15) and root (40.09) over control. Among all the treatments, combined application of *G. fasciculatum* +carbofuran @ 1 kg a.i./ha was found to be the best in respect of plant growth parameters of brinjal.

Das & Sinha (2006) studied the efficacy of *P. lilacinus*, carbofuran (as seed treatment) poultry manure and FYM alone and in combination for the management of *M. incognita* on okra and found that maximum increase in growth parameters viz., plant height, root length, fresh and dry weight of shoot and root including yield and decrease in number of galls, eggmases per root system and final nematode population in soil in the combined treatment with *P. lilacinus* @ 4 g/kg soil, carbosulfan 25 EC @ 0.2 per cent, poultry manaure @ 2.5 t/ha and FYM @ 2.5 t/ha. All the treatments were found to be significantly different from the control in increasing growth parameters and reducing in nematode population in soil. Studies on comparative efficacy of *Trichoderma harzianum* and *Pseudomonas fluorescens* against *M. incognita* and *R. solanacearum* complex on brinjal revealed that, combined application of *T. harzianum* and *P. fluorescens* was found to be the best in suppression of both nematode and bacteria along with increasing plant growth parameters (Baruah, 2006). The population of *M. incognita* and *R. solanacearum* were recorded minimum when both the antagonists viz., *T. harzianum* and *P. fluorescens* were applied together against the disease complex. Kalita and Bora (2006) found that the application of organic amendments and nematicides alone or in combination improved the plant growth characters of tea seedlings and reduction in number of galls and egg masses, eggs per egg mass of *M. incognita* as compared to control. However, maximum seedlings height length of roots and number of laterals per root system were obtained in both doses of carbofuran 3G 1. 5g and 3g/seedling) tested. The treatment with neemcake @ 5 g per seedling +carbofuran 3 G @ 1. 5 g per seedling was not only found to be best treatment.

Borah and Phukan (2007) studied the effect of *G. fasciculatum* on nutritional and biochemical changes in brinjal infested by *M. incognita* and found that plants inoculated with VAM alone had significantly increased plant growth parameters, nitrogen, crude protein, phosphorus, potassium, total sugar, total protein and total free amino acids content compared to uninoculated plants and nematode inoculated plants. The plants inoculated with VAM + nematode after 10 days (V+N) significantly decrease the number of galls, eggmasses, final nematode population in soil; increased nitrogen, crude protein, phosphorus, potassium, total sugar and total phenol, total amino acids in the roots of brinjal compared to nematode alone, nematode prior to VAM (N+V) and VAM + nematode simultaneously (VN).

References

Ahuja, S. (1983). Comparative efficacy of some nematicides against root-knot nematode, *Meloidogyne incognita* in tomato, okra, brinjal and cauliflower. *Third Nematol. Symp.*, H. P. Agri. Univ., Solan, P. 43.

Alam, M. M., A. M. Khan and S. K. Saxena, (1973). Efficacy of 'Vydate' oxamyl for the control of root-knot nematode, *Meloidogyne incognita* on eggplant and okra. *Indian J. Nematol.*, 3 : 148-52.

Alam, M. M., A. M. Khan and S. K. Saxena, (1974). Reaction of some cultivated varieties of eggplant, pepper and okra to the root-knot nematode, *Meloidogyne incognita. Indian J. Nematol.*, 4 : 64-8.

Alam, M. M., S. K. Saxena and A. M. Khan, (1977). Influence of interculture of marigold and margosa with some vegetable crops on plant growth and nematode population. *Acta Bot. Indica.*, 5 : 33-9.

Alam, M. M., Z. A. Siddiqi, A. M. Khan and S. K. Saxena, (1980). Effect of different cropping sequences on the population of plant parasitic nematode. *Indian J. Nematol.*, 10 : 35-9.

Anon. (1983). Proceedings of the Fifth Workshop of All India Co-ordinated Research Project on Nematode Pests of Crops and their control. Himachal Pradesh Agri. Univ., Solan.

Anon. (1985a). Proceedings of the Fourth Biennial Workshop of All India Co-ordinated Research Project on Nematode Pests of Crops and their control. Mohanlal Sukhadia Univ., Udaipur. 129 pp.

Anon. (1985b). Kufri Swarna, a high yielding late blight and cyst nematode resistant potato cultivar suitable for cultivation in Tamil Nadu Hills. Extn. Bull. No. 21 (E), Central Potato Res. Inst., Shimla, 4 pp.

Anonymous, (1993 b). Final Technial report, ICAR Adhoc Research Project on Citrus nematode, *Tylenchulus semipenetrans* and its management, AAU, Jorhat : 19 pp.

Anonymous, (1996). Annual Report (1995-96) AICRP on Betelvine. Compiled and edited by the Project Co-ordinator Betelvine, 2nd Oct., 1996 : p 93.

Anonymous, (1997a). Biennial report (1995-1997) AICRP on Plant Parasitic Nematodes & Integrated Approaches for their Control, Jorhat Centre, AAU, Jorhat. 38 pp.

Anonymous, (1998). Annual report (1997-98) AICRP on Betelvine. Compiled & edited by the Project Co-ordinator Betelvine. Sept. 15, 1998 : P 122.

Anonymous, (1999a). Bienial Report (1997-1999), AICRP on Plant Parasitic Nematodes & Integrated Approaches for their Control, Jorhat Centre, AAU, Jorhat. 40 pp.

Anonymous, (1999 b). Annual report (1998-99) AICRP on Betelvine. Compiled & edited by the Project Co-ordinator Betelvine. Sept. 19, 1999 : P 107.

Anonymous, (2000). Annual progress report (1999-2000) Ad-hoc project on Integrated management of rice root-knot nematode, *Meloidogyne graminicola* on up land rice in Assam. 18 pp with 3 plates.

Arya, M. and B. Tiagi, (1982). Reaction of some carrot cultivars to root-knot nematode, *Meloidogyne incognita. Indian J. Nematol.*, 12 : 397.

Atwal, A. S. and A. Mangar, (1969). Repellent action of root exudates of *Sesamum orientale* against the root-knot nematode, *Meloidogyne incognita* (Heteroderidae:Nematoda). *Indian J. Ent.*, 31 : 286.

Bajaj, K. L. and R. Mahajan, (1977). Phenolic constituents of tomato resistant and susceptible to *Meloidogyne incognita* (Kofoid & White) Chitwood. *Nematol. Mediterr.*, 5 : 329-33.

Bhagel, P. P. S and D. S. Bhatti, (1983a). Evaluation of pesticides for the control of phytonematodes on citrus. *Third Nematol. Symp.*, H. P. Agri. Univ., Solan, pp. 38-9.

Borah, Aparajita, (2000). Effect of Vesicular arbuscular mycorrhizal fungus, *Glomus fasciculatum* as bio-control agent for the management of root-knot nematode, *Meloidogyne incognita* on brinjal, Ph. D. thesis, AAU, Jorhat.

Borah, Aparajita and Phukan, P. N. (2000). Effect of VAM fungus, *Glomus fasciculatum* and root-knot nematode, *Meloidogyne incognita* on brinjal. *J. Agric. Sci. Soc.* N. E. India. 13 (2) : 212-214.

Borah, Aparajita and Phukan, P. N. (2003). Effect of interation of *Glomus fasciculatum* and *Meloidogyne incognita* on the growth of brinjal. *Ann. Pl. Protect. Sci.* 11 (2) : 352-354.

Borah, Aparajita and Phukan, P. N. (2007). Effect of *Glomus fasciculatum* in nautritional and biochemical changes in Brinjal infested by *Meloidogyne incognita. Indian J. Nematol.* 36 (2) : 189-191 (2006 published in 2007).

Borkakaty, D. (1993). Management of root-knot nematode, *Meloidogyne incognita* (Kofoid & White, 1919) Chitwood, 1949 on brinjal by integrated approach. M. Sc. (Ag) thesis, AAU, Jorhat.

Charles, J. S. K., T. S. Venkitesan and Y. Thomas, (1983). Field screening of banana against the burrowing nematode, *Radopholus simliis* (Cobb) Thorne. *Third Nematol. Symp.*, H. P. Agri. Univ., Solan, p. 28.

Darekar, K. S. and B. D. Patil, (1981). Reaction of some cowpea varieties to root-knot nematode. *Indian J. Nematol.*, 11 : 83-4.

Darekar, K. S. and B. D. Patil, (1982). Reaction of some grapevine varieties to root-knot and citrus nematode. *Indian J. Nematol.*, 12 : 390-2.

Das, Nabanita and Sinha, A. K. (2006a). Reaction of okra varieties to root-knot nematode, *Meloidogyne incognita. Indian J. Nematol.* 35 (2) : 214 (2005 published in 2006).

Das, Monoj (1999). Management of citrus nematode *(Tylenchulus semipenetrans,* Cobb, 1913) on Khasi mandarin *(Citrus reticulata* Blanco). M. Sc. (Ag.) thesis, AAU, Jorhat.

Deka, B. C. (1997). Crop loss assessment in okra due to *Meloidogyne inccognita* (Kofoid & White, 1919) Chitwod, 1949 & its management. M. Sc. (Ag) thesis, AAU, Jorhat.

Deka, B. C. & Rahman, M. F. (1997a). Management of *Meloidogyne incognita* in okra through seed soaking and soil application with some neem products in field condition. Paper presented in the Symposium on Major disease of crop plants in Eastern India & their management held on 10-11th December, 1997 at AAU, Jorhat.

Deka, B. C. & Rahman, M. F. (1998). Management of *Meloidogyne incognita* infesting okra by using neem parts and products. *J. Agric. Sci. Soc. N. E. India.* 11 (1): 124-126.

Deka, Ratul (2000). Pathogenicity and management of *Tylenchulus* semipenetrans (Cobb, 1913) on *Citrus jambhiri* M. Sc. (Ag) thesis, AAU, Jorhat.

Deka, Ratul, Sinha, A. K. and Neog, P. P. (2003a). Effects of *Paecilomyces lilacinus* and botanicals against *Tylenchulus semipenetrans* on *Citrus jambhiri. Indian J. Nematol.* 32 (2) : 230-232 (2002 published in 2003).

Deka, U. (1994). Management of *Meloidogyne incognita* (Kofoid & White, 1919) Chitwood, 1949 on tomato *(Lycopersica esculentum* Mill). M. Sc. (Ag) thesis, AAU, Jorhat.

Deka, M. (1996). Crop loss assessment of pea *(Pisum sativum* L.) due to *Meloidogyne incognita* (Kofoid & White, 1919) Chitwood, 1949 and its management. M. Sc. (Ag) thesis, AAU, Jorhat.

Dutta, A. (1999). Crop loss assessment in carrot due to root-knot nematode, *Meloidogyne incognita* and its management. M. Sc. (Ag) thesis, AAU, Jorhat.

Doraiswamy, K. Sundaram and K. Manjan, (1979). Root-knot nematode control in scented geranium *(Pelargonium graveolens* L. Hort). *Madras Agric. J.,* 66 : 205-6.

Ganguly, S. and D. R. Dasgupta, (1981). Prootein pattern in resistant and susceptible tomato varieties inoculated with the root-knot nematode, *Meloidogyne incognita. Indian J. Nematol.,* 11 : 180-8.

Gill, J. S. (1981). Fighting the foliar nematode of chrysanthemum. *Indian Hort.,* 25: 21-2.

Gill, J. S. and R. K. Walia, (1980). Control of chrysanthemum foliar nematode, *Aphelenchoides ritzemabosi* on *Zinnia elegans*. *Indian J. Nematol.*, 10 : 86-8.

Gunasekaran, C. R. (1979). Control of nematodes associated with patchouli, *Indian J. Nematol.*, 9 : 87.

Gupta, D. C. and M. C. Mukhopadhyaya, (1971). Effect of N, P and K on the root-knot nematode, *Meloidogyne javanica* (Treub) Chitwood. *Sci. Cult.*, 37: 246-7.

Handa, D. K. and B. N. Mathur, (1981). Control of root-knot of chillies with nematicides. *Nematol. Soc. India Symp.*, T.N. Agri. Univ., Coimbatore, p. 48.

Hazarika, K. Dutta, P. K. and Saikia, L. (1998). Field efficacy of *Paecilomyes lilacinus* against root-knot nematode in Betelvine. *J. Phytol. Res.* 11 : 171-173.

Hazarika, K., Dutta, P. K. and Saikia, L. (1999b). Management of root-knot nematode with *Paecilomyes lilacinus* and carbofuran in betelvine baroj. *Indian J. Plant Pathol.* 17 : 82-84.

Hussain, Zakir (2004). Integrated management of root-knot nematode *Meloidogyne incognita* and *Ralstonia solanacearum* complex in brinjal. *M. Sc. (Agri.) thesis, AAU, Jorhat.*

Jain, B. K. and D. S. Bhatti, (1978). Bare root treatment with systemics for controlling root-knot nematode in tomato transplants. *Indian J. Nematol.*, 8 : 19-24.

Jain, R. K. and D. S. Bhatti, (1983). Chemical control of *Meloidogyne javanica* in tomato through nursery bed treatment. *Third Nematol. Symp.*, H.P. Agri. Univ., Solan, pp. 37-8.

Jain, R. K. and D. S. Bhatti, (1985). Effect of summer ploughing alone and in combination with some other effective practices on the incidence of root-knot nematode *(Meloidogyne javanica)* on tomato (Culativar HS 101). *Indian J. Nematol.*, 15 : 262.

Jain, R. K. and D. S. Bhatti, R. D. Bhutani and Kalloo, (1983). Screening of germplasm of some vegetable crops for resistance to root-knot nematode, *Meloidogyne incognita*. *Indian J. Nematol.*, 13 : 212-5.

Kalita, Santanu, (2000). Pathogenicity and management of *Meloidogyne incognita* on tea seedlings. M. Sc. (Ag) thesis, AAU, Jorhat.

Kalita, S. and Bora, B. C. (2006). Effect of nematicides and organic amendments in the management of *Meloidogyne incognita* in Nursery tea. *Indian J. Nematol.* 36 (1) : 148-149.

Kandasamy, K. and C. V. Sivakumar, (1981). Effect of *Meloidogyne arenaria* on the germination of chilli *(Capsicum annum* Linn.) seeds. *Nematol. Soc. India Symp.*, T. N. Agri. Univ., Coimbatore, p. 16.

Khan, A. M., S. K. Saxena and M. W. Khan, (1971). Interaction of *Rhizoctonia solani* Kuhn and *Tylenchorhynchus brassicae* Siddiqi, 1961 in pre-emergence damping-off of cauliflower seedlings. *Indian J. Nematol.*, 1: 85-6.

Khan, A. M., S. K. Saxena and Z. A. Siddiqi, (1971a). Efficacy of *Tagete serecta* in reducing root infesting nematodes of tomato and okra. *Indian Phytopath.,* 24 : 166-9.

Khan, A. M., S. K. Saxena and Z. A. Siddiqi and R. S. Upadhyay, (1975). Control of nematodes by crop rotation. *Indian J. Nematol.,* 5 : 214-21

Khan, F. A. and A. M. Khan, (1973). Studies on the reniform nematode, *Rotylenchulus reniformis.* I. Host range and population changes. *Indian J. Nematol.* 3 : 24-30.

Khan, M. W., A. M. Khan and S. K. Saxena, (1973). Influence of certain oilcake amendments on nematodes and fungi in tomato yield. *Acta Bot. Indica,* 1: 49-54.

Khan, R. M. (1981). Studies on the effect of certain combinations of stunt, root-knot and reniform nematodes on cauliflower, brinjal and tomato. *Ph. D. thesis,* Aligarh Muslim Univ. Aligarh, India.

Kumar, S. and K. Nanjan (1983). Control of spiral nematodes in patchouli. *Fungicide* and *Nematicide Tests,* 39 : 93.

Lakshmanan, P. L. and C. V. Sivakumar, (1981). Control of *Rotylenchulus reniformis* in Bhindi by integrated method. *Nematol. Soc. India Symp.,* T. N. Agri. Univ., Coimbatore, P. 54.

Masood, A. and S. I. Hussain, (1976). Phednolic and ortho-dihydroxy phenolic changes and their role in the resistance and susceptibility of three tomato varieties to *Meloidogyne incognita. Indian J. Nematol.,* 6 : 86-93.

Maznoor, Sultan (1997). Management of citrus nematode *(Tylenchulus semipenetrans* Cobb. 1913) on khasi mandarin *(Citrus reticulate* Blanco). M. Sc. (Agri.) thesis, AAU, Jorhat.

Maznoor, S., Sinha, A. K. and Bora, B. C. (2003). Management of citrus nematode, *Tylenchulus* semipenetrans on Khasi mandarin, by *Paecilomyces lilacinus. Indian J. Nematol.* 32 (2) :153-155 (2002 published in 2003).

Misra, R. P. and N. N. Padhi, (1985). Relative efficacy of some nematicides and organic materials in the control of reniform nematode on French bean. *Indian J. Nematol.*, 15 : 270-1.

Muthaiyan, M. C., A. K. Purkar, R. P. Mishra, P. K. Sanyal, L. Patel and D. K. Dutta, (1983). Interception of nematodes from plants and plant materials at Plant Quarantine and Fumigation Station, Calcutta. *Pl. Prot. Bull.,* 35:

Muthaiyan, M. C. and K. Sridevi, (1982). Interception of plant parasitic nematode. *Indian J. Pl. Prot.,* 10 : 95.

Neog, P. P. (1997). Management of *Meloidogyne incognita* in Tea nursery. M. Sc. (Ag.) thesis, AAU, Jorhat.

Nirula, K. K. and K. K. Bassi, (1965). Thermotherapy for root-knot nematode *(Meloidogyne incognita)* in potato tubers. *Indian Potato J.,* 7 : 9-11.

Parvatha Reddy, P. (1983). Reaction of selected citrus root-stocks to the citrus nematode. *Proc. Intern. Citrus Symp.,* Bangalore, pp. 379-81.

Parvatha Reddy, P. (1984). Efficacy of seed treatment with three nematicides for the control of *Meloidogyne incognita* infecting cowpea, French bean and peas. *Indian J. Nematol.,* 14 : 39-40.

Parvatha Reddy, P. (1985a). Chemical control of root-knot nematodes infecting peas. *Indian J. Nematol.,* 15 : 120-1.

Parvatha Reddy, P. , H. C. Govindu and K. G. H. Setty, (1975). Further studies on the action of DL-methionine on *Meloidogyne incognita. Indian J. Nematol.* , 5: 200-6.

Parvatha Reddy, P. , and D. B. Singh, (1970). Control of root-knot nematode *Meloidogyne incognita* on tomato by chemical bare-root dips. *Indian J. Nematol.,* 9: 40-3.

Parvatha Reddy, P. , and D. B. Singh, (1984). Reaction of selected cultivars/ rootstocks of grapevine to *Meloidogyne incognita. Indian J. Nematol.,* 14: 73-4.

Patel, G. J., H. M. Shah and D. J. Patel, (1979). Reaction of some tomato cultivars to root-knot nematode. *Indian J. Nematol.,* 9 : 60-1.

Phukan, B. C. (1992). Management of citrus nematode, *Tylenchulus semipenetrans.* M. Sc. (Ag) thesis, AAU, Jorhat.

Prasad, K. S. K. and K. Krishnapa, (1981). Effect of bare root-dip treatments with pesticides on *Rotylenchulus reniformis* effecting brinjal. *Indian J. Nematol.,* 11: 154-8.

Prasad, K. S. K., D. Raj and R. K. Sharma, (1983). Effect of planting time on *Meloidogyne incognita* infesting potato at Simla Hills, *Third Nematol. Symp.,* H. P. Agric. Univ., Solan, pp. 5-6.

Raj, D. and J. S. Gill, (1983). Reaction of some potato varieties to root-knot nematode *Meloidogyne incognita. Third Nematol. Symp.,* H. P. Agri. Univ., Solan, pp. 27-8.

Raj, S. and P. N. Patel (1978). Studies on resistance in crops to bacterial diseases in India. Inheritance of multiple disease resistance in cowpea. *Indian Phytopath.,* 31 : 294-9.

Rajendran, G. and T. G. Naganathan, (1981). Evaluation of systemic chemicals for the control of *Rotylenchulus reniformis* on papaya. *Indian J. Nematol.,* 11: 130.

Rajendran, G., T. G. Naganathan and V. Sivagami, (1979). Studies on banana nematodes. *Indian J. Nematol.,* 9 : 54.

Rajendran, G., R. S. Vadivelu and T. S. Muthukrishnan, (1976). Pathogenicity of *Meloidogyne incognita* on crossandra. *Indian J. Nematol.,* 6 : 115-6.

Rao, V. R. and D. B. Singh, (1978). Chemical control of root-knot nematodes in okra with the aid of nematicides. *Indian J. Hort.,* 35 : 151-4.

Rao, V. R. and D. B. Singh, ( ). Control of *Meloidogyne incognita* on tuberose with nematicides. *Indian J. Hort.*

Rao, V. R. and S. K. Tikoo, (1975). Resistance of tomato cultivars to the root-knot nematode, *Meloidogyne incognita* (Kofoid and White, 1919) Chitwood, 1949. *Curr. Sci.,* 44 : 282-3.

Reddy, D. D. R. and A. R. Seshadri, (1972). Studies on some systemic nematicides. II. Further studies on the action of thionazin and aldicarb on *Meloidogyne incognita* and *Rotylenchulus reniformis. Indian J. Nematol.,* 2 : 182-90.

Reddy, D. D. R. and A. R. Seshadri, (1975). Estimation of root-knot nematode infestation from tomato seedlings by chemical bare-root dips or soil application. *Indian J. Nematol.,* 5 : 170-5.

Renjhen, P. L. (1973). History of Plant Quarantine and its Functioning in India. Govt. of India, Minist. of Agri., Directorate of Pl. Prot. Quarantine and Storage, Faridabad.

Routarary, B. N. and H. Sahoo, (1985). Integrated control of root-knot nematode, *Meloidogyne incognita,* with neem cake and granular nematicides on tomato. *Indian J. Nematol.,* 15 : 261.

Saikia, B. (1992). Pathogenicity, crop loss assessment and biological control of *Meloidogyne incognita* (Kofoid & White, 1919) Chitwood, 1949 on betelvine *(Pipper betle* L). M. Sc. (Ag) thesis, AAU, Jorhat.

Saikia, Jayanta, (2005). Efficacy of certain bio resources against *Meloidogyne* incognita on brinjal. M. Sc. (Ag) thesis, AAU, Jorhat.

Saikia, M. K. and Roy, A. K. (1995). Efficacy of *Paecilomyes lilainus* on the reduction of attack of *Meloidogyne incognita* on okra. *Indian J. Nematol.* 24 (2) : 163-167.

Sarmah, B. (1995). Integrated management of root-knot nematode, *Meloidogyne incognita* on brinjal. M. Sc. (Ag) thesis, AAU, Jorhat.

Saxena, S. K. , M. M. Alan and A. M. Khan, (1974). Chematherapeutic control of nematodes with Vydate oxamyl, VC-13 and Dazomat on Chilli plants. *Indian J. Nematol.,* 4 : 235-8.

Sen, K. and M. K. Dasgupta, (1981). Effect of rice hull ash on root-knot nematodes to tomato. *Trop. Pest Management,* 27 : 518-9.

Sethi, C. L., R. P. Nath, V. K. Mathur and S. Ahuja, (91972). Interceptions of plant parasitic nematodes from imported seed/plant materials. *Indian J. Nematol.,* 2 : 89-93.

Sharma, N. K. and C. L. Sethi, (1976a). Reaction of certain cowpea varieties to *Meloidogyne incognita* and *Heterodera cajani*. *Indian J. Nematol.*, 6: 99-102.

Shetty, K. D. and D. D. R. Reddy, (1985). Control of root-knot nematodes on *Solanum khasianum* by chemical bare root dips. *Indian J. Nematol*, 15:227-9.

Siddiqi, Z. A. , A. M. Khan and S. K. Saxena, (1973). Studies on *Tylenchorhynchus brassicae* II- Effect of temperature and moisture on multiplication of nematode. *Indian Phytopath.*, 26 : 139-47.

Siddiqi, Z. A. , M. W. Khan and A. M. Khan, (1976). Nematode population and yield of certain vegetables as influenced by oil-cake amendments. *Indian J. Nematol.*, 6 : 179-82.

Singh, D. B. and P. Parvatha Reddy, (1981b). Note on chemical control of root-knot nematodes infesting French bean. *Indian J. Agric. Sci.* , 51:534-5.

Singh, D. B. and P. Parvatha Reddy, (1982b). Chemical control of *Meloidogyne incognita* infecting cowpea. *Indian J. Nematol.*, 12 : 196-7.

Singh, D. B. and P. Parvatha Reddy, (1986). Inheritance of root-knot nematode resistance in cowpea. *Indian J. Nematol.*, 16 : 284-5.

Singh, D. B. , P. Parvatha Reddy, V. R. Rao and R. Rajendran, (1981). Reaction of some varieties and selections of French bean to *Meloidogyne incognita. Indian J. Nematol.*, 11 : 81-3.

Singh, D. B. , V. R. Rao and P. Parvatha Reddy, (1978). Evaluation of nematicides for the control of root-knot nematodes on brinjal. *Indian J. Nematol.*, 8:64-6.

Singh, D. B., P. P. Reddy and Shyamsundar Joshi, (1984). Histological, histopathological and histochemical investigations on root-knot nematode resistant and susceptible lines of cowpea. *Nematologica Mediterranea.* 12 : 213-9.

Singh, R. S. and K. Sitaramaiah, (1966). Incidence of root-knot of okra and tomato in oil-cake amended soil., *Pl. Dis. Reptr.*, 50 : 668-72.

Singh, R. S. and K. Sitaramaiah, (1971). Control of root-knot through organic and inorganic amendments of soil: Effect of sawdust and inorganic nitrogen. *Indian J Nematol.*, 1 : 80-4.

Singh, R. S. and K. Sitaramaiah, (1973). Control of Plant parasitic Nematodes with Organic Amendments of Soil. Res. Bull., G. B. Pant Univ. Agric. Technol., Pantnagar, 289 pp.

Singh, S. P., V. Pant, A. M. Khan and S. K. Saxena, (1984). Efficacy of certain nematicides and oil cakes for the control of stunt nematode. *Tylenchorhynchus brassicae* and parasitic fungi in the rhizosphere of cauliflower. *Indian Phytopath.*, 37 : 648-51.

Sinha, A. K., and Neog, P. P. (2003a). Effect of organic amendments for the management of *Tylenchulus* semipenetrans on Khasi mandarin. *Indian J. Nemaatol.* 32 (2) : 232-233. (2002 Published in 2003).

Sitaramaiah, K. , R. S. Singh and R. G. Choudhury, (1976). Control of root-knot (*Meloidogyne javanica*) of vegetable crops. *Ann. Rept.* of 1974-75, G. B. Pant Univ. Agric. Technol., Pantnagar, pp. 113-4.

Sivakumar, C. V. and M. Meerzainuddin, (1974). Influence of N, P and K on the reniform nematode *Rotylenchulus reniformis* and its effect on the yield of okra. *Indian J. Nematol.*, 4 : 243-4.

Sivakumar, C. V., S. Palanisamy and M. Balasubramanian, (1977). Control of *Rotylenchulus reniformis* in tomato nursery. *Indian J. Nematol.*, 7 : 78-80.

Sivakumar, C. V., S. Palanisamy and T. G. Naganathan , (1976). Persistance of nematicidal activity in seed treatment of okra with carbofuran and aldicarb sulfone. *Indian J. Nematol.*, 6 : 106-8.

Thakur, N. A. and C. C. Patel, (1985). Control of root-knot nematode, *Meloidogyne javanica* on tomato by chemical bare-root dips. *Indian J. Nematol*, 15 : 110-1.

Vadivelu, S. and T. S. Muthukrishnan, (1979). Control of root lesion nematode, *Pratylenchus delatteri* on crossandra. *Indian J. Nematol.*, 9: 86.

Varma, M. K. , H. C. Sharma and V. N. Pathak, (1978). Evaluation of systemic granular nematicides against *Tylenchorhynchus* spp. Parasitic on cabbage. *Indian J. Nematol.*, 8 : 59-60.

Venkitesan, T. S. and J. S. K. Charles, (1983). Sucker dip treatment for banana nematodes control. *Third Nematol. Symp.* , H. P. Agri. Univ., Solan, P. 45.

Phyto-nematodes in Crops: Their Identification, Treatment and Management, 2011
G.L. Sharma (editor), pp. 115-132
New India Publishing Agency, New Delhi (India)
E-mail : info@nipabooks.com; Website : www.nipabooks.com

Chapter 5

Maize Cyst Nematode, *Heterodera Zeae* - A Potential Threat to Maize Production

A. N. SRIVASTAVA, GAUTAM CHAWLA AND R. K. JAISWAL

Introduction

Maize *(Zea mays* Linn) is the most widely distributed crop of the world. It adapts easily to a wide range of production environments and is grown in tropical, subtropical and temperate regions of the world. It holds a unique position in world agriculture ranking third after rice and wheat in terms of area as well as production. In India too, it is the third most important cereal crop and is cultivated in 8. 17 m ha (2007-08) area. It contributes nearly 8% to the national food basket and more than Rs 100 billion to agricultural gross domestic products with an annual production of 19. 73 million tons. It is used as a source of human food (24%) animal feed (11%), poultry feed (52%), starch (11%), brewery and seed (1% each). In India it is largely grown in Northern states where it serves as an important staple food for an economically vulnerable population of submountainarean and hilly regions. Poultry sector is the leading consumer of maize at per capita broiler meat consumption of nearly of 2 kg. National Center for Agricultural and Economics and Policy Research (NCAP) has estimated a total demand of 22.73 mt by the end of 11th five year plan. Maize productivity in India (2.4 t /ha) is still far less in comparison to the developed countries (Average world yield >4.2 t /ha).

The yield production of maize crop in India is greatly affected by several biotic and abiotic factors;prominent among them are disease, insect pests and nematodes. A report from CIMMYT suggests that in India the biotic production constraints reduce maize production levels by more than 50% (Joshi *et al.*, 2005)A large number of plant parasitic nematode species have been found to be associated with this crop in different parts of the world. However, the maize cyst nematode *(Heterodera zeae*), lesion nematodes *(Pratylenchus* spp.), sorghum cyst nematode *(Heterodera sorghi*), root-knot nematodes *(Meloidogyne* spp.) and stunt nematode *(Tylenchorhynchus vulgaris*) are considered major nematode pests and important limiting factors in causing unthrifty growth of the crop, thus resulting in considerable economic loss in yield (Srivastava, 1980; Srivastava and Sethi, 1984b; Srivastava *et al.*, 1995). However, the maize cyst nematode, *H. zeae* is considered to be the most economically important nematode pathogens and has been recognized as a key nematode pest of maize in India (Srivastava and Chawla, 2005).

The maize cyst nematode, *H. zeae* was first detected and reported from India by Koshy, Swarup and Sethi in 1970 from a village Chapli, Udaipur district of Rajasthan state parasiting maize *(Zea mays)*. Its presence was also recorded from Pusa (Bihar)and Ludhiana (Punjab) in the same year besides its type locality. Eventhough, the presence of this nematode outside India was first observed in 1978 in the samples from Nile vally (Egypt), the widespread occurrence of the nematode from other parts of Egypt was reported only in 1981. Subsequently the nematode was also reported from Pakistan and USA in the same year. With the introduction of this nematode pest in the American continent particularly from the eastern shore region, kent county, Maryland, the importance of this nematode in limiting maize/corn production in agricultural economy was more realized. Earlier Mc Gregor (1978) indicated that introduction to USA of *H. zeae* ranking 15th in the list of 49 top ranking exotic pathogens with an introduction probability of 1-15 percent would amount to an expected economic impact of 320 million dollars.

H. zeae is characterized by light brown small lemon shaped cysts with thin cuticle and prominent posterior protuberance. The cone top is ambifenestrate with vulval slit and wide vulval bridge large, prominent four fingerlike 'cross bullae' as also presence of other fingerlike bullae around fenestration is the important and most reliable features of this nematode (Fig. 1). Basin, a band of thick cuticle around the fenestra, is wide and quite prominent. The vulval bridge is long and slender. It has a medium sized underbridge, which is equal to the length of vulval bridge. The eggs, retained in the body of the cyst, are easily visible from the outside due to thin body cuticle of the cyst. *H. zeae* also secrete a small egg sac that contain a few or one third of the eggs produced.

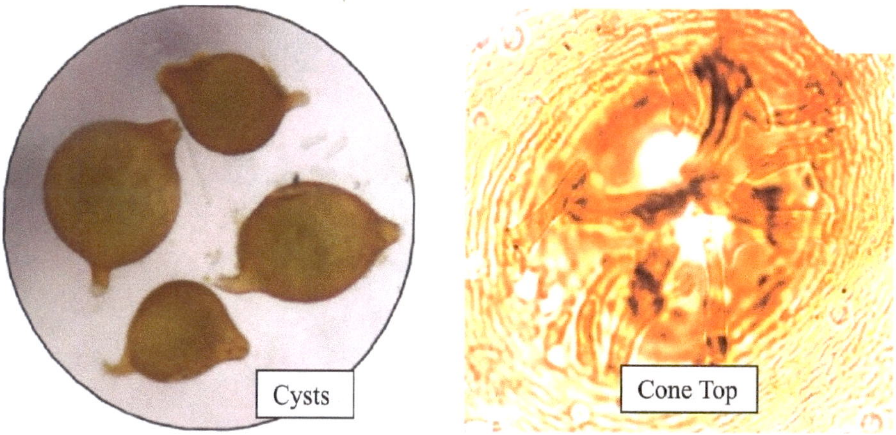

Fig. 1 : Newly formed cysts and cone top of Heterodera zeae

Distribution

H. zeae first reported by Koshy *et al.* (1970) from village Chapli, Udaipur (Rajasthan) is now widely distributed in major maize growing regions of northern, western, central and eastern parts of India. In India, maize growing areas of Himanchal Pradesh, Punjab, Haryana, Delhi, Gujarat, Rajasthan, Uttarakhand, Uttar Pradesh, Madhya Pradesh, Bihar and Maharastra. (Koshy and Swarup, 1971; Srivastava and Swarup, 1975; Darekar *et al.*, 1981; Sharma and swarup, 1984; Sharma *et al.*, 1984; Bajaj and Bhatti, 1984; Makadia *et al.*, 1988) are infested with maize cyst nematode (Fig. 2). The presence of this nematode at high altitudes in the states of Jammu & Kashmir (Srivastava and Kaushal, 1986; Khan *et al.*, 1989) and Himachal Pradesh (Srivastava and Kaushal, 1991) has also been reported. Besides, its widespread occurrence in India, the nematode has also been reported to occur in Egypt (Aboul-Eid and Khorab, 1981), Pakistan (Maqbool, 1981), USA (Sardenelli *et al.*, 1981), Thailand

(Chinnarsi *et al.*, 1995), Nepal (Sharma *et al.*, 2001), Portugal, Europe (Correia *et al.*, 2005).

Fig. 2 : Distribution of Maize Cyst Nematode, *Heterodera zeae* in India

Host Status

The maize cyst nematode, *H. zeae* is confined to the plants of Graminae family. Besides maize, important *kharif* and *rabi* cereals and millets namely, rice *(Oryza sativa)*, italian or foxtail millet *(Setaria italica)*, barnyard millet *(Echinochloa colona)*, fingermillet *(Eleucine coracona)*, barley *(Hordeum vulgare)*, wheat *(Triticum aestivum)*, rye *(Secale cereale)*, *Pennisetum americanum, Coix lachryma, Zea mexicana* and *Vetiveria zizanoides* are the hosts for *H. zeae* (Srivastava and Swarup, 1975, Lal and Mathur, 1982, Sharma and Swarup, 1984)). *H. zeae* has also been reported to occur in gram and garlic fields and citrus, pear and and mango orchards in Pakistan (Maqbool, 1981; Nasira and Shahina 2007) and fig *(Ficus* sp.) trees from Portugal (Correia and Abrantes, 2005). However their host status for *H. zeae* is still to be confirmed.

The common grasses and weed plants viz. *Archyranthes aspera, Cyperus rotundus, Digera muricata, Eclipta prostrata, Parthenium hysterophorus, Trianthema portulacastrum* and *Urochloa panicoidea* harbor the cyst populations of *H. zeae* in the fields, though in small numbers (Parihar *et al.*, 1991).

Biology

The maize cyst nematode, *H. zeae* reproduces parthenogenetically even though it is known to be a bisexual species. Males are generally very sporadic and rare. Temperature plays an important role in the biology of *H. zeae* (Srivastava and Sethi, 1985a). A temperature of 25°C is most favorable for juvenile emergence (91% emergence)from the cysts (Srivastava and Sethi, 1985b) while 25-30°C for embryonic development (Shahina and Maqbool, 1989). The emergence of second stage larvae starts even before the females turn yellow or light brown and extends over several weeks and no maturity period is required for hatching of eggs.

The development and reproduction of *H. zeae* is also greatly influenced by soil temperature. Hashmi *et al.*, 1993 also reported that reproduction of *H. zeae* increased directly with increase in temperature from 24 to 36°C. The duration of life cycle of the nematode is very short depending upon the maize cultivars and soil temperature. In India, the nematode *(H. zeae)* completes its one life cycle (J2 to J2) in 15 to 17 days on maize cultivars Shakti and Hi-starch respectively at a temperature range of 27 and 38°C. It has been speculated that *H. zeae* may complete 5-6 generations during one maize cropping season under Indian conditions (Srivastava and Sethi, 1985b).

Soil type is an important ecological factor which influences the biology including reproduction of the nematode (Srivastava and Sethi, 1984a). Moderately light soil (sandy loam- sand mixture in the ratio 2:3) is preferred the most for its reproduction and any addition of clay proportion to soil mixture results in a significant decline in the cyst production of *H. zeae*.

Population Dynamics and Survival

The population density of the nematode fluctuates greatly in maize- based cropping systems and attains high peak populations in October/November months coinciding with the maturity of the maize crop (Fig. 3) suggesting the host specific nature of the nematode (Srivastava and Sethi, 1986). The occurrence of peak population densities of the nematode in different years varied due to the change in the sowing time of the crop, prevailing soil temperature and moisture and cropping duration. The presence of very little populations of J2 surviving in the soil coincides with the crop growth period being at non- detectable level during December to June. The general course of the curve in both the years was similar to normal high and low population peaks coinciding with growth of the host crop. Cysts with infective juveniles apparently persist in fallow soil for about two years.

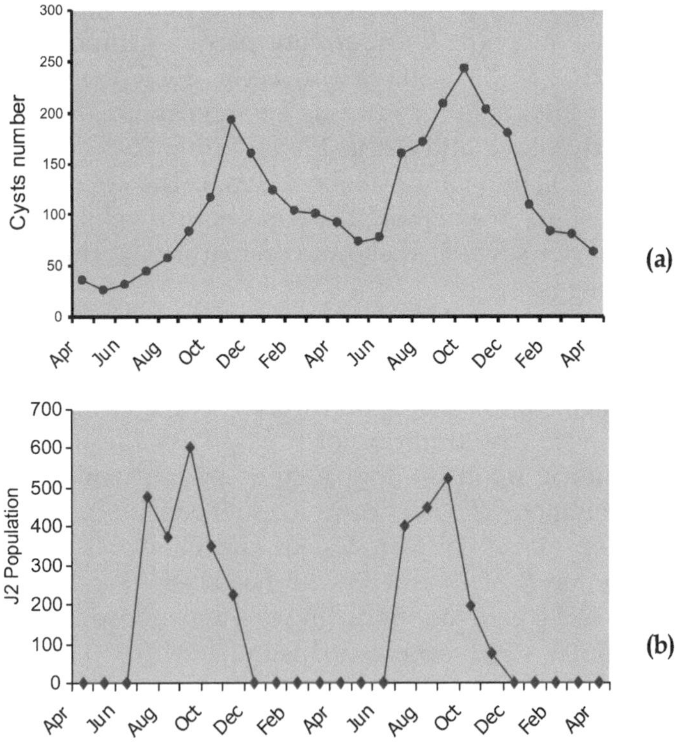

Fig. 3 : Population dynamics of **a)** Cysts and **b)** juveniles (J_2) in soil under maize based cropping system.

In Maryland, USA, the eggs and/or second-stage juveniles inside cysts of *H. zeae* have been reported to survive during winter in the field with no detectable mortality. It has been observed that the entire contents of the cyst do not empty out even in one year (Krusberg and Sardanelli, 1989). So, this survival strategy of the nematode, coupled with the number of generations (5-6) during the crop season gives an indication of the damaging potential of this nematode. Being multivoltine, it may quickly attain very high population densities.

Economic Losses

The pathogenic potential of *H. zeae* has been established and demonstrated on maize in India by Srivasatava, 1980 and on corn in USA by Krusberg, 1988. Srivastava and Sethi (1984b) showed that plant growth reductions are directly correlated with initial nematode population density. The damage threshold level was found to be 2-4 J_2 per g soil. Hashmi *et al.* (1993b) found consistent reductions in maize plant weights in the presence of *H. zeae* at optimum temperature regimes for the nematode.

In India, *H. zeae* could cause an economic losses of about 21 % in maize cv. Deccan-103 (Delhi) at 4 J_2/cm^3 soil, 29% at 6 J_2/cm^3 soil in maize cv. Ganga Safed - 2 (Udaipur, Rajasthan) and 17% at 4 J_2/cm^3 soil in Local maize cv. Pusa (Bihar) (Fig. 4) in *kharif* seasons (Anon, 1987, 1989; Srivastava and Chawla, 2005). Krusberg *et al.* (1997) reported that maize growth (dry weight) and yield were suppressed by 13 to 73% in the presence of *H. zeae.* These plant responses to *H. zeae* were greater in coarse-textured than in fine-textured soil. Fertilizer amendments did not alleviate suppression of plant growth by *H. zeae.* The nematode caused more damage to maize plants in hot, dry rather than cool, wet seasons.

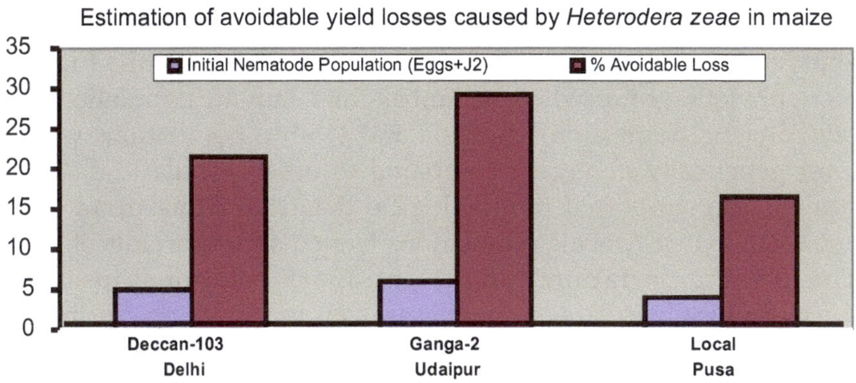

Fig. 4 : Economic Yield losses on Maize due to infection by Heterodera zeae

The infected maize crop with *H. zeae* exhibits poor, patchy and unthrifty growth with stunting and pale yellow and narrow foliage. Plant growth reduction is directly correlated with the initial population density of nematode. The infected root appears bushy and poorly developed. The diseased plants tassel earlier and bear smaller cobs with relatively fewer grains. The presence of cysts on the rootsurface is the most important characteristic used in the identification of *H. zeae*. In addition to causing direct damage, *H. zeae* also synergises other pathogens. Wilt pathogen, *Cephalosporium maydis* caused severe damage to maize crop in presence of *H. zeae* (Singh and Sirdhana, 1988). Similarly synergistic interaction between *H. zeae* and *Fusarium pallidorosem* occurred leading to significantly greater damage to maize roots (Anon., 1993).

Biotypes/Races

Physiological variations have been recorded in *H. zeae* populations obtained from different geographical areas or within a region of the country. Srivastava and Sethi (1984b) compared 3 geographically isolated populations of *H. zeae* collected from Pusa (Bihar), Delhi and Udaipur (Rajastan) for their virulence and ability to reproduce and multiply on different cultivars of maize and found that they varied significantly in their ability to reproduce and multiply on these hosts. They also found the population from Pusa Bihar to be highly virulent.

Samstipur population earlier reported to be more virulent strain of *H. zeae* (Srivastava and Sethi, 1984b) differed from other populations of *H. zeae* which were more variable and distinct compared to Delhi, Himachal Pradesh, Punjab and Haryana populations by PCR-RFLP analysis of internal transcribed spacers, ITS1 and ITS2 (Rao *et al.* 2008). Gavas *et al.* (2009a) found that out of six populations of *H. zeae* analyzed Samastipur, Wellington, Udaipur and Kulu formed a monophyletic group whereas Ambala and Jammu populations were polyphyletic. Samastipur, Ambala and Jammu populations were the most genetically divergent compared to other populations. Genetic studies also supported morphological data that Samastipur, Jammu and to an extent Ambala populations were distinct variants of *H. zeae*. The genetic variation and evolutionary relationships among 7 geographically isolated populations of *H. zeae* from India using DNA sequencing of the ITS region of rDNA suggested strong isolation among different populations of *H. zeae* (Gavas *et al.*, 2009b).

Three races of *H. zeae* have been reported from Haryana (Ambala, Hisar and Sonepat), India using maize and vetiver as host differentials (Bajaj and Gupta, 1994). The host races are i) multiplying on maize only (Ambala population), ii) multiplying on vetiver only (Sonepat population), and iii) multiplying on both maize and vetiver (Hisar population). These three races of *H. zeae* also differed in catalase phenotype (Singh *et al.*, 1998a). The isoenzyme patterns of esterase and malate dehydrogenase of *H. zeae* races differed in number, concentration and relative mobility (Singh *et al.*, 1998b).

The population of *H. zeae* from Egypt, India and USA might differ in their abilities to reproduce on certain plants. Three biotypes of *H. zeae* have also been reported from Egypt (Khair *et al.*, 1989). Ringer *et al.* (1987) reported that population of *H. zeae* from USA, India and Egypt differed in their ability to reproduce on certain hosts.

Management Strategies

Chemical

Several chemical control trials conducted at IARI research farm, New Delhi under field conditions and also at farmer's fields in different parts of the country especially at Udaipur (Rajasthan) and Pusa (Bihar) using organophosphate and carbamate nematicides for managing *H zeae* populations through soil, seed and foliar spray treatments of nematicides have been reported to be very effective against populations of this nematode (Srivastava, 1980; Sethi and Srivastava, 1986; Srivastava and Sethi, 1989a, 1989b; Srivastava and Jagan Lal, 1997, 2004, 2007). Seed dressing with different organophosphate and carbamate nematicides both under pot and field conditions indicated enhanced emergence of juveniles of *H. zeae* while their penetration into maize roots was inhibited to considerable extent (Srivastava, 1980, Sethi and Srivastava, 1986). Aldicarb used at 0. 75 and 1. 5 per cent and fensulfothion @ 0. 25 and 0. 5 per cent were found to be more effective in reducing the populations of *H. zeae* (Sethi and Srivastava, 1986). Similarly Kaul and Sethi (1987a) also reported the effect of some systemic nematicides on the emergence of the larvae from the cysts of *H. zeae* and found that cysts when exposed to five different concentrations (62. 5 to 500 mg/ml) of aldicarb, carbofuran, oxamyl and phenamiphos resulted in decreased emergence of larvae from the cyst. However, phenamiphos was proved to be most effective nematicide in reducing juvenile emergence *in vitro.*

In yet another study Sethi and Kaul (1987) reported that root invasion was least when juveniles of *H. zeae* were exposed to carbofuran @ 10 mg/ml concentrations for 6 hours. Foliar spray of maize plants with phenamiphos @ 250 or 500 mg/ml 3 days prior to nematode inoculation was found more effective than post inoculation sprays (Kaul and Sethi, 1987b)

Carbosulfan (25 ST)used as seed treatment @2-3% (w/w)and soil application of carbofuran (3G), phorate (l0G)and sebuphos (l0G) @ 2Kg a. i/ha have also been found to be very effective in reducing nematode populations and subsequently increasing the crop yield (Srivastava and Sethi, 1989a, 1989b). The pre-sowing soil applications of phenamiphos resulted in higher reduction of cyst production of *H. zeae* than other chemicals (Kaul and Sethi, 1988a, 1988b). The carbosulfan seed treatment accompanied with carbosulfan foliar spray treatment resulted in additional control of the nematode (Srivastava and Sethi, 1989b;Srivastava and Lal, 1997, 2004). The potential of carbosulfan seed treatment proved to be very effective and economical management strategy for reducing the total requirement (80 to 100 g carbosulfan required for treating 1 kg of maize seeds)as also the cost of the chemicals against *H. zeae* on maize. Srivastava and Lal (2004)found that carbosulfan seed treatment at 2% w/w combined with 1 kg carbofuran/ha was effective in increasing the growth of maize infected with *H. zeae.* Carbofuran and carbosulfan significantly reduced the cyst population of *H. zeae* compared to the inoculated control. Carbosulfan seed treatment at 2% w/w alone or combined with foliar spray treatment at 250 micro g/ml reduced the cyst population to a considerable extent (Srivastava and Lal, 2004). The treatment of neem based products/formulations namely neem seed kernel powder and achook powder as seed dressing @ 5 and 10 % also reduced the population of *H. zeae* to some extent but enhanced the crop growth of maize considerably.

The fecundity of the nematode (eggs and nematode juvenile per cyst) reduced maximum when cadusafos was applied at 1 kg a. i. /ha at the time of sowing compared to carbofuran application (Srivastava and Lal, 2007). Their results further indicated that the split dose of either of the chemicals given at 21 days interval was comparatively less effective compared to single application. Cadusafos soil application at 1 kg a. i. /ha when applied either at the time of sowing or in split dose was more effective than carbofuran in enhancing maize (cv. Deccan-103) growth parameters (i. e. shoot length, shoot and root fresh and dry weights)and in reducing the final population of maize cyst nematode (Srivastava and Lal, 2007).

Even though, carbofuran, phorate, sebuphos and carbosulfan are quite effective and capable in containing the population densities of *H. zeae* thereby enhancing plant growth and subsequently increasing the crop yield, their use in crops like maize may not be favoured by the farmers, particularly since maize is not as remunerative crop as other cereal crops such as rice and wheat and hence nematode management strategies for maize must be based on some other approaches preferably non-chemical methods like crop rotation and nematode resistant cultivars which have been proved to be quite effective against other cyst nematodes under experimental conditions. Perhaps a combination/inte-gration of different approaches including soil organic amendments and biocontrol agents may be useful for developing the integrated management schedule for this nematode.

Non-chemical strategies

Cultural

Crop rotation : It is an effective and economical approach for managing several important species of cyst nematodes. There is a good scope for adopting/using crop rotation as a control strategy for *H. zeae* as host range of this nematode is restricted to graminaceous plant (Graminae family)only. Since *H. zeae* is host specific, monoculturing of maize crop in the same field should be avoided, as it is likely that under continuous maize cropping systems, nematode populations may increase considerably, ultimately resulting in significant yield loss. Hence, two years rotation with non-host crops (preferably non-cereals like vegetables, pulses and oilseeds) can be fruitful, as it would bring down the nematode populations below economic threshold levels. Haroon *et al.* 2009 reported that in microplots medicinal plants viz. *Ambrosia maritima, Calendula officinalis, Hyoscyamus muticus, Linum usitatissimum*, and *Origanum vulgare*, significantly reduced the number of *H. zeae*. Survival of hatched juveniles was significantly reduced after 24 h in root exudates of *A. maritima, C. officinalis* and *O. vulgare* when compared to the controls. Juveniles, inoculated onto root explant cultures with added exudates, failed to penetrate and/or develop. None of the five medicinal plants allowed complete development of the cyst nematode *H. zeae* inside the roots system. Non-poaceous crops in a cropping sequence system during the winter season in infested soil may reduce the nematode population on the following summer crops.

Ismail (2009) also observed that populations of *H. zeae* are strongly influenced by the cropping sequence. He found that when broad bean-corn or Egyptian clover-corn rotations were practiced, the population increase was insufficient to cause economic damage to corn; while when barley-corn or wheat-corn rotations were practiced, the nematode increased strongly and could cause considerable damage.

Deep/Summer ploughing

Summer ploughing and keeping the fields free from weeds can also help in checking the survival and perpetuation of *H. zeae* population in off seasons. Two to three deep ploughing at 10-15 days interval during April/May months in hot summer also reduces the nematode populations to a considerable extent.

Resistant / Tolerant varieties

The use of resistant and tolerant varieties is yet another most effective and economical management strategy for plant parasitic nematodes but cyst nematode resistant cultivars are not always available. Resistant source in maize crop against *H. zeae* is also not available;however, a few cultivars provide some degree of resistance against this nematode. Maize cultivars Ageti-76 and Karnal-1 have been found to be moderately resistant against *H. zeae* in Haryana, India. Similarly Shahina *et al.*, (1989)have also reported Sarhad White, Gauhar, Azam and Composite-15 maize varieties as moderately resistant against *H. zeae* in Pakistan.

Soil amendments

The incorporation of organic substances from plants and amending soil with non-edible deoiled cakes (e. g. neem etc.) which are reported to improve the soil fertility and organic matter status have also been proved effective in managing the populations of *H. zeae*. Field trials using various soil organic amendments against this nematode have clearly indicated that combination of mustard cake and tobacco dust @ 2. 5 q/ha, a practice adopted by farmers in the eastern parts of the country, is as efficient as carbofuran soil treatment @2 kg a. i. /ha in terms of increased crop yield and containment of *H. zeae* populations (Anon, 1987). The practice which serves as one of the important components of integrated approach for this nematode can easily be adopted by the farmers of other regions of the country.

Biocontrol agents

Several biocontrol agents have been exploited for management of cyst nematodes, including *H. zeae* and few of these have shown good promise. Although there is not much work on use of biocontrol agents against maize cyst nematode *(H. zeae)*however, certain nematophagus fungi such as *Cataneria, Verticillium* and *Gleocladium* have been isolated from cysts of *H. zeae.* Also *Arthobotrytis coincides* and *Monacrosporium salinum* have shown some potential to kill second-stage juveniles of *H. zeae in vitro* (Anon, 1987). *H. zeae* isolate of *Pasteuria* (Bajaj *et al.,* 2001)may give a good biological control of maize cyst nematode. In Haryana, *H. cajani* is attacked by a wider spectrum of antagonists as compared to *H. avenae* or *H. zeae* (Walia, 2009). The real potentialities of biocontrol agents for the management of this nematode are still to be exploited and explored.

Conclusion

The maize cyst nematode, *Heterodera zeae* is the most important nematode pathogen and is considered as the key nematode pest of maize in India. The nematode is widely distributed in major growing areas of northern, central, western and eastern states of India. Its presence at higher altitudes in northern hills, especially in Jammu & Kashmir and Himachal Pradesh needs immediate attention to study the role played by this nematode in limiting maize production in the country. So far, the nematode has not been recorded from the two southern states, Karnataka and Andhra Pradesh, even though maize productivity there is very high as compared to other states. But it is possible that with ongoing extensive survey programme, we may encounter it in the near future. Further, maize growing areas of north-eastern states of the country such as Assam, Meghalaya, Manipur and Sikkim also need to be explored for the presence of nematode species.

Results on various chemical trials have clearly shown that carbosulfan seed treatment and carbofuran and sebuphos soil application enhances the crop yield with considerable reduction in the nematode population of *H. zeae.* Further, work on this aspect with reference to cost-benefit ratio is needed. However, recommendations of high cost technology involving use of nematicides, may not find favour with the farmers particularly because maize is not a high value crop as rice and wheat and hence nematode management technology for maize must be based on non-chemical methods particularly crop rotation and nematode resistant cultivars. It is likely that under continuous maize cropping nematode populations of *H. zeae* may increase considerably, ultimately resulting in significant yield losses therefore, monoculture should be avoided. Two years rotations with non-host crops preferably non-cereal like vegetables, pulses and oil seeds can be fruitful, as it would bring down the nematode population below economic threshold level appreciably. Further, maize genotypes screening work should largely be taken up to identify the sources of resistance/tolerance against *H. zeae* and also other nematodes. Investigations are also needed to work out the influence of various soil organic amendments generally adopted by farmers for their incorporation in economically viable integrated management of the nematodes. Also, possibilities for exploitation of bioagents should be explored for managing the nematodes.

This virulence in one population of the same species is of special interest. This may be explained by looking at the genotypic and phenotypic variability in these populations.

References

Aboul-Eid, H. Z. & Ghorab, A. I. (1981). Occurrence of *Heterodera zeae* in maize fields in Egypt. *Egyptian Journal of Phytopathology*. 13: 51–61.

Anonymous (1987). First Annual Technical Report, *PL-480 Project: FG-In-673 (IN-ARS-326)*, pp. 31.

Anonymous (1989), Final Technical Report, *PL-480 Project: FG-In-673 (IN-ARS-326)*, 48pp.

Anonymous (1993). Final Technical Report, *I. C. A. R. Cess Fund Scheme: F. No. 2-5/85-PP*, 32pp.

Bajaj, H. K. & Bhatti, D. S. (1984). Occurrence of *Heterodera zeae* in Haryana State, India. *Haryana Agricultural University Journal of Research*. 14: 551.

Bajaj, H. K. & Gupta, D. C. (1994). Existence of host races in *Heterodera zeae* Koshy *et al.*, *Fundamental & Applied Nematology*. 17: 389–390.

Bajaj, H. K. , Mehta, S. K. & Walia, R. K. (2001). Host preference of *Pasteuria* isolates from *Heterodera cajani, H. zeae* and *Xiphinema insigne*. *International Journal of Nematology*. 11: 226-228.

Chinnasri, B. , Tangchitsomkid, N. & Toida, Y. (1995). *Heterodera zeae* on maize in Thailand. *Japan Journal of Nematology*. 24: 35–38.

Correia, F. J. S. & Abrantes, I. M. de O. (2005). Characterization of *Heterodera zeae* populations from Portugal. *Journal of Nematology*. 37: 328-335.

Darekar, K. S. , Kadam, M V. , Ayri, D. S. & Dorge, S. K. (1981). Occurrence of maize cyst nematode, *Heterodera zeae* in Maharashtra State. *Journal of Maharashtra Agricultural University*. 6: 156.

Haroon, S. A. , Othman, E. & Youssef, R. M. (2009). The effect of root exudates from certain Egyptian medicinal plants on cyst nematode, *Heterodera zeae*. In'Cereal cyst nematodes:status, research and outlook.' (Eds IT Riley, JM Nicol, AA Dababat)pp. 227-232. (CIMMYT: Ankara, Turkey).

Hashmi, S. , Krusberg, L. R. & Sardanelli, S. (1993). Reproduction of *Heterodera zeae* and its suppression of corn plant growth as affected by temperature. *Journal of Nematology*. 25: 55-58.

Ismail, A. E. (2009). Impact of winter wheat, barley, broad bean and clover as preceding crops on population densities of corn cyst nematode, *Heterodera zeae* on corn in Egypt. In 'Cereal cyst nematodes: status, research and outlook' (Eds IT Riley, JM Nicol, AA Dababat) pp. 237-241. (CIMMYT: Ankara, Turkey).

Joshi, P. K, Singh, N. P. , Singh, N. N. , Gerpacio, R. V. & Pingali, P. L. (2005). Maize in India: Production Systems, Constraints, and Research Priorities. Mexico, D. F. : CIMMYT.

Kaul R. K. & Sethi C. L. (1987b) Penetration and development of *Heterodera zeae* in *Zea mays* as influenced by foliar spray of phenamiphos and oxamyl. *Indian Journal of Nematology* 17: 237-240.

Kaul, R. K. & Sethi, C. L. (1988b) Effect of time of application of certain systemic nematicides on final cyst production and fecundity of *Heterodera zeae*. *Indian Journal of Nematology*. 18: 134-135.

Kaul, R. K. & Sethi, C. L. (1988a) Soil application of systemic nematicides in relation to emergence, penetration and development of *Heterodera zeae*. *Indian Journal of Nematology*. 18: 1-5.

Kaul, R. K. , Sethi C. L. (1987a). Effect of some systemic nematicides on the *in vitro* emergence of the larvae from the cysts of *Heterodera zeae*. *Indian Journal of Nematology*. 17: 214-217.

Kaushal K. K. , Srivastava A. N. , Pankaj, Chawla Gautam & Singh Khajan (2007). Cyst forming nematodes in India - A Review. *Indian Journal of Nematology* 37: 1-7.

Khair, A. M. , Farahat, A. A. & Abadir, S. K. (1989). Studies on the corn cyst nematode (CCN) *Heterodera zeae* in Egypt; IV variation in development and reproduction of four different populations of some corn cultivars. *Pakistan Journal of Nematology*. 7: 69–73.

Koshy, P. K. , Swarup, G. & Sethi, C. L. (1970). *Heterodera zeae* n. sp. (Nematoda: Heteroderidae), a cyst-forming nematode on *Zea mays*. *Nematologica*. 16: 511–516.

Krusberg, L. R. (1988). The corn cyst nematode, *Heterodera zeae* in the United States. *In: Advances in Plant Nematology*. 171-175.

Krusberg, L. R. & Sardanelli, S. (1989) Survival of *Heterodera zeae* in soil in the field and in the laboratory. *Journal of Nematology*. 21: 347-355.

Krusberg, L. R. , S. Sardanelli, & Grybauskas, A. P. 1997. Damage potential of *Heterodera zeae* to *Zea mays* as affected by edaphic factors. *Fundamental & Applied Nematology*. 20: 593-599.

Lal, A. & Mathur, V. K. (1982). Occurrence of *Heterodera zeae* on *Vetiveria zizanioides*. *Indian Journal of Nematology*. 12: 405-407.

Makadia, B. M. , Jagani, D. K. & Patel, D. J. (1988). Occurrence of *Heterodera zeae* on maize in Gujarat. *Indian Journal of Nematology*. 18: 130.

Mani, A. & Prakash, K. S. (1992). Distribution of plant parasitic nematodes associated with certain principal crops in Andhra Pradesh. *Current Nematology*. 3: 21-26.

Maqbool, M. A. (1981). Occurrence of root-knot and cyst nematodes in Pakistan. *Journal of Nematology*. 13: 448-449.

Mc Gregor, R. C. (1978). People placed pathogens. The emigrant pests. In: Plant Disease. An advance treatise Vol II. How disease develops in population. Edited by Horsefall, J. C. and Cowling, E. B.

Nasira, K. & Shahina, F. (2007). Nematode investigation in some cereals, fruits and vegetables of Pakistan. *Pakistan Journal of Nematology*. 25: 15-27.

Parihar, Aruna; Yadav, B. S. & Siddiqui, A. U. (1991). Reaction of some cultivated and weed plants to the corn cyst nematode, *Heterodera zeae*. *Indian Journal of Nematology* 21: 94-95.

Qasim, M. & Ghaffar, A. (1986). *Heterodera zeae* in rhizosphere of declining almond trees, *Prunus amygdalus* in Baluchistan, Pakistan. *Nematology Mediterranea*. 14: 159.

Ragesh, G. , Antony, G. & Kaushal, K. K. (2009a). Phylogenetic analysis of ITS region of r-DNA reveals strong genetic divergence among geographically isolated populations of *Heterodera zeae*. *International Journal of Nematology*. 19: 196-202.

Ragesh, G. , Grace, T. , Rao, U. , Kaushal, K. K. , Vangapandu, S. (2009b). Genetic differentiation among populations of *Heterodera zeae* as revealed by Restriction Fragment Length Polymorphism. *International Journal of Nematology*. 19: 189-195.

Rao, U. , Grace, T. , Kaushal, K. K. & Vangapandu, S. (2009). Molecular characterization of Indian populations of maize cyst nematode, *Heterodera zeae* using PCR-RFLP of ribosomal DNA. *Indian Journal of Nematology*. 38: 81-85.

Ringer, C. E. , Sardanelli, S. & Krusberg, L. R. (1987). Investigations of the host range of the corn cyst nematode, *Heterodera zeae* from Maryland. *Annals of Applied Nematology*. 1: 97-106.

Sardanelli, S. , Krusberg, L. R. & Golden, A. M. (1981). Corn cyst nematode, *Heterodera zeae* in the United States. *Plant Disease*. 65: 622.

Sethi C. L. & Kaul, R. K. (1987). Nematicidal exposure of larvae of *Heterodera zeae* and their development in maize root. *Indian Journal of Nematology*. 17: 261-263.

Sethi, C. L. &Srivastava, A. N. (1986). Management of *Heterodera zeae* on maize through pesticidal seed dressing. *Indian Journal of Nematology*. 16:82-86.

Shahzad, S. & Ghaffar, A. (1986). Tomato a natural host of *Heterodera zeae* in Pakistan. *International Nematological Network Newsletter*. 3: 38.

Sharma, N. K. , Damanjeet & Sharma, S. B. (1984). Incidence of maize cyst nematode, *Heterodera zeae* in Himachal Pradesh. *Indian Journal of Nematology*. 14:198.

Singh, S. , Kumar, S. & Bajaj, H. K. (1998a). Catalase and Glutamate oxaloacetate transaminase isozymes patterns of races of *Heterodera cajani* and *H. zeae*. *Haryana Agricultural University Journal of Research*. 28: 175-177.

Singh, S. , Kumar, S. , Bajaj, H. K. & Ganguly, A. K. (1998b). Esterase and malate dehydrogenase patterns of races of *Heterodera cajani* and *H. zeae*. *Indian Journal of Nematology*. 28: 105-109.

Srivastava, A. N. (1980) Studies on the maize cyst nematode, *Heterodera zeae*. Ph. D. Thesis, IARI, New Delhi. pp. 99.

Srivastava, A. N. &Chawla Gautam (2005) Maize cyst nematode, *Heterodera zeae*– a key nematode pest of maize and its management. *Technical Bulletin* (TB-ICN: 39/2005), Division of Nematology, IARI, New Delhi. 18pp.

Srivastava, A. N. & Kaushal, K. K. (1991). *Heterodera zeae,* at high altitudes in Himachal Pradesh. *Indian Journal of Nematology*. 21: 163.

Srivastava, A. N. & Lal, J. (1997). Effect of carbosulfan and sebuphos applied singly and in combination on reproduction of *Heterodera zeae* and growth of *Zea mays*. *Annals of Plant Protection Sciences*. 5: 94-97.

Srivastava, A. N. & Lal, J. (2004). Effect of carbofuran and carbosulfan following soil, seed and foliar application on plant growth of maize and reproduction of maize cyst nematode, *Heterodera zeae*. *Indian Journal of Nematology*. 34: 200-203.

Srivastava, A. N. & Lal, J. (2007). Effect of split application of cadusafos and carbofuran on reproduction and fecundity of maize cyst nematode, *Heterodera zeae* and maize growth. *Indian Journal of Nematology*. 37:78-80.

Srivastava, A. N. & Sethi C. L. (1986). Seasonal population fluctuations of *Heterodera zeae, Rotylenchulus reniformis* and *Tylenchorhynchus vulgaris* in maize-cowpea-wheat rotation. *Indian Journal of Nematology.* 16: 16-18.

Srivastava, A. N. & Sethi, C. L. (1984a). Effect of soil type on reproduction of *Heterodera zeae. Indian Journal of Nematology.* 14: 72

Srivastava, A. N. & Sethi, C. L. (1984b). Relationship of initial populations of *Heterodera zeae* with plant growth of maize and nematode reproduction. *Indian Journal of Nematology.* 14: 110–114.

Srivastava, A. N. & Sethi, C. L. (1985a). Effect of temperature on larval emergence on *Heterodera zeae. Indian Journal of Nematology.* 15: 242-243.

Srivastava, A. N. &Sethi, C. L. (1985b). On the larval penetration and biology of *Heterodera zeae* on maize. *Indian Journal of Nematology.* 15: 18-20.

Srivastava, A. N. & Sethi, C. L. (1989a). Effectiveness of nematicidal spray as a method to manage maize cyst nematode, *Heterodera zeae. Indian Phytopathology.* 42: 290.

Srivastava, A. N. &Sethi, C. L. (1989b). Seed treatment as a method in control of maize cyst nematode, *Heterodera zeae. Indian Phytopathology.* 42: 303.

Srivastava, A. N. & Swarup, G. (1975). Preliminary studies on some graminacious plants for their susceptibility to the maize cyst nematode, *Heterodera zeae. Indian Journal of Nematology.* 5: 257–259

Srivastava, A. N., Walia, R. K. & Singh, H. (1995). Nematode problems in maize. In:Nematode pest management-An Appraisal of Eco-friendly Approaches (Gopal Swarup, D. R. Dasgupta and J. S. Gill eds.), Nematological Society of India. pp 134-139.

Walia, R. K. (2009). Biological Control - Recapitulating the Basic Principles. *Indian Journal of Nematology.* 39: 129-137.

Phyto-nematodes in Crops: Their Identification, Treatment and Management, 2011
G.L. Sharma (editor), pp. 133-153
New India Publishing Agency, New Delhi (India)
E-mail : info@nipabooks.com; Website : www.nipabooks.com

Chapter 6

Phytonematodes: A Constraint to Rice Production and their Management

A. N. SRIVASTAVA, R. K. JAISWAL AND G. CHAWLA

Introduction

Rice (*Oryza sativa* L.), one of the most important cereal crops of India, occupies 23.3 per cent of total cropped area of the country. Rice contributes 43 per cent of total food grain production and 46 per cent of total cereal production. It continues to play a vital role in the nation's food grain supply. It is also a staple food of nearly half of the world population. Rice is primarily a high caloric food having protein content (6-17%) with high biological value and a low amount of fat (2-2.5%). B-group vitamins content in rice grains are at par with wheat, however, calcium content in rice is generally low. It ranks third after wheat and maize in terms of worldwide production. Asia accounts for 90 and 92% of the world's rice area and production, respectively. Thus, rice production, consumption and trade are concentrated in Asia. Among the rice growing countries in the world, India has the largest areas under rice crop and ranks second in production next to China. Rice is cultivated in India in about 45 million hectares under irrigated (46%), rainfed lowland (28%), rainfed upland (12%), and flood-prone (14%) ecosystems (Sarla and Mallikarjuna, 2005). In India, rice occupies first position among the cereals in respect of both area and production. West Bengal is the highest rice producing state while Tamil Nadu ranks first place in productivity.

Plant parasitic nematodes, the hidden enemy, constitute one of the most devastating pests group and are responsible for causing huge yield losses to almost all kinds of crops. Across the world, damages attributed to plant parasitic nematodes are estimated to $ 157 billion each year (Abad *et al.*, 2008), however, the National loss in 24 different crops is estimated to the tune of 21 billion rupees per year (Jain *et al.*, 2007). Such a colossal loss becomes unbearable for a poor country like India in particular and any other country of the world in general. Rice has a number of important nematode pests that can seriously affect growth and yield of the crop. More than 200 species of plant parasitic nematodes have been reported to be associated with the rice crop world wide (Prot *et al.*, 1994). Some of the nematodes have a restricted distribution, while others occur throughout the rice-growing regions of the world. These nematodes are either foliar or root parasites and produce a wide range of symptoms on rice crop. In the present manuscript, only economically important plant parasitic nematodes of rice crop are briefly discussed.

Key nematode pests of rice in India

A. Root parasites

i) Rice root-knot nematode, *Meloidogyne graminicola*

ii) Rice-root nematodes, *Hirschmanniella* spp.

iii) Rice cyst nematodes, *Heterodera* spp.

B. Foliar Parasites

iv) White tip nematode, *Aphelenchoides besseyi*

v) Rice stem nematode, *Ditylenchus angustus*

1. Rice root-knot nematode

Rice root-knot nematode, *Meloidogyne graminicola* (Golden and Birchfield, 1968) is a well established nematode pest of rice. This nematode is most severe in nurseries, direct sown and transplanted rice in upland and medium low land irrigated fields (Manser, 1968, 1971; Rao and Israel, 1971, 1972). The root-knot disease caused by *M.*

graminicola is characterized by abnormal swelling of roots (known as root-knots or galls), yellowing, stunting and wilting of plants depending on the initial population density (Pi) of the pathogenic nematode in soil and nematode-to-root biomass ratio (Singh *et al.*, 2006). Plowright and Bridge (1990) reported that a high initial population densities of *M. graminicola* caused wilting of seedlings along with severe reduction in growth parameters while low population caused only reduction in growth parameters.

Distribution

This nematode is widely distributed in rice growing areas of the world including India (Israel *et al.*, 1963), U.S.A. (Golden and Birchfield, 1968), Laos (Manser, 1968), Thailand (Buangsuwon *et al.*, 1971), Bangladesh (Hoque and Talukdar, 1971), Nepal (Pokhrel, 2001) and Pakistan (Munir and Bridge, 2003) but occurs in more severe form in Asia (Rao *et al.*, 1986; Bridge *et al.*, 1990; Jairajpuri and Baqri, 1991; Prot *et al.*, 1994; Prot and Matias, 1995; Soriano *et al.*, 2000; Soriano and Reversat, 2003; Padgham *et al.*, 2004). In India, this disease was first reported from Orissa by Israel *et al.*, (1963). Subsequently, occurrence of the disease was also reported from different states of India by several workers viz., West Bengal (Pal and Jayaprakash, 1983), Assam (Subramanian and Velayuthan, 1983), Tripura (Sahu and Chawla, 1986), Andhra Pradesh (Sharma and Prasad, 1995), Haryana (Gaur *et al.*, 1996), Bihar (Kumar and Sinha, 1996), Jammu (Singh *et al.*, 2007c) and Jharkhand (Kumar *et al.*, 2010).

Symptoms

The above ground symptoms are nonspecific in nature as yellowing, stunting, drying of leaves, wilting, delayed flowering by 10-15 days and reduced number of tillers. The presence of characteristic 'hook shaped' or 'ring-like' root galls on the root tip of growing rice seedlings is the confirmatory evidence for the association of this nematode. The root systems of infected seedlings are usually heavily laden with root galls. Such galls are usually smaller in size. The degree of symptom manifestation differs with time of infection, age of the plants and load of nematode inoculum, etc.

Biology

The first and second stage juveniles of *M. graminicola* develop inside the eggs. Most juveniles hatch inside the galls and reinfect the same roots but a few emerge into the soil. In the soil, the second stage juveniles are attracted to the roots and usually penetrate the roots closely behind the root tip by inserting their stylets (Buangsuwon *et al.*, 1971; Rao and Israel, 1973). The juveniles then migrate first towards the root tip and reach to the vascular cylinder. This migration happens intercellularly by mechanical and possibly enzymatic softening of the middle lamella. The parasites finally start feeding on three to ten cells, which are rapidly turned into multinucleated giant cells, by endomitosis and cell hypertrophy. At the same time as the giant cells are formed, the cells of the neighbouring pericycle start to divide, giving rise to a typical gall or root-knot. Rice seedlings shows first visible symptoms in the form of swelling of roots 40 hours after inoculation and usually one or two root galls appear 48 hours after being exposed to nematodes (Singh *et al.*, 2006). Female lays egg in a gelatinous matrix with or without fertilization. However, parthenogenesis is often encountered in root-knot nematodes. The newly hatched second stage juveniles remain in the maternal gall or migrate intercellularly through the aerenchymatous tissues of the cortex to new feeding sites within the same root (Bridge and Page, 1982). This behaviour appears to be an adaptation by *M. graminicola* to flooded conditions enabling it to continue multiplying within the host tissues even when roots are deeply covered by water. *M. graminicola* complete its life cycle within 19-51 days depending on the temperature and season (Rao and Israel, 1973; Yik and Birchfield, 1979; Bridge and Page, 1982; Halbrendt, 1997 and Dabur *et al.*, 2004). *M. graminicola* is well adapted to flooded conditions and can survive in waterlogged soil as eggs in egg masses or as juveniles for long periods. Numbers of *M. graminicola* decline rapidly after 4 months but some egg masses can remain viable for at least 14 months in waterlogged soil. *M. graminicola* can survive in soil flooded to a depth of 1 m for at least 5 months (Bridge and Page, 1982), it cannot invade rice in flooded conditions but quickly invades when infested soils are drained (Manser, 1968). *M. graminicola* can be spread in soil and on seedlings of other crop hosts planted in infested field. Since *M. graminicola* occurs in flooded rice too, there is additional possibility of dissemination in irrigation and run-off water.

Economic Losses

The rice root-knot nematode is one of the constraints to rice production that causes significant yield losses in upland and rainfed lowland rice production (Jairajpuri and Baqri, 1991; Prot and Matias,

1995; Soriano *et al.*, 2000). It is likely that *M. graminicola* will further contribute to the decline of rice yield as the trend towards intensification of production will support increased nematode population densities. The losses caused by *M. graminicola* may vary from negligible to heavy depending on the severity of disease. Biswas and Rao (1971) reported a loss of 15 to 17% in rice yields by *M. graminicola* in pot condition. However, Rao and Biswas, (1973) further reported 21 to 64% yield loss due to this nematode. Jairajpuri and Baqri (1991) also reported grain yield losses from 16 to 32%. Netscher and Erlan (1993) reported that *M. graminicola* caused 28 to 87% yield loss in upland rice in Indonesia. The extent of yield loss caused by *M. graminicola* also varied with agro-ecosystems as 20% losses have been reported under intermittent flooding (Tandingan *et al.*, 1996), 30% in semi deep water (Prasad *et al.*, 1985) and 70% under saturated condition when the nematode infects young susceptible seedlings at transplanting stage (Prot *et al.*, 1994). Soriano *et al.* (2000) recorded 11 to 73% yield losses by this nematode under simulation of intermittently flooded rice, whereas under simulated upland conditions, yield loss varied between 20 and 98% (Plowright and Bridge, 1990; Prot and Matias 1995; Tandingan *et al.*, 1996). Padgham *et al.* (2004) also reported 16 to 20% yield loss caused by *M. graminicola* in low land rainfed rice in Bangladesh. After harvesting of rice, the nematode may survive in egg stage in soil or continue to reproduce on various weeds. Female often remains concealed within root tissues and eggs are laid in cortical tissues and hatched juveniles reinfect the same roots.

Management Strategies

The use of resistant cultivars, controlled flooding, crop rotation, weeding, biocontrol agents and chemicals play important role in the management of the rice root-knot nematode, *M. graminicola*.

Use of Resistant Cultivars, Flooding, Soil Solarization and Crop Rotation

Although resistance would provide the most sustainable means of control,only relatively few rice lines are truly resistant.Cultivars of the African rice,*Oryza glaberrima*,are resistant to *Meloidogyne* spp.and progeny being derived from a cross between *O.sativa* and *O.glaberrima* can provide improved acceptable cultivars with resistance to these nematodes.Continuous flooding is highly effective for the control of *M. graminicola* (Kinh *et al.*,1982). *M.graminicola* survives normal flooding; however, crop damage can be avoided by raising seedlings in flooded soils,thus preventing root invasion by the nematodes(Bridge and Page, 1982).Similarly,yield losses due to *M.graminicola* may be prevented or minimized when the rice crop is flooded early and kept flooded until a late stage of development(Soriano *et al.*,2000).Soil solarization is also effective to reduce the effect of *M.graminicola* on nursery bed. It uses clear polyethylene sheets which are laid on the surface of beds for a period of 3-4 weeks in sunny weather.It reduces the population of *Meloidogyne* spp.in rice beds over 80% and improves seedling growth (Ganguly *et al.*,1996).Some crops are reported as poor hosts of *M. graminicola* such as castor,cowpea,sweet potato, peanut, maize, sunflower, sesame, turnip, and okra may be useful in a rotational cropping system for managing this nematode.

Weeding

Some of the prominent weeds viz., *Echinochloa* spp, *Phalaris minor, Cynodon dactylon, Cyperus rotundus, Eleucine indica*, and *Paspala mscrobiculatum* etc. serve as reservoirs for *M. graminicola*. The presence of such weeds helps in increasing the population of *M. graminicola* in the soil. The weed hosts are usually infected by second stage juveniles which grow within root galls. The developed females lay eggs in larger number, which are mostly retained within galls even after hatching. Thus a large number of infective J_2 s from soil is trapped within roots of these weeds. Weeding of such hosts, therefore, reduces the population of nematode and thereby a good proportion of the inoculums are declined in the soil. This will obviously reduce the build up of inocula of *M. graminicola* in rice fields.

Biological Control

Several biocontrol agents are reported for capturing and killing of plant parasitic nematodes *in vitro* or *in vivo*.Singh *et al.* (2007a) reported that *Catenaria anguillulae* parasitized and killed the second stage juveniles of *M.graminicola* under natural condition.Thus natural enemies like *C.anguillulae* can modulate the population of *M. graminicola* in the soil. Further, Singh *et al.* (2007b) reported association of nematophagous fungi with root galls of rice caused by *M.graminicola* and its biological control by *Arthrobotrys dactyloides* and *Dactylaria brochopaga*. They also reported that mass culture of *A.dactyloides* and *D. brochopaga* reduced the number of root galls,number of females and number of eggs and juveniles and consequently improved the growth parameters of rice in pots.Stirling *et al.*(1998),Stirling and Smith (1998)and Kumar and Singh(2006)also reported the control of root-knot nematode of tomato by application of *A.dactyloides*.Padgham and Sikora (2007)reported that inoculation of rice root with bacterium, *Bacillus megaterium*, resulted in 40% reduction in nematode penetration and gall formation.In addition,colonization of rice roots with *B.megaterium* also decreased migration of *M.graminicola* to the root zone by 60%.

Furthermore,exposure of *M.graminicola* eggs to secondary metabolites of *B. megaterium* reduced hatching by 60%. Lee *et al.*(2009)reported that *Fusarium* and *Trichoderma* species are potential biological control agents against *M. graminicola* in rice depending on the initial nematode intensity in the soil.They showed that inoculation of rice root seedlings with conidia of endophytic and rhizosphere *Fusarium* isolates, root-galling was reduced by 29-42% while root weight was increased by 33%. *Trichoderma* species also, showed similar biological control effects by reduced galling by 38%. A Field application of chemical nematicides is not recommended, but their use in seedling nurseries could possibly prove economically effective.

2. Rice-root nematodes

The rice-root nematodes,*Hirschmanniella* spp.is a migratory endo-parasites of roots which infest most of the flooded rice fields in subtropical and tropical areas,especially in Southeast Asia(Bridge *et al.*, 1990; Cuc and Prot, 1992; Prot *et al.*, 1994). Because flooded rice agro ecosystems contribute 95% of the total rice production in Southeast Asia,rice-root nematodes have the greatest potential economic impact in rice(Prot,1993).There are several species of *Hirschmanniella* that found on a wide variety of crops,and over a dozen of them,known as the rice root nematodes,occur on rice.The most well known species are *H.oryzae*, *H.belli*, *H.gracilis*, *H.imamuri*, *H.mucronata* and *H.spinicaudata*.

Distribution

Rice-root nematode, *H. oryzae* was first reported as *Tylenchus oryzae* on rice *(Oryza sativa* L.) in Java, Indonesia (Haan and Van, 1902) Subsequently, this species was reported from several other countries and areas viz. Japan, Thailand, India, Pakistan, Malaysia, Madagascar, SriLanka, Nigeria, Venezuela, U. S. A., and Ivory-Coast (Sher, 1968; Ichinohe, 1972) Species of *Hirschmanniella* are found in all rice growing areas of the world and comprise one of the few groups of nematodes that can withstand anaerobic conditions such as those occurring in flooded rice fields. In India, rice-root nematodes are prevalent in all the rice growing tracts in varying proportions (Rao and Israel, 1971;Venkitesan *et al*, 1979; Kuriyan *et al*, 1985; Randhawa *et al*, 1990 and Sharma *et al.*, 1992).

Symptoms and Biology

The above ground symptoms are non-specific as stunted growth, leaf chlorosis, reduced tillering and delayed flowering. The juveniles and adult stages penetrate through entire length of roots and feed on cortical cells leading to the formation of channels or cavities in the roots. Its feeding sometimes extends to central vascular regions. The infected roots exhibit water-soaked brown lesion which are of mostly spindle shaped. The physiological function of infected plants is disrupted and plants growth reduced. Nematodes commonly found in the aerenchyma tissues and, since aerenchyma cells are adaptations to aquatic life, permitting the movement of oxygen from leaves to roots (Raalte, 1943), the damage to these tissues is likely to restrict the passage of oxygen, thus placing the roots under oxygen stress. The feeding activities of the *Hirschmanniella* species have deleterious effects on the rice plant in terms of mechanical damage to the stele and parenchyma tissues, disruption of growth, and predisposition of the rice root to secondary infection through the feeding and invasion sites. All of these would result in poor grain yield. The population of *Hirschmanniella* species was found maximum during active growth phages of rice. The populations build up of this nematode increases after transplanting up to 80 days of rice (Singh and Jain, 1995) and declines when rice roots starts degenerating. Rice-root nematode survives better in poorly drained clay and heavy soils. It can survive even in high temperature of May-June (35-45^{0}C) as well as low temperature of December-January (8-12^{0}C) in the North Indian conditions (Mathur and Prasad, 1973). Their survival in soil is much longer than in roots in flooded soils. *H. oryzae* can survive more than 12 months in wet soils (Muthukrishnan *et al.*, 1977). The weeds viz. *Cynodon dactylon, Cyperus rotundus, Echinochloa* spp., *Oryzae nivara*, and *Panicum miliacleum* etc. found in rice fields serve as alternative hosts and account for build up of nematode population. Under rice-wheat cropping systems, the nematode maintains a very high population, though wheat is not a host for this nematode. Its perpetuation occurs mainly through irrigation water, flood water, soil adhering to farm implements, field workers and also rice seedlings.

Economic Losses

Haan and Van (1902) and Goodey (1933) established that *H. oryzae* was injurious to rice plants. Yield reductions of 56% and 87% have been associated with *H. oryzae* in Indonesia and India, respectively (Vecht *et al.*, 1952;Mathur and Prasad, 1972). Panda and Rao (1971) also recorded up to 70% yield reduction in rice due to *H. mucronata*. Ahmad *et al.* (1984) reported 19% yield reduction in rice due to this nematode from West Bengal. Jonathon and Velayutham (1987) have recorded 27 to 39. 4% loss in grain yield and 23 to 42% loss in straw yield due to *H. oryzae*. A yield loss of 30% caused by rice-root nematode, *H. spinicaudata* is recorded in pot experiment (Thio, 1992).

Management Strategies

Absence of rice and other susceptible hosts cause a severe decline in nematode population. In sequential cropping of flooded lowland rice with dry land crops, population is reduced but only if the intervals between susceptible crops are more than one year. The time of transplanting of rice in the field has been found to affect the population build up of *H. oryzae*. Early planting i. e. in June or middle of July favors higher build up of nematode populations than late planting (Randhawa *etal.* 1991) hence, early planting should be avoided. Where green manure legumes are used to increase soil fertility there has been marked control of *Hirschmanniella* spp., especially with *Sesbania rostrata* and *Sphenoclea zeylanica*, which either act as trap crops or produce toxic root exudates. Use of organic amendments such as pressmud, FYM, neem cake, coir waste, poultry manure, neem and *Calotropis giganta* leaves have also proved effective against rice-root nematode (Jonathon and Pandirajan, 1991). A number of weeds, which are known to serve as host plants for this nematode should be removed from the nematode infested fields both in the absence of rice and during crop growth to keep the nematode population under check. The use of various nematicides for the management of *Hirschmaniella* spp. has been attempted by their application in fields and nurseries, as root dip for transplanting seedlings and soaking of seeds.

3. Rice-cyst nematodes

Four species of cyst nematodes are known pests of rice in different parts of the world:*Heterodera sacchari* in West Africa and Trinidad,

H. oryzicola in India, *H. elachista* in Japan, and *H. oryzae* in West Africa and Bangladesh. In India, *H. oryzicola* is a serious root infesting nematode of rice. This nematode is potentially important since suitable conditions for its growth and development are present in rice growing belts areas.

Distribution

Rice-cyst nematode, *H. oryzicola* (Raoa & Jayaprakash, 1978) has been reported as a serious nematode pest of rice in Kerala. Charles and Venkitesan (1984) reported the occurrence of the *H. oryzicola* on banana *(Musa* AAB) cv. Nendran from the state of Kerala, India. It has also been known to occur in Burdwan and Bankura districts of West Bengal. This nematode is one of the serious pests of rice and banana in Kerala (Usha and Kuriyan, 1981). Its occurrence has also been recorded from Karnataka and Goa.

Symptoms

The browning of roots and chlorosis of leaves, retardation in growth, early flowering of plants by 10 to 15 days and partial filling of grains are the typical symptoms of infection of *H. oryzicola*. The presence of brown cyst on rice root is the confirmatory evidence of infestation of this nematode. One life cycle is completed in 30 days and 12 generations of this nematode may occur in a year. The nematode spread through infested seedlings, irrigation water or farm implements.

Economic Losses

Infestation of rice by *H.* or*yzicola* caused reduction in nitrogen, phosphorus, iron, and reducing sugars in shoots and roots. Total sugars, protein, IAA, cytokinin and thiamine were reduced in the cyst nematode infested plants (Rao *et al.*, 1988). Rice cyst nematode can cause up to 38% yield loss in rainfed or well drained soils throughout the country (Prasad *et al.*, 1987). Usha and Kuriyan (1981) also reported 41.9 per cent grain yield loss in rice due to this nematode.

Management Strategies

Soaking of seeds with phenamiphos @ 0.02% for 6 hours has resulted good control of this nematode. Soil application of carbofuran or phorate 1 kg a.i./ha at 7 and 50 days after treatment has also shown promising result. Rotation with non-host plants is an effective means of reducing soil populations. Non cereal crops are non-hosts of this nematode. Resistance to *H. sacchari* has been identified in crosses between *Oryza glaberrima* and *O. sativa*, which could prove useful in the future. Spread of this nematode occurs mainly through transportation of cysts through seeds, farm implements, etc. hence, quarantine measures should be adopted. Regulatory measures on the movement of banana rhizomes as well as rice seedlings from infested areas have also to be adopted to check the spread of this nematode.

4. White tip nematode

White tip nematode, *Aphelenchoides besseyi* is a specialized parasite attacking aerial parts of its natural host, rice. Though rice is the most suitable host of this nematode, it also infects tuberose, onion, soybean, sugarcane, oat, millets, orchids etc.

Distribution

This nematode is disseminated to most rice growing areas of Africa, North, Central, and South America, Asia, Eastern Europe and Pacific Islands. In India, it has been recorded in serious form from Gujarat, Tamil Nadu, Madhya Pradesh, Andhra Pradesh and West Bengal.

Symptoms and Biology

The most characteristic disease symptom is'white tip'in rice leaf produced by this nematode due to which the common name of the nematode is'white tip nematode'. Affected plants shows a characteristic symptoms on leaf tips;which become chlorotic or whitened for a distance of upto 5 cm. Eventually the infected leaf becomes dry and shreds while flag leaf becomes twisted and panicles may not emerge or if emerged may have high sterility, distorted glumes, and small and distorted kernels. It is easy to detect the presence of nematode within the rice seeds. In field, the initial appearance of symptoms as the leaf tip up to 5cm becomes pale yellow or whitish at tillering stage and subsequently leaves get dry. These symptoms are found for a short period in the plant. The tip of the flag leaf are often twisted which may obstruct the emergence of panicles. Infested panicles are shorter and lighter in weight as compared to healthy panicles. *A. besseyi* survives as pre-adults as well as in adult stages (quiescent state) beneath the hull of rice kernel and does not survive in soil after harvesting rice plants. Infected seeds or presence of other alternative hosts help nematode to survive up to next crop. They usually remain in coiled anhydrobiotic state in rice between lemma and palea up to 3 years. The presence of *A. besseyi* is detected mainly by soaking the seeds in a dish of water and examining directly using a dissecting microscope. The infected rice seed is the only means for rapid spread of *A. besseyi*. It also spread through irrigation water or flood water. Female lays eggs on rice plants. All the developmental stages occur on rice plant. On rice, infested seed is the primary inoculum source. When the crop is sown, the nematode becomes active and moves to the growing points of leaves and stems where feed ectoparasitically. *A. besseyi* may reproduce parthenogenetically. The optimum temperature for development is 21-25°C, the life cycle takes 10 days at 21°C and 8 days at 23°C, and several generations are completed in a cropping season. Upto 14 nematodes, mostly pre-adults, have been found on a single seed, where they remain coiled up inside the palea. The grain itself is never invaded. *A. besseyi* can withstand desiccation, retaining viability for 2-3 years on dry grain, but dies in 4 months on grain left in the field;the nematode is not reported to survive long periods in the soil between crops.

Economic Losses

The importance and severity of *A. besseyi* can vary with country, locality, and rice environment. It is considered to be important in India and Africa, but causes little damage in the USA, Thailand, or Japan. In Bangladesh, more than 50% of deepwater rice fields have been shown to be infested with this nematode. Yield loss in susceptible rice depends mainly on the number of nematodes in infected seed and the percentage of nematode infected seed sown;300 live nematodes/100 seeds have been suggested as an economic damage threshold density. Since early in the 20th century, *A. besseyi* has been reported causing serious losses to rice in Japan and parts of the USA (especially from 1935 to 1945). However, nematode damage has become less important with the use of resistant cultivars and other control measures (Vuong and Rodriguez, 1970). Nematode infestations results in poorly filled or unfilled, chaffy grains;the proportion of the latter may be as high as 40% in some cultivars, but early-ripening cultivars are relatively less affected. Yield reductions in rice of 17-54% and 0-24% have been recorded for susceptible and resistant cultivars, respectively.

Management Strategies

Prevention by use of clean seed or eliminating nematodes from seed prior to planting is the most effective means of preventing yield loss. Hot water treatment of seed is probably the most useful and cheapest method of reducing crop losses when seed is possibly contaminated.

Seeds are pre-soaked in cool water for 18–24 hours followed by immersion in water at 51-53°C for 15 minutes. Higher temperatures (55–61°C) for 10–15 minutes are required if the seed is not pre-soaked (Todd and Atkins, 1959). The duration of the exposure depends on the degree of infestation. However, along exposure may impair seed germination. Other methods which can contribute to a reduction of nematode populations include the use of resistant or tolerant cultivars, early planting when the rice season is preceded by a cooler period and low seed-bed planting densities. Nematicides applied either as foliar sprays or as soil treatments, can reduce the nematode population sufficiently to prevent symptom expression. Chemicals as seed disinfectants have been tried with promising results. Carbofuran, aldicarbsulfone and methomyl at 0.1% have shown high percentage of nematode mortality but at the same time impaired 11-29% seed germination.

5. Rice stem nematode

The rice stem nematode, *Ditylenchus angustus*, is the causal agent of 'Ufra' disease in Bangladesh, India, Myanmar and Vietnam (Ou, 1985). Being an ectoparasite, it feeds on the younger tissues of unemerged leaves, leaf sheaths, peduncles and developing spikelets.

Distribution

Rice stem nematode, *D. angustus* originated as an important pest of deepwater rice in South and Southeast Asia and, therefore, has a restricted distribution due to deepwater rice being found in only a few areas. It is a parasite of both wild and cultivated species of *Oryza* in the major river deltas of Bangladesh, Assam, Burma, and Vietnam that are prone to deep and very deep flooding in the rainy season. However, it is now widespread in Bangladesh on lowland rice systems, and may spread into the lowland crop in other countries. Rice stem nematode is usually problematic in rice grown in deep water situations.

Symptoms

The name 'stem nematode' is derived from the stem inhabiting nature of the nematode. It is an obligate parasite and serious pest of rice causing popular disease symptom referred to as 'Ufra disease'.

The symptoms produced by rice stem nematode are popularly known as Ufra or dakpora disease. The Ufra symptoms appear in patches and subsequently spread to the entire field. The nematode attacks at vegetative stage that results in yellowing or whitish pattern on the leaf sheath and margin becomes corrugated. In due course of time, the splash pattern turns brownish stains and stem and internodes become black. Twisting of leaf and leaf sheath are commonly found symptoms. Sometimes infested nodes give bushy appearance due to branching. The Ufra symptoms may be grouped as:

Swollen or Thor Ufra: Panicle does not come out, it remains enclosed within the leaf sheath and infected portion tending to branch.

Pucca or Ripe Ufra: Panicle emerges partially and panicle bears filled grains at tip only. The primary sources of infection of *D. angustus* are rice stubbles, straw, wild rice and weeds found in rice fields. The nematode can over winter through quiescent state (fourth stage juvenile) which remains viable up to 15 months. They live in coiled anhydrobiotic state in grains, dried plants parts left in the fields and reinfect the crop in the next season.

Biology

Rice stem nematode, *D. angustus* occurs in the leaves, inflorescences, young seeds, and rolled stems of growing plants and also in crop residues but it is not seed-borne. The nematode survives in plant residues, the stem tissues, and partially or fully enclosed panicles between seasons in a desiccated state. Nematodes can migrate from diseased plants or plant residues to healthy plants in water, and by stem and leaf contact under high humidity. The nematode is spread primarily through irrigation water, although most nematodes die after a few days in water. *D. angustus* needs at least 75% humidity to migrate on the foliage and is more damaging in wet areas. Nematodes in water can invade young rice plants within 1 hour. Greatest infection occurs at temperatures of 27–30°C and the nematode has a short life cycle of 10–20 days. The presence of *D. angustus* can be confirmed by cutting pieces of about 5 mm long from the rolled leaf stems and placing in a small dish of water. Stem pieces are cut longitudinally and left in the water for 24 hours. The rolled leaves or young inflorescence can be teased apart in a Petri dish with water and observed directly. Numerous nematodes (hundreds to thousands) will be active from fresh material, but they require some time to resume activity from dried panicles.

Economic Losses

Wherever *D. angustus* does occur, it can cause total losses in individual fields. In Bangladesh, annual yield losses of 4% have been estimated while in Assam and west Bengal (India), losses are estimated at 10–30% in some areas. It is also recognized as a serious problem in the Mekong delta in Vietnam. Substantial yield losses can occur when transplanted seedlings are infected, even at low percentage infection. In Bangladesh and some parts of India, 100% yield loss has been recorded due to severe attack of *D. angustus*.

Management Strategies

Incorporation of crop residues into the soil by ploughing can reduce ufra as nematodes decline quickly in soil. Growing non-host crops such as jute or mustard in rotation with rice can also reduce ufra incidence. Extending the period of time between growing rice crops and lengthening the overwinter period (by delaying sowing, transplanting after flooding, and using early maturing cultivars) can reduce the populations of nematodes and severity of ufra disease on successive rice crops. Destruction of weeds and ratoon rice and burning of stubble and straw are efficient and have long been suggested to control ufra disease (Butler, 1919). However, it is sometimes difficult to burn the stubble and straw because of standing water in the field, or because a large proportion of the straw is removed for cattle feeds and fuel, which leaves an insufficient amount for effective burning. The nematode is spread by water flow; therefore, improvement in river bunds and the construction of irrigation canals to control water flow may prevent spread of the nematode (Sein and Zan, 1977). Application of zinc (Miah *et al.*, 1984) and calcium silicate reduce ufra incidence and increase crop yield. Some resistance to *D. angustus* has been found in deepwater rice cultivars and will be effective if commercially available. Chemicals such as benomyl, carbofuran, fensulfothion, hexadrin, monocrotophos, and phenazine have been tested to control *D. angustus* (Miah and Bakr, 1977; Sein 1977; Cox and Rahman, 1979; Rahman *et al.*, 1981; Miah and Rahman, 1985). Chemical treatments efficiently control the nematode but they are uneconomical and not feasible.

References

Abad, P., Gouzy, J., Marc, M. & Jean (2008). Genome sequence of the metazoan plant parasitic nematode, *Meloidogyne incognita*. *Nature Biotechnology*. 26: 909-915.

Ahmad, N., Das, P. K. & Baqri, Q. H. (1984). Evaluation of yield losses in rice due to *Hirschmanniella gracilis* (de Man, 1988) Luc & Goodey, 1963 (Tylenchida:Nematoda) at Hooghly (West Bengal). *Bulletin Zoological Survey of India.* 5:85-91.

Biswas, H., & Rao, Y. S. (1971). Toxicological assays of nematicides with rice nematodes. *Journal of Phytopathology.* 24:159-165.

Bridge, J., Luc, M. &Plowright, R. A. (1990). Nematode parasites of rice. In: Luc, M., Sikora, R. A. & Bridge, J. (Eds). Plant parasitic nematodes in sub-tropical and tropical agriculture. Wallingford, UK, CAB International, pp. 69-108.

Bridge, J. &Page, S. L. J. (1982). The rice root-knot nematode, *Meloidogyne graminicola,* on deep water rice *(Oryza sativa* sub. sp. *indica*). *Revue de Nématologie.* 5: 225-232.

Buangsuwon, D., Tonboon-Ek, P., Rajirachoon, G., Braun, A. J. & Taylor, A. L. (1971). Nematodes. In: Rice diseases and pests of Thailand. English edition, 61-67. Rice Protection Research Centre, Rice Department, Ministry of Agriculture, Thailand.

Butler, E. J. (1919). The rice worm *(Tylenchus angustus*) and its control. Mem. Dep. Agric. India: Bot. Ser. 10: 1-37.

Charles, J. S. K. & Venkitesan, T. S. (1984). New hosts of *Heterodera oryzicola* (Rao and Jayaprakash 1978) in Kerala, India. *Indian Journal of Nematology.* 14: 181-182.

Cox, P. G. & Rahman, L. (1979). Synergy between benomyl and carbofuran in the control of ufra. *International Rice Research Newsletter.* 4: 11.

Cuc, N. T. T. & Prot J. C. (1992). Root parasitic nematodes of deep-water rice in the Mekong Delta of Vietnam. *Fundamental and Applied Nematology.* 15: 575-577.

Dabur, A. S., Taya, A. S. &Bajaj, H. K. (2004). Life cycle of *Meloidogyne graminicola* on paddy and its host range studies. *Indian Journal of Nematology.* 34: 80-84.

Ganguly, A. K., Pankaj, A. and Sirohi, A. (1996). Effect of soil solarization of rice nursery beds to suppress plant parasitic nematodes. *International Rice Research Notes.* Los Baños, Philippines, pp. 2-3.

Gaur, H. S., Singh, J., Sharma, S. N. & Chandel, S. T. (1996). Distribution of community analysis of plant parasitic nematodes in rice growing areas of Haryana, India. *Annals of Plant Protection Sciences.* 4: 115-121.

Golden, A. M. & Birchfield, W. (1968). Rice root-knot nematode *(Meloidogyne graminicola*) as a new pest of rice. *Plant Disease Reporter.* 52: 423.

Goodey, T. (1933). Plant parasitic nematodes- *Anguillnlina oryzae* (van Breda de Haan) Goodey 1932, a nematode parasite of the roots of rice, *Oryza sativa* L. *Journal of Helminthology.* 14:107-112.

Halbrendt, J. (1997). Root-knot Nematodes: An IPM CRSP pest profile. In: Update Integrated Pest Management, Collaborative Research Support Program. Virginia Polytechnic Institute and State University. USA. 3:2.

Haan, de B. & Van, J. (1902). Een aaltjesziekte der rijst, omo mentek of omo bambang. Mededeelingen uit's Lands Plantentium. L. I 11:1-65.

Hoque, M. O. and Talukdar, M. J. (1971). A new disease of rice caused by the nematode *Meloidogyne* sp. Proceedings of Pakistan Science Conference, Peshawar.

Ichinohe, M. (1972). Nematode disease of rice. Economic Nematology. J. M. Webster, ed. Academic Press, London, UK. pp. 127-143.

Israel, P., Rao, Y. S. & Rao, Y. R. V. J. (1963). Investigations on nematodes in rice and rice soils. *Oryza* 1: 125-127.

Jain, R. K., Mathur, K. N. & Singh R. V. (2007). Estimation of losses due to plant parasitic nematodes on different crops in India. *Indian Journal of Nematology*. 37: 219-221.

Jairajpuri, M. S., & Baqri, Q. H. (1991). Nematode Pests of Rice. New Delhi, India, IBH Publishing Co. pp. 66.

Jonathon, E. I. &Pandirajan, P. (1991). Effect of organic amendments in controlling rice root nematodes. *International Rice Research Newsletter*. 16: 25

Jonathon, E. I & Velayutham, B. (1987). Evaluation of yield loss due to rice root nematode, *Hirschmaniella oryzae*. *International Nematological Network Newsletter*. 4: 8-9

Kinh, D. N., Huong, N. M. & Ut, N. U. (1982). Root-knot disease of rice in the Mekong Delta, Vietnam. *International Rice Research Newsletter*. 7: 15.

Kumar, A., Lal, H. C. &Akhtar, J. (2010). First report of root-knot nematode of rice from Jharkhand, India. *Journal of Mycology and Plant Pathology*. 40: 296-297.

Kumar, D. and Singh, K. P. (2006). Assessment of Predacity and effect of *Arthrobotrys dactyloides* for biological control of root-knot disease of tomato. *Journal of Phytopathology*. 154: 1-5.

Kumar, K. & Sinha, R. K. P. (1996). First record of *Meloidogyne graminicola* on rice from Bihar. *Oryza* 33: 289.

Kuriyan, K., Bai, H. & Ushakumari, R. (1985). Rice root nematode, *Hirschmaniella oryzae* and burrowing nematode, *Radopholus similis* in the soils of Kerala state. *Indian Journal of Nematology*. 15: 280.

Le, T. T. H., Padgham, J. L. &Sikora, R. A. (2009). Biological control of the rice root-knot nematode, *Meloidogyne graminicola* on rice, using endophytic and rhizosphere fungi. *International Journal of Pest Management*. 55: 31-36.

Manser, P. D. (1968). *Meloidogyne graminicola* - a cause of root-knot of rice. *Plant Protection Bulletin* F. A. O. 16: 11.

Manser, P. D. (1971). Notes on the rice root-knot nematode in Laos. *Plant Protection Bulletin.* F. A. O. 19: 138-139.

Mathur, V. K. & Prasad, S. K. (1972). Role of the rice root nematode, *Hirschmanniella oryzae* in rice culture. *Indian Journal of Nematology*. 2: 158-167.

Mathur, V. K. & Prasad, S. K. (1973). Survival and host range of the rice root nematode, *Hirshcmaniella oryzae. Indian Journal of Nematology*. 3: 88-93.

Miah, S. A. & Bakr, M. A. (1977). Chemical control of ufra disease of rice. PANS 23 (4) : 412-413.

Miah, S. A. & Rahman, M. L. (1985). Severe ufra outbreak in transplanted rice in Bangladesh. *International Rice Research Newsletter*. 10: 24.

Miah, S. A., Shahjahan, A. K. M., Sharma, N. R., Shah, A. L. & Bhuyian, N. (1984). Interaction of zinc deficiency and ufra disease on rice in Bangladesh. *Bangladesh Journal of Agriculture*. 9: 61-67.

Munir, A. & Bridge, J. (2003). Rice root-knot nematode *Meloidogyne graminicola* (Golden and Birchfield 1965) from rice in Pakistan. *Pakistan Journal of Nematology*. 21: 133-135.

Muthukrishnan, T. S., Rajendran, G., Ramamurthy, V. V. & Chandrasekaran (1977). Pathogenicity and control of *Hirschmaniella oryzae. Indian Journal of Nematology*. 7: 8-16.

Netscher C, Erlan. (1993). A root-knot nematode, *Meloidogyne graminicola*, parasitic on rice in Indonesia. *Afro-Asian Journal of Nematology*. 3:90-95

Ou, S. H. (1985). Rice Diseases. 2nd ed. Commonwealth Mycological Institute, Kew, England, UK.

Padgham, J. L., Duxbury, J. M., Mazid, A. M. & Abawi Hossain, M. (2004). Yield loss caused by *Meloidogyne graminicola* on lowland rainfed rice in Bangladesh. *Journal of Nematology*. 36: 42-48.

Padgham, J. L. & Sikora, R. A. (2007). Biological control potential and modes of action of *Bacillus megaterium* against *Meloidogyne graminicola* on rice. 26: 971-977.

Pal, A. K. & Jayaprakash, A. (1983). Root-knot nematode damage to rice in West Bengal, India. *IPR Newsletter*. 8: 14-15.

Panda, M. &Rao, Y. S. (1971). Evaluation of losses caused by the rice root-nematode *(Hirchmaniella mucronata). Indian Journal of Agricultural Sciences*. 41: 611-614.

Plowright, R., & Bridge, J. (1990). Effect of *Meloidogyne graminicola* (Nematoda) on the establishment, growth and yield of rice cv. IR 36. *Nematologica*. 36: 81-89.

Pokhrel, R. R. (2001). Rice cropping patterns and plant parasitic nematodes. *Journal of Institute of Agriculture and Animal Science.* 21-22:187-196.

Prasad, J. S., Panwar, M. S., & Rao Y. S. (1985). Occurrence of root-knot nematode, *Meloidogyne graminicola* in semi-deepwater rice. *Current Science.* 54: 387-388.

Prasad, J. S., Panwar, M. S. & Rao, Y. S. (1987). Nematode problems of rice in India. *International Journal of Pest management.* 33:127 – 136.

Prot, J. C. (1993). Monetary value estimates of nematode problems, research proposal and priorities: The rice example in South and Southeast Asia. *Fundamentals and Applied Nematology.* 16: 385-388.

Prot, J. C. & Matias, D. M. (1995). Effect of water regime on the distribution of *Meloidogyne graminicola* and other root- parasitic nematodes in a rice field toposequence and Pathogenicity of *M. graminicola on* rice cultivar UPL Ri 5. *Nematologica.* 11: 219-228.

Prot, J. C; Soriano, I. R. S. & Matias, D. (1994). Major root parasitic nematodes associated with irrigated rice in the Philippines. *Fundamentals and Applied Nematology.* 17: 75-78.

Raalte, M. H. van. (1943). On the oxidation of the environment by the roots of rice *(Oryza sativa* L). Syokubutu-Iho 1:15-33.

Rahman, M. L., Sharma, N. R. & Miah, S. A. (1981). Incidence and chemical control of ufra in boro fields. *International Rice Research Newsletter.* 6: 12.

Randhawa, N., Khera, S., Singh, I. & Sakhuja, P. K. (1990). Plant parasitic nematodes associated with rice in Punjab. *Plant Disease Research.* 5:59-63.

Randhawa, N., Singh, I., Sakhuja, P. K. & Malhi, S. S. (1991). Effect of date of transplanting of Basmati rice cultivars and spacing on the population build up of rice root nematode, *Hirschmaniella oryzae. Indian Journal of Nematology.* 21: 131-135.

Rao, Y. S. & Biswas, H. (1973). Evaluation of yield losses in rice due to the root-knot nematode, *Meloidogyne graminicola. International Journal of Nematology.* 3: 74.

Rao, Y. S. & Israel, P. (1971). Studies on nematodes of rice and rice soils. Influence of soil chemical properties on the activity of *Meloidogyne graminicola,* the rice root-knot nematode. *Oryza.* 8: 33-38.

Rao, Y. S. & Israel, P. (1972). Effect of temperature on hatching of eggs of the rice root-knot nematode, *Meloidogyne graminicola. Oryza.* 9: 73-75.

Rao, Y. S. & Israel, P. (1973). Life history and bionomics of *Meloidogyne graminicola* the rice root-knot nematode. *Indian Phytopathology.* 26: 333-337.

Rao, Y. S. & Jayaprakash, A. (1978). *Heterodera oryzicola* n. sp. (Nematoda: Heteroderidae) a cyst nematode on rice *(Oryza sativa* L.) from Kerala state, India. *Nematologica.* 24: 341-346.

Rao, Y. S., Jayaprakash, A. & Mohanti, J. (1988). Nutritional disorders in rice due to infestation by *Heterodera oryzicola* and *Meloidogyne graminicola. Revue de Nématology*. 1: 375-380.

Rao, Y. S., Prasad, J. S. & Panwar, M. S. (1986). Nematode problem in rice: crop losses, symptomatology and management. In: Plant parasitic nematodes of India. Problems and progress, pp. 279-299. Eds G. Swarup and D. R. Dasgupta. New Delhi: Indian Agricultural Research Institute.

Sahu, S. C. & Chawla, M. L. (1986). A new virulent strain of rice root-knot nematode, *Meloidogyne graminicola* from Agartala, India. *International Rice Research Newsletter*. 11: 40.

Sarla, N. & Mallikarjuna, S. B. P. (2005). *Oryza glaberrima*: A source for the improvement of *Oryza sativa. Current Science*. 89: 955-963.

Sein, T. (1977). Varietal resistance to ufra disease in Burma. *International Rice Research Newsletter*. 2: 3.

Sein, T. & Zan, K. (1977). Ufra disease spread by water flow. *International Rice Research Newsletter*. 2: 5.

Sharma, N. N., Edward, J. C., Mishra, S. L. & Chandrasekhar, M. (1992). Studies on Phytonematodes associated with important rice growing tracts of India. *Current Nematology*. 3: 107-109.

Sharma, R. & Prasad, J. S. (1995). First record of *Meloidogyne graminicola* on rice in Andhra Pradesh. *Oryza*. 32: 59.

Sher, S. A. (1968). Revision of the genus *Hirschmaniella* Luc and Goodey, 1963. *Nematologica*. 14: 243-275.

Singh, I. & Jain, R. K. (1995). Nematode problems of rice crop in India. I: Rice root nematode. Nematode Pest Management: An appraisal of eco-friendly approaches. pp. 140-150.

Singh, K. P., Jaiswal, R. K. & Kumar, N. (2007a). *Catenaria anguillulae* Sorokin: A natural biocontrol agent of *Meloidogyne graminicola* causing root-knot disease of rice *(Oryza sativa* L.). *World Journal of Microbiology and Biotechnology*. 23: 291-294.

Singh, K. P., Jaiswal, R. K., Kumar, N. & Kumar, D. (2007b). Nematophagous fungi associated with root galls of rice caused by *Meloidogyne graminicola* and its control by *Arthrobotrys dactyloides* and *Dactylaria brochopaga. Journal of Phytopathology*. 155: 193-197.

Singh, K. P., Jaiswal, R. K., Kumar, N. &Kumar, D. (2006). Biomassof nematodes & associated roots:A determinant of symptom production in root-knot disease of rice *(Oryza sativa*). *Journal of Phytopathology*. 154: 676-682.

Singh, V. K., Kalia, C. S. & Kaul, V. (2007c). New record of root-knot nematode, *Meloidogyne graminicola* infecting rice in Jammu. *Indian Journal of Nematology*. 37: 94.

Soriano, I. R. S., Prot, J. C. & Matias, D. M. (2000). Expression of tolerance for *Meloidogyne graminicola* in rice cultivars as affected by soil type and flooding. *Journal of Nematology*. 32: 309-317.

Soriano, I. R. S. &Reversat, G. (2003). Management of *Meloidogyne graminicola* and yield of upland rice in SouthLuzon, Philippines. *Nematology*. 5: 879-884 .

Stirling, G. R. & Smith, L. J. (1998). Field test of formulated products containing either *Verticillium chlamydosporium* or *Arthrobotrys dactyloides* for biological control of root-knot nematodes. *Biological Control*. 11: 231-239.

Stirling, G. R., Smith, L. J., Licastro, K. A. & Eden, L. M. (1998). Control of root-knot nematode with formulation of nematode trapping fungus *Arthrobotrys dactyloides*. *Biological Control*. 11: 224-230.

Subramanian, P. & Velayuthan (1983). Control of rice root nematode. *International Rice Research Newsletter*. 8: 15-16.

Tandingan, I. C., Prot, J. C., & Davide, R. G. (1996). Influence of water management on tolerance of rice cultivars for *Meloidogyne graminicola*. *Fundamentals of Applied Nematology*. 19: 189-192.

Thio, B. (1992). Laboratory study on actions of *Orseolia oryzivora* (Diptera:Cecidomyiidae) and *Hirschmaniella spinicaudata* (Tylenchina: Pratylenchidae) on rice. Dissertation thesis, Université de Ouagadougou, Burkina Faso.

Todd, E. H. & Atkins, J. G. (1959) White tip disease of rice. II. Seed treatment studies. *Phytopathology* 49: 184-188.

Usha, K andKuriyan, K. J. (1981). Cyst nematode, *Heterodera oryzicola* on rice in Kerala. 1. Estimation of loss in rice due to *Heterodera oryzicola* infestation in field conditions. *Indian Journal of Nematology*. 11: 106.

Vecht, J., Vander&Bergman, B. H. H. (1952). Studies on the nematode, *Radopholus similis* (van Breda de Haan) Thorne and its influence on growth on the rice plant. Pemberitan Balai Besai Penjelidikan Pertanian Bogor No. 131:1-75.

Venkitesan, T. S., Nalinakumari, T. & Satyakumar, J. (1979). Distribution of rice root nematode in Kerala. *Indian Journal of Nematology*. 9:55.

Vuong, H. H.; Rodriguez, H. (1970) Lutte contre les nématodes du riz à Madagascar (résultats d'expérimentation 1968-1969). *Agronomie Tropicale*. 25: 52-66.

Yik, C. P. and Birchfield W. (1979). Host studies and reactions of cultivars to *Meloidogyne graminicola*. *Phytopathology*. 69: 497-499.

Phyto-nematodes in Crops: Their Identification, Treatment and Management, 2011
G.L. Sharma (editor), pp. 155-168
New India Publishing Agency, New Delhi (India)
E-mail : info@nipabooks.com; Website : www.nipabooks.com

Chapter 7

Nematode Problems of Seed Spice Crops and their Management

I. J. PARUTHI

Introduction

Seed spices have prominent place in India's national economy because of their very high domestic consumption and growing demand for export. This group of spices, being consisting of mostly annual crops, is grown extensively in rotation with food and cash crops and also as inter crops under rain fed as well as irrigated conditions.

There are about 20 seed spice crops grown in India. Out of which nine are important in terms of production and export. The most important ones are coriander, cumin, fennel and fenugreek (major seed spices) and ajowain, aniseed, black cumin, celery and dill (minor seed spice crops). Although these crops are grown all over India but Haryana, Rajasthan, Madhya Pradesh, Uttar Pradesh, Orissa, Tamil Nadu and Andhra Pradesh are the main seed spice growing states.

In Haryana, fenugreek and coriander are the important seed spice crops grown on a large scale whereas fennel and cumin are cultivated only on a very small scale. In Rajasthan, fenugreek, coriander, cumin and fennel are grown on a large scale throughout the state whereas minor seed spices like Ajowain, celery and dill are cultivated in some pockets only. Several biotic and abiotic stresses affect the successful cultivation of these crops. Besides insect pests and other disease causing agents, several species of nematodes also attack these crops and cause significant losses. However, very little information on these organisms is available in literature because of their microscopic size, hidden mode of life and production of non diagnostic symptoms on the above ground parts of the plants.

Although a large number of nematodes like *Hoplolaimus, Tylenchorhynchus, Helicotylenchus, Psilenchus, Coslenchus, Basiria, Meloidogyne, Paratylenchus, Pratylenchus, Heterodera, Ditylenchus, Aphelenchus, Rotylenchus, Rotylenchulus reniformis*, have been found associated with seed spice crops of Haryana, Rajasthan and other states of India (Anon, 2008), but root knot nematode (*Meloidogyne* spp.) lesion nematode (*Pratylenchus* spp.) and reniform nematode (*Rotylenchulus reniformis*) are considered as the major nematode pests of these crops due to their endo and semiendoparasitic nature. This chapter deals with various aspects of important nematode pests of seed spice crops.

A. Major Nematodes

1. Root-knot nematode (Meloidogyne spp.*)*

Root-knot nematode (*Meloidogyne* spp.) is considered as the most important nematode of seed spice crops because of its widespread occurrence, endoparasitic nature and wide host range. *M. incognita* and *M. javanica* have been most frequency found associated with these crops in India. Other species, *M. hapla* and *M. arenaria* have also been reported to parasitize cumin, fennel, coriander, ginger and turmeric from Himachal Pradesh, Maharastra and Orissa states.

Historical back ground: Occurrence of *M. incognita* on some seed spice crops from Haryana and Rajasthan was reported by Paruthi *et. al*. (1987) and Midha & Trivedi (1988b) respectively. Kumar and Paruthi (2007) reported the common occurrence of Meloidogyne spp. on coriander, cumin, fennel and fenugreek crops in Haryana and Rajasthan. Jain and Paruthi (2005) reported that fenugreek and coriander as good hosts whereas fennel and cumin shared poor and moderate host response, respectively.

Host range: Fenugreek, coriander, cumin and fennel crops are considered as hosts of this nematodes but their susceptibility varies according to nematode species, crop plant and nematode inoculum level. Jain and Paruthi (2005) reported that fenugreek and coriander showed good response to *M. incogntia* whereas cumin and fennel showed moderate and poor response, respectively with a root knot index at 2.6 and 1.5. However, Patel *et.al.* (2002a) from Gujarat reported fennel as a good host of *M. javanica*.

Anon (2006) studied response of commonly available species of root knot nematodes (*M. incognita, M. javanica* and *M. graminicola*) of Haryana on fenugreek, coriander, cumin and fennel). *M. incognita* and *M. javanica* showed almost similar responses on all the seed spice crops but the response of *M. incognita* was comparatively more than that of *M. javanica*. All the crops showed non host response to *M. graminicola*.

Distribution: Occurrence of *M. incognita* and *M. javanica* is found in all spice growing areas of the country whereas incidence of *M. arenaria* and *M. hapla* has been reported from Himachal Pradesh, Gujarat, Maharashtra and Orissa.

Nature of damage: General above ground symptoms caused by this nematode are stunted growth, yellowish leaves with reduced branches/ tillering resulting in poor and fewer flowers, pods and seeds.

The roots of such infested plants are normally deformed and show some bead like structures known as galls or knots. The size and number of galls depends upon host response, nematode species and nematode inoculum present in the soil. On fennel and cumin, generally, small sized galls are formed. Upon teasing, in such galls or knots, large number of different stages of nematodes can be seen inside the roots. *M. arenaria* and *M. hapla* produce comparatively small sized knots than *M. incognita* and *M. javanica*.

Pathogenicty: Pathogenic response of *Meloidogyne spp.* on spice crops has been reported to vary from crop to crop and nematode species (Kumar and Pathak, 2004; Paruthi *et al.*, 2008).

Paruthi *et al.* (2008) reported that 100 J2 of *M. incognita*/kg soil to be pathogenic to coriander, cumin and kasuri fenugreek whereas pathogenic level in desi fenugreek and fennel was 1000 J2/kg soil (Fig. 1, 2&3). Midha and Trivedi (1988a) also reported 100 J2 of *M. incognita* to be pathogenic to coriander crop from Rajasthan. Kumar and Pathak (2004) reported 500J2 of *M. incognita* as pathogenic level in fenugreek.

Histopathology: Histopathological studies revealed that giant cells induced by nematode are produced in the vascular tissues which have cell wall invagination to draw nutrients from adjoining cells. These giant cells are multinucleate cells produced by mitosis without cytokinesis (Dropkin, 1969). Anon (2006) studied histopathology of *M. incognita* in fenugreek, coriander, cumin and fennel (Fig. 4 A-D). The study revealed well developed adult females and conspicuous giant cells in fenugreek (Fig. 4A) and coriander (Fig. 4B) indicating their susceptible response while in cumin (Fig. 4C) and fennel (Fig. 4D) moderate development of nematode as well as giant cells indicated their less response to the feeding of nematode.

Sharma and Trivedi (1993) reported that infestation of *M. incognita* in fenugreek plants showed increase in all metabolites in galls compared to healthy roots. Insoluble polysaccharides and proteins were abundant in giant cell walls and nematode bodies. Lignin and ascorbic acid were mainly present in the xylem and giant cell cytoplasm, respectively. Starch and lipids occurred in the nematode body

Losses: Avoidable losses due to *M. incognita* have been reported as 51.6% in coriander, 43.3% in cumin and 41.5% in fennel by using carbofuran @ 2kg a.i./ ha (Midha and Trivedi, 1991). Patel *et al*. (2002a) reported 21 per cent loss in grain yield of fennel due to *M. javanica* pathotype-2 from Gujarat.

Life Cycle: Root knot nematodes (*Meloidogyne spp*.) are sedentary endoparasites. The second stage juvenile is the infective stage. The larvae penetrate the root system and start swelling and develop into third and fourth stages very quickly. The females attain sac like shape and lay eggs on root surface in a gelatinous matrix known as egg mass. Each egg mass contains hundreds of eggs of this nematode. Reproduction in these nematodes is generally by parthenogenesis.

The life cycle of *M.incognita* was studied in two cultivars of *Trigonella foenum gracium* Sharma and Trivedi (1992a). They reported that J2 entered the roots within 34 hours of inoculation.Egg formation was seen in the female body 36 days after inoculation in susceptible cv. UM-84 and 44 days after inoculation in resistant cv.UM-34. The life cycle was completed in 42 to 44 days in UM-84 and in 48-50 days in UM34

Anon (2008) studied the life cycle development of *M. incognita* on fennel from Haryana. The nematode did not complete its generation on fennel under pot conditions. Although various stages of development, except males, were observed but more than 70 per cent of the females were disintegrated and did not produce eggs and egg masses on the infested roots.

Management

Cultural practices: Addition of leaf powder of *Aegle marmelos, Withania somnifer, Hibiscus rosasinensis, Murraya koenigi* and stem of *Cuscuta reflexa* significantly enhanced the plant growth of fenugreek and reduced galling by *M. incognita* (Sharma and Trivedi, 1992b). Application of neem and mustard cake @ 5g/kg soil has been found effective against *M. incognita* in fenugreek, coriander and cumin (Anon, 2006 & 2007). Paruthi and Nandal(2008a)reported the efficacy of neemate (neem based pesticides)against *M.incognita* in cumin. Patel *et al.*(2002b) tested soil solarization with 25LLDPE film, rabbing with castor @ 7 kg/ m2, mustard cake @ 5t/ ha and carbofuran @ 3 kg a.i./ ha as nursery bed treatment against *M. javanica* pathotype 2 in fennel. Among all treatments, soil solarization and rabbing were found to be the most effective in improving seedling growth and transplantable seedlings by reducing root knot disease and weeds.

Use of resistant varieties: Planting nematode resistant varieties is of utmost importance to check the nematode spread and also to avoid crop losses due to phytonematodes. A large number of varieties of seed spices found resistant against *Meloidogyne spp.* have been listed in Table 1.

Biological control : Midha *et al.* (2001) reported the efficacy of *Paecilomyces lilacinus* @ 2g powder/ kg soil against *M. incognita* in coriander. RKB-68, RKB-91 strains of *Bacillus* spp. have been found effective against *M. incognita* in fenugreek, coriander and cumin (Anon, 2007, and Paruthi *et al.*, 2009).

Chemical Control

Seed treatment: Use of carbosulfan, as seed soaking @ 1000 ppm concentration on a.i. basis for 6 hrs. or as seed dressing @ 2-3% (w/w on a.i. basis) has been found effective against *M.incognita* in fenugreek, coriander and cumin. However performance of seed dressing was better than the seed soaking treatment (Anon, 2006 and 2007).

Soil treatment: Application of carbofuran or phorate @ 1kg a.i./ha has been found effective against *M. incognita* in fenugreek and coriander (Midha and Sharma, 2008). Patel *et al.* (2002b) reported the efficacy of carbofuran as nursery bed treatment @ 3kg per hectare against *M. javanica* pathotype-2 in fennel.

2. *Lesion nematode* (*Pratylenchus* spp.)

This is an important nematode pest of seed spice crops due to its migratory endoparasitic nature and wide host range. Anon (2006) while surveying of Haryana and Rajasthan recorded its incidence in 5.6 to 34.9% samples of fenugreek, coriander, cumin and fennel crops from Rajasthan and Haryana states.

Host range: This nematode is a polyphagous pest and has also been reported from spice growing areas of Rajasthan and Haryana on fenugreek,fennel coriander and cumin,dill celery etc. (Anon,2006 & 2007)

Symptoms: Not much is reported about symptomatology on attack of lesion nematode on seed spices. However, stunted growth, yellowing of leaves and dark brown lesions or discoloration on the roots are the general symptoms of its attack.

Biology: *Pratylenchus spp.* are migratory endo parasitic nematode. The nematode enters roots and moves intercellularly and feeds on cortical cells. The nematode larvae and adults both have the capacity to leave and enter roots. Females lay eggs singly within root system or in soil. The second, third and fourth stages feed on roots. The nematodes with in the roots move freely through the channels between radial lamellae of the cortical cells forming cavities and eggs are deposited within the roots.

Management: Not much information is available on the management of this nematode seed spice crops.

3. *Reniform nematode (Rotylenchulus reniformis)*

This is a semi-endoparasitic sedentary nematode. In Haryana and Rajasthan it is found associated with the rhizosphere of fenugreek, fennel and cumin crops in 5-15% of the samples (Anon, 2006 & 2007).

Hosts: This is also a polyphagous pest and has been found to infect various seed spice crops like fenugreek fennel and cumin in Haryana and Rajasthan (Anon, 2006 & 2007).

Symptoms: This nematode is reported to cause stunted growth, yellowish leaves and poor branching.

Biology and life cycle: Although not much information is available on life cycle and biology of this nematode in seed spice crops. However Siva Kumar and Seshadri (1971) have studied its life cycle in detail. Normally young females enter the roots near the root tip. During feeding, their head position remains embedded in the roots cortex while the posterior part of the nematode remains outside in the soil. Near the head region 2-3 cells in the cortex are stained deeper than the surrounding ones known as nurse cells. The female starts feeding on these cells and assumes reniform shape with swollen posterior part and egg sac. This nematode is known to complete its life cycle with in 21 days at the optimum temperature.

Management: Not much information is available on management aspects of this nematode in spice.

B. Minor Nematodes

Besides above mentioned nematodes, some other nematodes like stunt nematodes (*Tylenchorhynchus spp*), lance nematode (*Hoplolaimus spp.*) and *Helicotylenchus* spp. (Spiral Nematode) have also been reported to infect seed spice crops in Haryana and Rajasthan (Anon, 2006 and 2007). Very meager information is available on these nematodes in spite of their occurrence in large number in some of the samples of seed spices. These nematodes are considered of minor importance because of their ectoparasitic nature.

Conclusion and Futuristic Approaches

Survey of seed spice crops of Haryana and Rajasthan though revealed the presence of several phytonematodes but on the basis of frequency of occurrence and their nature of damage most attention has been paid on root knot nematode (*Meloidogyne* spp.). The other nematodes like *Pratylenchus* spp. and *R. reniformis* have not been properly attended in spite of their high frequency of occurrence in seed spice crops. The work on the ecto-parasitic nematodes like- *Tylenchorhynchus* spp., *Helicotylenchus* spp. and *Hoplolaimus* is also needed to be initiated due to their availability in large number in some of the samples of seed spices. Seeing the importance of these crops in the economy of our country, complete intensive survey of seed spice growing areas of various important states like Haryana, Rajasthan, Gujarat, Madhya Pradesh and Uttar Pradesh is needed to be under taken for preparation of maps showing the incidence, spread and prevalence of key nematode pests of these crops

Studies on avoidable losses due to these nematodes at farmer's fields need to be undertaken in major seed spice crops in order to understand the economic implications of the damages caused by these pests.

Fig. 1 : Pathogenicity of *M.Incognita* on seed spice crops. Inoculum levels (L-R); A-C, 0, 10, 100, 1000, 10000 J_2 per plant

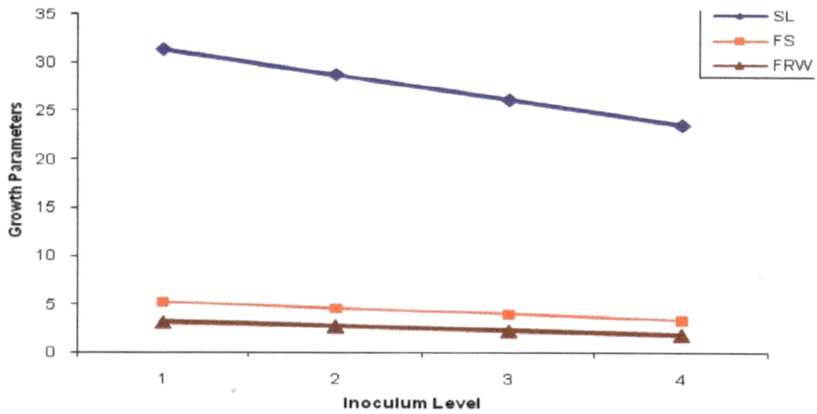

Fennel

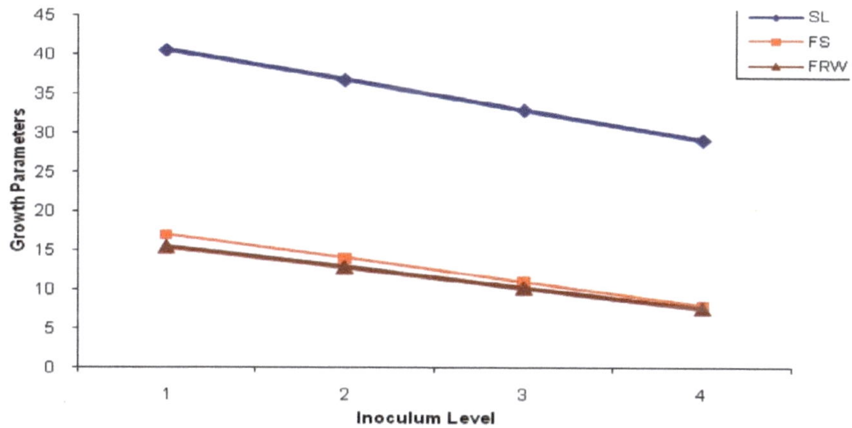

Fenugreek

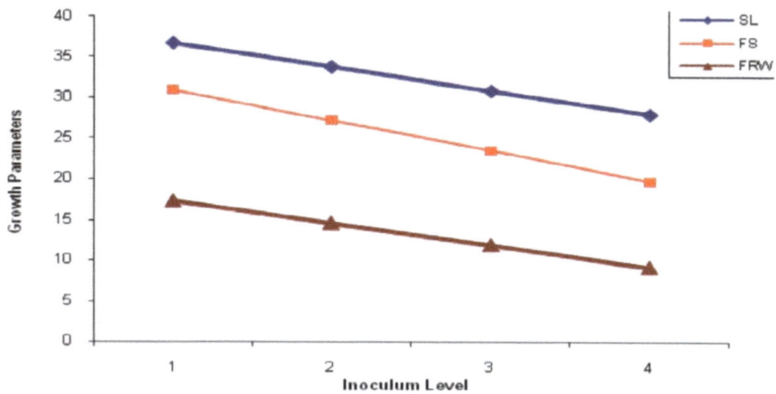

Fenugreek (Kasuri)

Fig. 2 : Fitted regression line (y = a + b log x) of *Meloidogyne incognita* on seed spice crops.

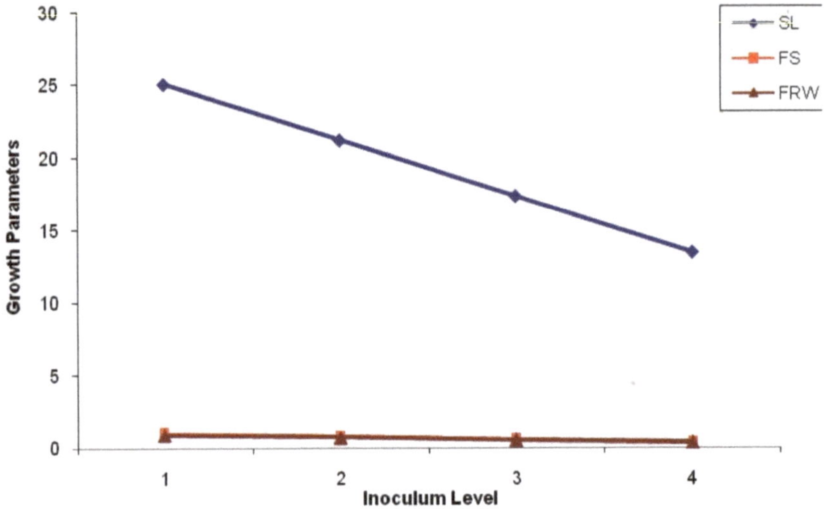

Cumin

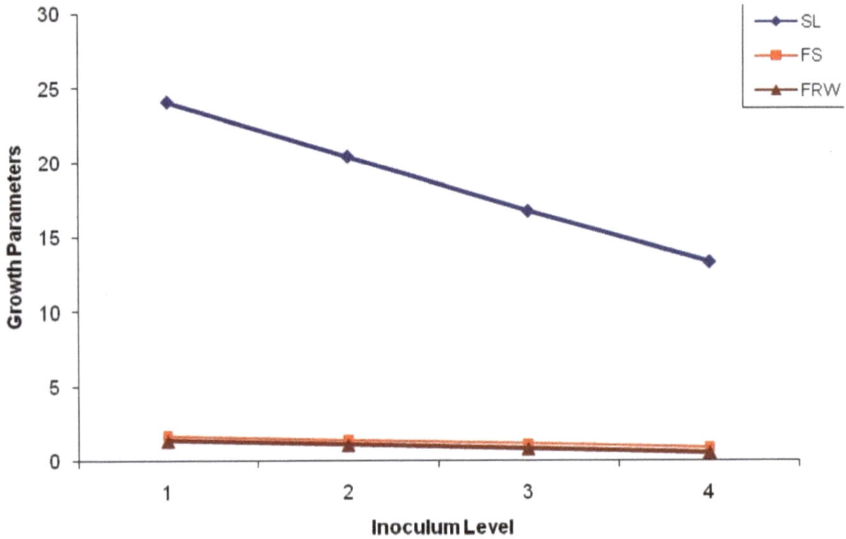

Coriander

Fig. 3 : Fitted regression line (y = a + b log x) of *Meloidogyne incognita* on seed spice crops.

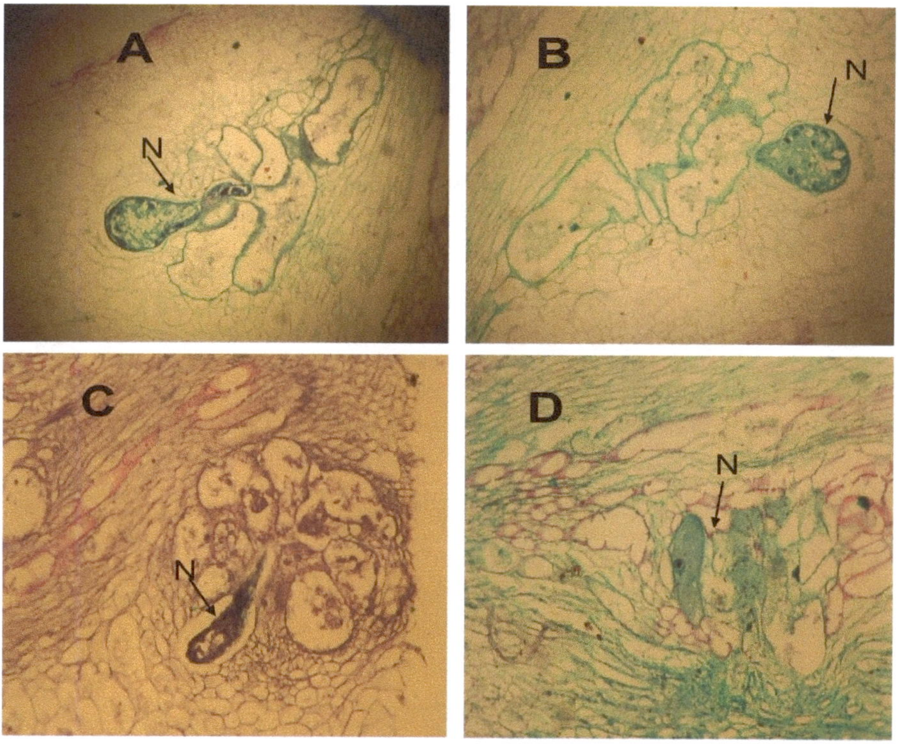

Fig. 4 : Histopathological changes due to *M. Incognita* in of seed crops (**A**-Fenugreek, **B**-Coriander, **C**-Fennel, **D**-Cumin).N- Nematode Feeding of Giant cells.

Table 1 : Germplasm of seed spice crops showing resistant response against *Meloidogyne* spp.

Crop	Lines/varieties	Response	Reference
Fenugreek	VLM-9, 24, 67, 112, 113 & 127	Resistant	Sharma & Trivedi (1988)
	NLM, TG 2336, VM34 & 35	Resistant	Sharma *et al.* (1988)
	Selection-15	Moderately	Thakar *et al.* (1987)
	UM32, 34, 117, 118, 122, 128, 143, 144	Resistant	Yadav & Chand (2001)
	HM335 and 357	Resistant	Anon (2005)
Coriander	Co-1, Co-2	Resistant	Midha & Trivedi (1988b)
	CS-4, CS-6, UD-20, 21, 124, 239, 241, 296, 375 and 475	Resistant	Yadav & Chand (2001)
Cumin	CVT RS-1, CVT JC-1, CVT JC-2, CVT2C-3, CVT VC-43	Resistant	Midha & Trivedi (1989)
	BTC-19	M.R.	Patel *et al.* (1986)

Work on host-parasite relationship (i.e. host status, pathogenicity, life history, *histopathology* etc) of other associated nematodes like *Pratylenchus spp* and *R.* reniformis which are not much knocked so far may also be under taken on priority.

Further work on screening and confirmation of resistance in germplasm of seed spice crops against key nematode pests also needs to be undertaken.

Work on use of rhizobacterial strains for managing nematode pests of these crops needs further strengthening.

Field evaluation of different effective management strategies should be tested at multi locations at the farmers fields on a large scale to find out the economically viable long lasting and effective practices integrated for management of key nematodes pests of seed spices.

References

Anon, 2005. First annual report of 'Studies on nematode problems of seed spice crops', Department of Nematology, CCS HAU, Hisar, 7pp.

Anon, 2006. Second annual report of 'Studies on nematode problems of seed spice crops', Department of Nematology, CCS HAU, Hisar, 41pp.

Anon, 2007. Third annual report of 'Studies on nematode problems of seed spice crops', Department of Nematology, CCS HAU, Hisar, 23pp.

Anon, 2008.Annual report of department of Nematology, CCS HAU Hisar

Dropkin, V. H.1969.Cellular response of plants to nematode infestation. *Ann.Rev. Phytopath.*7:101-122.

Jain, R.K. and Paruthi, I.J. 2005. Host status of different seed spices against root-knot nematode (*M. incognita*). National symposium on recent advances and research priorities in Indian Nematology. Held at IARI, New Delhi from 9-10 Dec. 2005: 21.

Kumar, A. and Paruthi, I.J. 2007. Prevalence and incidence of root-knot nematode (*Meloidogyne* spp.) in seed spice crops of Haryana and Rajasthan. National Symposium on Nematology in 21st Century: Emerging paradigms (22-23 Nov. 2007) held at Assam Agricultural University, Jorhat, p. 60.

Kumar, M. and Pathak, K.N.2004.Advers effect of root-knot nematode, *Meloidogyne incognita* on spinach beet, *Beta vulgaris bengalensis* and Fenugreek. *Trigonella foenum graecum*. *Indian J. Nematol.* 34: 123-128.

Midha, R.L. and Sharma, G.L.2008.Management of *Meloidogyne* on coriander on cumin. National seminar on Recent trends in research on spices and aromatic plants(ed .by.Arora, S.K., Partap, P.S., Batra, V.K, Bhatia, A.K.Malik,T.P. and Telhan,S.K.) held at CCS HAU, Hisar, Haryana.p.82.

Midha, R.L. and Trivedi, P.C. 1988a. Effect of different inoculum levels of *M. incognita* on coriander. *Indian J. Nematol.* 18: 372-373.

Midha, R.L. and Trivedi, P.C. 1988b. Evaluation of coriander varieties against *Meloidogyne incognita. Indian J. Nematol.* 18: 375.

Midha, R.L. and Trivedi,P.C.1989. Evaluation of cumin varieties against root-knot nematodes. *Meloidogyne incognita. Indian J. Nematol.* 19: 264.

Midha, R.L. and Trivedi, P.C.1991.Estimation of losses by Meloidogyne incognita on coriander, cumin and fennel. *Current Nematology* 2:159-162.

Midha, R.L., Trivedi, P.C. and Yadav, B.D. 2001. Biocontrol agents against *Meloidogyne incognita* on coriander. National congress on centenary of Nematology in India appraisal and future plans. Held at IARI, New Delhi on 5-7 Dec., 2001: 146-147.

Paruthi, I.J.,Jain,R.K.and Gupta, D.C. 1987. A note on reaction of some spices to root-knot nematode (*Meloidogyne javanica).Haryana J.Hort. Sci.* 16: 154-155.

Paruthi, I.J., Kumar, A. and Nandal, S.N. 2008. Pathogenic response of root-knot nematode on seed spice crops of Haryana. National seminar on Recent trends in research on spices and aromatic plants(ed .by.,Arora, S.K., Partap, P.S., Batra, V.K, Bhatia, A.K. Malik, T.P. and Telhan, S.K.,) held at CCS HAU, Hisar, Haryana.p.83.

Paruthi, I.J. and Nandal, S.N. 2008a. Efficacy of neemate (Neem based pesticide) as soil treatment against root-knot nematode in seed spices. IBID.p.84.

Paruthi, I.J. and Nandal S.N. 2008b. Nematode problems of spice crops of northern India and their management. National seminar on Recent trends in research on spices and aromatic plants(ed .by. Telhan, S.K., Malik, T.P. Bhatia, A.K. Arora, S.K. and Batra V.K.) held at CCS HAU, Hisar, Haryana.pp.143-149.

Paruthi, I .J, Bansal R k and Nandal, S.N 2009.Evaluation of antagonistic efficacy of various rhizobacterial strains against root-knot nematode (*Meloidogyne incognita*) in fenugreek. Proceedings of workshop on prospects and problems in production, processing and marketing of medicinal and aromatic plants in Haryana. Held at CCSHAU, Hisar, March, 20-21, 2009, p85.

Patel, D.J., Patel, H.R. and Patel, C.C. 1989. Susceptibility of spices to root-knot nematodes. *In: Proc. First National Seminar on Seed Spice*, 24-25 October, 1989, Jaipur, India, pp. 256-265.

Patel, D.J., Patel, S.T. and Jogani, D.K. 1986. Preliminary screening of cumin varieties against root-knot nematodes.*Indian J.Nematol.*16:130.

Patel, R.G., Patel, M.R., Vyas, R.V., Patel, B.A. and Patel, B.N. 2002 a. Assessment of avoidable yield losses due to root knot nematode (*Meloidogyne javanica*) pathotype 2 in fennel field. National symposium on biodiversity and management of nematodes in cropping systems for sustainable agriculture 11-13 Nov. 2002 at Department of Nematology, Agriculture Research Station, Durgapura, Rajasthan., p 36

Patel, B. A., Patel R.G., Patel, H.R, Vyas, R.V., Patel, B.N and Patel, J. G 2002 b Management of root knot nematode, *Meloidogyne javanica* pathotype 2 in fennel nursery. National symposium on biodiversity and management of nematodes in cropping systems for sustainable agriculture 11-13 Nov. 2002 at Department of Nematology, Agriculture Research Station, Durgapura, Rajasthan., p 50

Sharma, A. and Trivedi, P.C. 1988. Screening of *Trigonella foenum-graecum* varieties/cultivars against root-knot nematode *Meloidogyne incognita. Int. Nematol. Network Newsl.* 5: 30-31.

Sharma, A. and Trivedi, P.C. 1992a. Studies on the life cycle of *Meloidogyne incognita* in the cultivars of *Trigonella foenum-graecum. Nematol. medit.* 20: 215-216.

Sharma, A. and Trivedi, P.C. 1992b. Control of root-knot disease of *Trigonella foenum*-graecum using decomposed leaf powder. *In: Proc. 3rd International Conference on Plant Protection in the Tropics* 6: 25-28. Malaysian Plant Protection Society, Kuala Lumpur, Malaysia.

Sharma, A. and Trivedi, P.C. 1993. Histochemical alterations in various cell constituents induced by *Meloidogyne incognita* in *Trigonella foenum-graecum* L. *Curr. Nematol.* 4: 71-76.

Sharma, G.L., Mathur, B.N. and Bhargava, L.P. 1988. Reaction of some fenugreek (*Trigonella foenum-graecum* L.) lines against *Meloidogyne incognita. Indian J. Nematol.* 18: 356.

Sharma, G.L., Mathur, B.N. and Jain, M.P. 1989. Screening of fenugreek and coriander genotypes against *Meloidogyne incognita. In Proc. First National Seminar on Seed Spices, Jaipur, India* 266p.

Sivakumar,C.V.and Seshadri,A.R.1971. Life history of reniform nematode, *Rotylenchulus reniformis*.Linford and Oliverira.*Indian J.Nematol.*1:7-20.

Thakar, N.A., Patel, H.R. and Patel, C.C. 1987. Reactions of chickpea and fenugreek varieties/lines to root-knot nematodes. *Indian J. Nematol.* 17: 143-144.

Yadav, B.D. and Chand, R.2001.Evaluation of tomato, pea, cowpea, coriander, fenugreek, cluster bean and bottle gourd germplasm against root knot nematode, *Meloidogyne incognita*. National congress on centenary of Nematology in India appraisal and future plans. Held at IARI, New Delhi on 5-7Dec.2001: p 134.

Phyto-nematodes in Crops: Their Identification, Treatment and Management, 2011
G.L. Sharma (editor), pp. 169-202
New India Publishing Agency, New Delhi (India)
E-mail : info@nipabooks.com; Website : www.nipabooks.com

Chapter 8

Phytosanitary Measures to Check Dispersal of Nematodes through Banana Suckers

RAJAN

Introduction

Banana is the forth ranked agricultural crop in the world and first among fruits. A large number of pests and diseases threaten banana production. The major constraints in production are diseases and pests transmitted through the conventional planting materials. Although the average yield for bananas worldwide is over 40 t per hectare, production in India averages only 20 t per hectare. The nematode parasites are among the most important pests on commercial and subsistence banana production and their control remains difficult, as the nematodes are involved in the destruction of primary roots that disrupt the anchorage system resulting in toppling or uprooting of plants. The most widespread and important are burrowing nematode, *Radopholus similes*, some species of root lesion nematode, *Pratylenchus coffeae* and *P. goodeyi*, species of root knot nematode, *Meloidogyne incognita* and *M. javanica*; the spiral nematode, *Helicotylenchus multicinctus*, *H. mucronatus* and semi endo-parasitic nematode *Rotytlenchus reniformis*. In addition to these five important genera, there are more than 175 other less known species belonging to 47 genera of nematodes associated with *Musa* sp. through out the world. Crop losses due to nematodes range from 20% (Sasser and Freckman, 1987; Kumar *et al.*, 2008) to 50% as reported from Uganda (Speijer *et al.*, 1994, 1998, 1999a, 1999b, 2001).

During the past 500 years the production of many crops expanded worldwide coupled with exploration and expansion of travel and trade. Inadvertently hitchhiking pests were introduced to new areas of production. This is especially true for vegetative propagated crops. As an example nematodes and other pests were carried with banana plants when they were introduced to different geographical regions throughout the world (Kashaija *et al.*, 1999, Sikora *et al.*, 2007, Jones 2007). The burrowing nematode has a wide geographical distribution in banana growing regions because this nematode has spread with infected rhizomes from Southeast Asia (Myanmar, Thailand, Indonesia) to new banana growing areas in the world (Marin *et al.*, 1998). As a well documented example - in 1950s a susceptible planting material was distributed by the United Fruit Company to their plantations in at least seven countries in Central America, the Caribbean and South America. Unfortunately the suckers grown and distributed from Panama were infested with burrowing nematode, thus making sure that a serious nematode problem could spread through out the world (Marin *et al.*, 1998). As a well known example, in the late 1800 and early 1900s scientists in the USA imported soil from China and Japan to ensure adequate inoculum of nitrogen fixing bacteria for newly introduced legumes, that acted as a pathway for many pests/diseases.

The burrowing nematode is also associated with another important pathogen on banana: root rot disease caused by *Fusarium oxysporum* f.sp. *cubense*. The fungus invades intact banana roots and colonizes parenchyma cells especially when the roots are wounded mechanically by the nematode (Pocasangre, 2007). The existing practice of chemical control of nematodes has mitigated the crop loss to some extent but leave behind lot of residues causing environmental threat (Luc *et al.*, 2005, Binks and Gowen, 1996). The development of aseptic shoot-tip culture methods has enabled selected banana germplasm to be multiplied quickly and has reduced many of the risks associated with the international movement of germplasm (Dochez *et al.* 2009). A major drawback to breeding for resistance has been the sterility of the popular Cavendish clones. However genetic engineering and other biotechnology-based approaches to improvement are viable alternative methods. Chemical control has been employed successfully in the past to counter pests and pathogens, but resistance to fungicides and pesticides is becoming a major issue. Biological control methods are being employed more often in integrated pest management programms. The uses of tissue cultured banana have been found to be more susceptible to nematodes as compared to conventional planting material, especially in the first crop cycle (Niere *et al.*, 1999).

The WTO (1995) has given a new dimension to international trade. It has removed quantitative restrictions in import which implies the likelihood of a huger inflow of agricultural commodities therefore, increasing the risk of introduction of exotic pests associated with them. Because of their endoparasitic behavior, some of the nematodes can be disseminated easily via seed (*Anguina, Aphelenchoides,* and *Ditylenchus*), plants or vegetative plant parts (e.g., nursery stock, tubers, bulbs, cuttings; *Ditylenchus, Radopholus, Pratylenchus,* and *Aphelenchoides*), or plant products (e.g., timber; *Bursaphelenchus xylophilus*). As a consequence, several of these nematodes are regulated through international phytosanitary regulations. Lehman (Lehman, 2004) compiled a list of plant-parasitic nematodes most frequently regulated in international trade. Out of the 15 regulated species on his list, 8 species are migratory endoparasitic nematodes: *Aphelenchoides besseyi, Ditylenchus dipsaci, Radopholus similis, Ditylenchus destructor, Aphelenchoides fragariae, Bursaphelenchus xylophilus, Anguina tritici,* and *Bursaphelenchus cocophilus.* Specific restrictions against other migratory endoparasitic root feeders are imposed by Argentina, Brazil, Chile, Egypt, Mexico, Morocco, South Africa, Syria, and Uruguay (Hockland *et al.*, 2006). The listed root-lesion nematode pests include mainly tropical and subtropical species such as *Pratylenchus coffeae, P. loosi, P. scribneri,* and *P. zeae* and the temperate species *P. penetrans.*

The Sanitary and Phytosanitary Agreement of WTO suggests that member countries should update their quarantine systems, incorporate the element of Pest Risk Analysis (PRA) for making regulatory decisions. Risk is assessed based on biological parameters of the pest. At the same time the other equally important responsibilities of plant protection organizations are to identify foreign pests of crops and access the potential damage that those pests could cause if introduced in a country. The analysis is done, based on information available about distribution, biology, survival, entry, spread, establishment and economic importance of each pest. It is not sufficient to just identify that something might go wrong,there is need to focus on how likely this is going to happen.The risk analysis of certain important nematode pests of banana suckers has been carried out; the results are being presented in the chapter.

In addition some pests got introduced accidentally through cargo, baggage, ship, stores, domesticated animals and ship ballast; unfortunately many non native nematodes were inadvertey introduced with some of these new crops and the contaminated soil that accompanied them. In 1916, it was recognized that seed and soil inoculations had serious drawbacks since cotton wilt, pea wilt, melon wilt and nematodes were being distributed. It has been speculated that soybean cyst nematode may have been introduced in many soybean importing and growing countries with the soil.

Nematode Invasions

Fortunately nematodes have a number of characteristics that limit their natural long distance dispersal, including (i) their restricted movement (ii) obligate parasitism (iii) narrow host ranges of certain species (iv) sharp population declines in the absence of hosts (v) survival depending on environmental conditions and (vi) crop management practices. The long-distance dispersal of these plant parasites is largely passive through infected planting material. Key means of dissemination include movement of soil on equipment and plant parts, crop transplants, water, animals and contaminated packaging materials.

The process of invasion of an exotic nematode pest species has four phases (i) arrival (ii) establishment (iii) integration and (iv) spread. Arrival is the introduction of a new population into a community of established species, typically by some factors or pathway other than locomotion. Association of plant parasitic nematodes with agricultural planting material and commodities suggest that the probability of arrival is high. Nematodes with stages that are resistant to desiccation may spread widely. Nematodes are ready and rapidly spread throughout a field and among fields by irrigation systems and land leveling. In general those climatic and environmental conditions that favors host also favor the nematode pests. The nematodes and their hosts have evolved over many thousand of years, resulting in specific recognition signals between host and parasite.

The fact that nematodes are small, spatially aggregated, no specific symptoms and difficult to identify increases the probability of their chances of escape from detection on arrival. The difficulty of detecting them in plant and soil material increases the probability that they will be successfully introduced. Establishment occurs when the newly arrived species maintains population through multiplication/ reproduction; not continuous immigration. Integration is not a rapid process; it may proceed over a long period of time. Inactive dispersal, spread of plant parasitic nematodes through, soil, water, wind and agricultural equipments occurs in the nearby areas.

Nematodes are generally excellent invaders of new habitats. They have evolved strategies for exploiting favorable environments and with stand harsh conditions. The feeding of nematodes within host tissues and their survival capabilities contribute to the ease with which they are disseminated with plants. Once introduced the generally nonspecific nature of symptoms increases the probability that their presence may go undetected or unrecognized for considerable periods. Their dispersal in and around plant tissues and throughout the soil contributes to the difficulty of targeting them in management or eradication programs.

The introduction of invasive species, propagation of others already present and the spread of so called emerging species is harmful to the plantation crops. In spite of control measures available and existence of plant health certificates, transport of plant products and contaminated soil open up ways in which accidental or intentional introduction can occur. The consequences can be catastrophic. For instance, introduction of the invasive species *Radopholus similis*, originally from South-East Asia, into all areas of banana cultivation began in the early XVth Century. Not only did it cause severe damage to plantations but also eventually prompted the massive long term use of pesticides, worldwide. Another example concerns the introduction of the nematode *Scutellonema bradys*, originally from Africa into the Caribbean. The *S. bradys* holds back the development of yam cultivation. Some verities of the crop are severely affected and have even died out because of the nematode.

The introduction of "physiological races" morphologically indistingui-shable from nematode species already present, raises the question of the risk of increasing the genetic reservoir of these populations. Such races have different parasitic capacities, in their host range, aggressiveness or virulence to face the resistant gene in the host crop. If the strain "race citrophilus" of *R. similes* spread beyond its present occurrence, which is limited to Florida, the consequences could be disastrous for farmers and plant protection agencies.

The most important determinant of rate of spread in agriculture is the movement of infested plants and planting material. Especially important is material that is to be propagated and distributed as nursery stock. Sale and movement of infested nursery stock, seed or turf immediately spreads the nematode pest to uninfested areas and distributes it throughout the planting site. The nematodes *Radopholus similis* and *Meloidogyne* sp. have long been present in Philippine soils. Their existence however, was not a problem until large volumes of nematode-infected 'Giant Cavendish' banana planting materials were imported from Central America and commercially raised by farmers in the early 70's. The nematodes spread to the native banana varieties and became a prominent production problem throughout the country. Plant-pathogenic nematodes are not only a limiting factor in the production of banana in India, Sri Lanka and the Philippines,they also cause heavy yield losses in most other Southeast Asian countries where,bananas provide an essential source of food and income for millions of(mostly-poor)people.

Burrowing Nematode (*Radopholus similes*) : The burrowing nematode is the number one pest of banana, being wide spread in tropical and subtropical regions of the world except Israel, the Canary Islands, Cape Verde Islands, Cyprus, Crete, Mauritius and Taiwan. It also appears to be absent from some of the important areas of production in the highlands of Eastern Africa. In India, its first occurrence was reported from Palghat District of Kerala (Nair *et al.* 1966). Subsequently the nematode was reported from all banana growing states of the country including Andaman and Lakshadweep Islands. Economic loss due to the nematode infestations was estimated at 50% in Central America (Volkers and Gamboa, 1988, Vilardebo *et al.* 1988). In India, *R. similes* is responsible for 31-41% yield loss in banana (Nair, 1979). The survival and multiplication of the nematode, deep in the cortex of rhizomes or suckers helped in its faster spread to clean areas.

The burrowing nematode is a migratory endoparasite of roots of over 200 woody and herbaceous perennials including important commercial crops such as banana and citrus.All life stages are easily transported within plant tissues (roots and suckers) and associated soil. With wide host range it is one of the most economically important nematode in tropical and subtropical regions of the world.Burrowing nematode has been intercepted on vegetative propagating material of several crops.Its worldwide distribution is relatively recent(beginning of the 19th century)and is due to the transfer of infected plant material, particularly banana sets,from country to country for commercial propagation. The wide distribution of *R.similis* is correlated with the areas where banana sets of the sub-group Cavendish (AAA) were imported. Adaptation may cause the development of a wider host range as it spreads on different AAA,AAB and ABB clones in Africa,and on ornamental plants which increasingly are being exported to regions outside the tropics.

Recently, several *Radopholus* species of local importance have been described; however, the knowledge about their host range is limited. *Radopholus citri* and *R.bridgei* are pathogens of citrus (Machon and Bridge, 1996) and turmeric (Siddiqi and Hahn,1995),respectively,in Java. *R. musicola* (Stanton *et al.*, 2001) was described from banana in northern Australia,and *R.duriophilus* (Nguyen *et al.*, 2003)and *R.arabocoffeae* (Trinh *et al.*, 2004) were described from Vietnam, where they were found on durium and/or coffee. By now more than 30 *Radopholus* species had been described,one third of these during the past two decade all associated with plants in Africa,Asia,or New Zealand. The genus *Radopholus* contains a small number of species whose geographical distribution is restricted in some countries, at present but are potentially destructive pathogens because they can cause heavy crop losses. *R. citri* was described from roots of citrus seedlings in Indonesia. In pathogenicity experiments, populations of only 1000 nematodes per plant caused severe root degradation and significantly reduced plant growth. *R. duriophilus* was described in 2003 from roots of durian in the Western Highlands of Vietnam, where there was tree decline and death of young trees.

Lesion Nematode (*Pratylenchus coffeae*)

In India the crop loss due to the lesion nematode has been reported to be 44% (Suderaraju *et al.*, 2002; 2003). The nematode is reported to have spread through out banana growing areas through suckers (Koshy *et al.*, 1978; Sudha and Sundaraju, 1996). Another species of lesion nematode, *P. thornei* is reported from Assam only. There is direct relationship between nematode population and physico-chemical properties of the soil (Sundaraju and Cannayane, 2002). The *P. coffeae*, quite prevalent in India is probably a native of the Pacific countries but now has a world wide distribution almost equal to that of *R. similes*. It is most likely that it has spread due to either non existence of quarantine machinery or due to quarantine failure, through commercial planting material.

The *P. coffeae* is widespread and variable species, which may represent a species complex. Molecular and morphological studies on isolates of *P. coffeae* and closely related species have demonstrated the difficulties involved in identification (Duncan *et al.*, 1999). The complex taxonomic status is yet to be satisfactorily resolved. *P. coffeae* is a migratory endoparasite of the root cortex; also of the corms of banana, plantain and *Musa textiles*, where it feeds and multiplies. It causes lesions and is often associated with primary and secondary rot in the root cortex, corms, tubers and rhizomes. The developmental stages of *P. coffeae* infesting coffee in Java are as follows: the first moult takes place within the egg, three moults occur outside, and the eggs hatch in 6-8 days at 28-30°C.

The *P. coffeae* is of particular importance on bananas in the Pacific island countries (Bridge, 1988a,1988b) and parts of Africa (Sarah 1989), and on bananas, plantains and abaca (*Musa textilis*) in Cuba, Central and South America (Koenning and Schmitt 1986, Gowen and Quénéhervé 1990). *P. coffeae* rarely occurs as the only nematode pest on the roots of *Musa* species. Although localized in South Africa, where it is found as the main nematode pest, the estimated crop losses can be as high as 80% (Sarah, 1989). In Cuba, high population densities cause toppling of the plants and limit production of the crop; economic threshold densities of the nematode have been estimated at 5000-10,000 nematodes per 100 g roots in good soils but as low as 1000-5000 nematodes per 100 g roots in soils with low fertility (Fernández and Ortega, 1998).

P. coffeae can be detected by extraction from root, corm, tuber and rhizome tissues of plants showing damage symptoms. It can also be found in soil extracts. Nematodes are extracted from plant tissues and soils using standard nematological extraction techniques. In bananas and related crops, nematode damage is sometimes assessed by recording the incidence of uprooting per hectare per month. This may also be correlated with assessments of necrosis on primary roots and corms. These techniques can be used by those who are familiar with nematode symptoms but care should be taken not to confuse lesions caused by plant parasitic nematodes with those resulting from other root-infesting pests and fungal pathogens.

Increasing work is being done on assessing varietal resistance to *P. coffeae*.Price *et al.* (1996) reviewed techniques for field screening of *Musa* germplasm.Binks and Gowen (1996) also did field evaluation of *Musa* germplasm against *P.coffeae* and *R.similis*. Banana cultivars were screened by Collingborn and Gowen (1997), the varieties Kunnan and Paka being considered highly resistant whereas Pisang Sipulu, Pisang Tongat and Pisang Mas were not considered to be resistant to the nematode. Collingborne *et al.* (1998) screened Indian cultivars of *Musa* for resistance or tolerance to *P.coffeae* and found that Yamgambi km 5, Kunnan and Paka were most resistant.Stoffelen *et al.* (1999) investigated the host plant response of 13 Musa genotypes from Malaysia and Vietnam. Although differences in susceptibility between cultivars were observed, no resistance was detected. Stoffelen *et al.* (1999) also screened 32 Papua New Guinea banana varieties against *P. coffeae*, but none showed resistance to *P. coffeae*.

The Root Knot Nematode, *Meloidogyne incognita*

The root knot nematode, *Meloidogyne incognita* infestation is quite wide spread in India (Tiwai and Dave, 1985), where as *M. javanica* is confined to higher temperature zones of the country. The nematode is highly pathogenic and causes a yield loss to the tune of 30% to the banana crop in Tamil Nadu (Jonathan and Rajendran, 2000).

The *M.incognita* is probably the most widely distributed and economically important species of plant parasitic nematode in tropical and subtropical regions of the world.Two-thirds of the root-knot nematode samples obtained from a number of tropical countries were of *M.incognita*.In India alone,232 plant genera have been reported as hosts to *M. incognita* (Krishnappa, 1985) and worldwide the species is a parasite of a wide range of crop plants. *M. incognita* is widely endemic in subtropical and tropical regions, thus quarantine regulations have become superfluous and many countries have removed it from the list of regulated pests. Identification of species within the genus *Meloidogyne* requires the services of specialized taxonomists (Cofcewicz *et al.*, 2005). In addition to classical taxonomic techniques, biochemical methods have also been developed (Esbenshade and Triantaphyllou, 1985). The situation regarding *M. incognita* is complicated by the existence of morphologically indistinguishable races differentiated by their ability to reproduce on a range of test host plants (Hartman and Sasser, 1985).

Field symptoms are typically of stunted, poorly growing plants with yellowing leaves. Infected root systems show characteristic knots or galls, the severity of which varies with the degree of nematode infection and species and variety of *Musa* spp. plant parasitized (Dropkin, 1989). *M. incognita* is similar to other species of *Meloidogyne*. Confusion within this group is common as the characters used tend to be variable. Above-ground symptoms are non-specific (see Symptoms): examination of roots can reveal the presence of swollen roots and galls. Closer examination, for example with a hand lens, can show the presence of egg masses on the root surface.

The Spiral Nematode, *Helicotylenchus multicintus*

The *Helicotylenchus multicinctus* is the major nematode problem in tropical and subtropical regions. The nematode is a problem even in the areas such as Israel and Taiwan where *R. similes* is absent. The other nematode species recorded on banana are *H. dihystera, H. africanus* and *H. erythrinae*. Under Indian conditions, plant growth is severely affected by the nematode (Rajendran and Sivakumar, 1996).

The *H. multicinctus* is regarded as an important parasite of bananas in almost every banana-growing area of the world. It has also been recorded to have a wide host range (Goodey *et al.*, 1965). Gowen and Quénéhervé (1990) found it in Côte d'Ivoire, in association with 12 other hosts. In Cuba, the nematode was detected on 32 host plants (Stoyanov, 1967). The nematode attacks suckers, primary and secondary roots. Quénéhervé and Cadet (1985) also reported lesions in the corm.

According to the reports of Minz *et al.* (1960) and Zuckerman and Strich-Harari (1963), *H. multicinctus* is able to complete its life cycle within the cortical layer of the roots where all eggs, juveniles and adult stages can be found. Therefore it is regarded as an endoparasite, as there is no evidence of nematode migration through the cortical tissues (Blake, 1966). While studying the host-parasite relationships of this species on banana roots, Blake (1966) observed that adults inoculated fed within 36 hours upon parenchymatous cells partly embedded in the root; after 4 days they were wholly inside the cortical layer to a depth of about 4-6 cells. Surrounding tissues showed cellular damage consisting of contracted cytoplasm, ruptured walls and enlarged nuclei. Discoloration and necroses often also occurred near the infection points.

The introduction of the nematode into new areas occurs by suckers, usually brought from old infested plantations; infected planting material is the main means of dissemination. Strich-Harari *et al.* (1966) studied the spread of nematodes in a growing plant, whereas Stoyanov (1967) observed that the nematode can also survive for 4 months without the host plant. Minz et al. (1960) first reported a decline of banana (Musa AAA, Cavendish) plantations in the Jordan Valley, with substantial losses in yield within 3 years. Their data are the most reliable because the spiral nematode occurred in the absence of *Radopholus similis* and *Pratylenchus coffeae*, the other two major pests of banana. Colbran and Saunders (1961) and Edmunds (1969) reported that *H. multicinctus*, together with other nematodes, caused considerable reduction in banana production in Australia and the Windward Islands,respectively. *H. multicinctus* seems also to be a major nematological problem on banana and plantains in several locations, such as Florida, Argentina, Cuba, South Africa, Cyprus, India and Pakistan (McSorley and Parrado, 1986). Vovlas *et al.* (1994) reported it as the most important nematode associated with banana in Sao Tomé.Although *H.multicinctus* has been reported to cause considerable root damage in many banana-growing areas of the world, estimates of crop losses are not available because of the close association between species, and their interactions with environmental factors.

The damage symptoms are similar to those caused by other serious root endoparasites of the family Pratylenchidae, both on banana plants and plantains. The nematode attacks the outer cortical part of the root causing characteristic necrotic lesions, which are at first yellow in colour and then turn reddish-brown to black. Minz *et al.* (1960) and Luc and Vilardebó (1961) reported brown root flecks in discoloured tissues, that they indicated were symptomatic of *H. multicinctus* infestation; later, Zuckerman and Strich-Harari (1963) noticed that these were associated with egg laying within the root cortex. Rarely, in heavy infestations, the root lesions can coalesce and the necrosis becomes extensive, causing distortion followed by decay of the root. Infected plants are stunted and may topple. Because of the endoparasitic habits of the species, populations of *H. multicinctus* can be detected by collecting soil, especially samples from around the root system, with appropriate tools (trowel, soil sampling tube, spade, shovel).

Nematicides are one of the most important means of controlling *H. multicinctus* and other plant parasitic nematodes on bananas, and are generally widely used by growers producing fruit for the international export trade. Minz *et al.* (1960), Strich-Harari *et al.* (1966), Luc and Vilardebó (1961) and Philis (1971) reported nematicidal trials on banana using DBCP; different degrees of efficacy were found due to the influence of local conditions and environmental factors. For toxicological reasons this fumigant has recently been replaced in many countries of the world. For a review see Siddiqi (1973) and McSorley and Parrado (1986); a more comprehensive review, including other major damaging endoparasitic nematodes, is by Gowen and Quénéhervé (1990).

The Cyst Nematode, *Heterodera oryzicola*

The cyst nematode, *Heterodera oryzicola* is an important pest of banana in Kerala state (Charles and Venkitesan, 1993) and Tamil Nadu (Sundararaju *et al.*, 2002) of India. Only the roots of host plants are attacked by *H. oryzae* and secondary symptoms are produced. Damage symptoms are similar for most root-parasitic nematodes of rice including *H. oryzae*. The symptoms are similar to those of nutrient deficiency, including reduced tillering, stunting and leaf chlorosis (yellowing) (Fortuner and Merny, 1979). Cysts of *H. oryzae* are detected by close inspection of the roots, preferably with a hand lens. The examination of stained roots under a microscope can reveal the presence of the endoparasitic stages of the nematode in roots (Southey, 1986). Accurate identification of the species requires the extraction of cysts and second-stage juveniles from soil and roots (Hooper, 1990) and close examination (Luc and Taylor, 1977). There has been very little work on the control of *H. oryzae*. Several crop species, are poor hosts or non-hosts to *H. oryzae* in pot experiments (Merny and Cadet, 1978).

The *H. oryzae* is generally found in low-lying, rainfed areas in Côte d'Ivoire and Senegal, West Africa, and in Bangladesh. There are no phytosanitary measures in force for *H. oryzae* in any country.

Intervention strategies

The first step in preventing the establishment of exotic nematodes is by excluding their entry into pest free regions. Most of the spread is through tourism and trade between countries and continents, that facilitate introduction of planting material/suckers, that contain nematode pests. Such a spread can have dreadful consequences for crop-growing, in countries where the agricultural economy is most often fragile. Improved knowledge of plant parasitic nematodes should lead to better awareness of the risks linked to their introduction or their spread in order to be able to devise and implement well-targeted control methods. The burrowing nematode nursery certification program serves as another means of protection.

Exclusion may be done through chemical,physical or cultural methods, developing pest free areas. If eradication is not feasible, then containment efforts may be undertaken to prevent further spread.If an exotic nematode becomes established,growers would have costly but a variety of control measures. Nematode management is mainly based on crop rotation and chemical control (Gowen and Quénéhervé 1990, Edmunds 1971). However in those areas where bananas are grown continuously crop rotation can not be practiced,while at the same time the price of chemical nematicides is often prohibitive for small farmers. It is also important that most nematicides are extremely toxic for the environment.

Exclusion of Host Plants by Adding Declaration on Phytosanitary Certificate

Declaration on Phytosanitary Certificate may require that specific pests do not occur in the country of export/ region. In many cases nematodes are indirectly excluded by default because the country or origin can not meet all the requirements for other types of pests and pathogens listed on the Phytosanitary Certificate (Table 1). The presence of *R. similis* can be detected by inspecting and extracting nematodes from roots and bases of pseudostems (corms) in bananas and plantains; from roots and surrounding soil in crops such as black pepper, coconut, betel vine and sugarcane; and mainly from rhizomes, corms and tubers in ginger, turmeric, taros and yams. Samples taken near to the base of the stem of the mother plant will contain roots of different ages and vigour, and consist predominantly of primary roots with relatively smaller quantities of secondary and perhaps no tertiary roots. It is in this region that roots will contain highest populations of *R. similis* in the root cortex. Generally, the greatest numbers of nematodes occur in the roots of the most actively growing suckers. Samples are best collected within a horizontal distance of up to 30 cm from the base of the plant and down to a depth of 30 cm from the soil surface (Araya *et al.*, 1999).

In quarantine the techniques used to extract *R. similis* from bananas may depend on the available laboratory facilities and assistance, and use may be made of non-standard methodology availabe locally. This should not prevent or discourage nematologists from adapting a technique which can be used routinely by different operators to give reproducible results. Whatever extraction procedure is used it is important to obtain a representative root sample which should be chopped in 0.5 cm lengths, mixed thoroughly, and a 25-g subsample taken for processing. A 24-hour period of incubation is sufficient for macerated root samples. Chopped roots should be incubated for 2-4 days and mist extractions may be run for up to 14 days in some laboratories. It is customary to report nematode populations per 100 g of fresh roots although this quantity is seldom used for extractions (Gowen and Quénéhervé, 1990, Quénéhervé and Berg 2005).

Table 1: Exclusion of nematodes by indirect quarantine regulations which prohibit entry of crops or genera

Crop/ Genera	No. of countries prohibiting	Percentage of countries prohibiting		
		Plants	Seeds	Plants and Seeds
Fruit crops				
Citrus	62	55	0	45
Cocoa	28	29	7	64
Fragaria	20	65	0	35
Musa	39	54	0	46
Pome	37	85	0	15
Prunus	37	85	0	15
Ribes	16	69	0	31
Vitis	41	90	0	10
Vegetable crops				
Ipomoea	23	61	0	39
Solanum	48	77	0	23
Forest crops				
Acer	14	100	0	0
Castanea	34	76	0	24
Conifers	27	100	0	0
Crategus	14	100	0	0
Juglans	21	100	0	0
Populus	27	93	0	7
Quercus	25	92	0	8

Contd...

Salix	22	100	0	0
Sorbus	24	96	0	4
Unimus	32	84	0	16
Other crops				
Coffee	20	100	0	0
Gossypium	35	100	0	0
Helianthus	15	20	0	80
Hevea	28	29	0	71
Nicotiana	26	35	7	58
Palm	16	56	0	44
Rice	33	21	18	61
Rosa	22	100	0	0
Theobroma	43	19	2	79
Tea	20	45	0	55
Saccharum	40	63	0	37

Data from survey of regulations of 125 countries

Regulatory Strategies

Avoiding infestations by exotic nematode is the highest priority. The use of certified nematode free planting stock is critical input. The movement of soil from infested fields should be avoided. Entry is restricted from all areas under quarantine of soil and potting media, plants and plant parts with roots, parts of plants, products below ground or at soil level and rhizomes for propagation.

Control and Eradication

An eradication program through soil removal and nematicide treatment may be attempted. For larger infestation a chemical eradication may be tried. For larger infestation a chemical eradication program may be tried. Eradication in large areas is difficult due to absence of fumigants.An alternative is to remove or destroy all roots and other plant material,treat the soil with nematicides and maintain it plant free for 2-3 years.Due to the wide host ranges of most of the plant feeding nematodes considered in this study, growing a non host crop would require elimination of host weeds. This approach was attempted in Florida, it failed both as an eradication strategy and as containment strategy.

In bananas, planting 'nematode-disease-free' suckers is recommended. Some growers or organizations maintain 'disease-free' nurseries from which new planting material, usually sword suckers, is collected. More commonly, planting material is taken from existing banana fields and these are more likely to be infested with nematodes. If the external tissue of the corm has purple or reddish-brown lesions, these, together with root stumps and adhering soil should be removed with a machete (pared) until only white corm tissue is exposed. Paring suckers should take place away from the field, and corms with severe lesions should be discarded. Although useful, the paring technique may never be totally effective in removing all nematode infection; this treatment is often complemented by dipping suckers in a nematicidal solution or, more effectively, by coating them with a nematicidal mud. The most reliable means of avoiding the introduction of the nematode is to use completely nematode-free banana plantlets grown from meristem cultures (Quénéhervé and Cadet, 1985; Gowen and Quénéhervé, 1990).

The immersion of banana suckers in water held at a constant 55°C for periods of 15-25 minutes has been a commercial practice in Australia and Central and South America (Stover, 1972). However, the technique is difficult to manage because of the critical balance that is required to achieve a temperature that is lethal to nematodes in the corm tissue but does not cause permanent damage to the plant. This factor can also be important if suckers are not of uniform size. Nematicides are widely used by commercial banana growers producing fruit for the international export trade. Less specialized production, serving local markets, may not justify the high cost of chemical treatment. A number of organophosphate, oxime carbamate and carbamate nematicides are used on bananas either as granular or emulsifiable concentrate formulations. Effective control of *P. coffeae* can be achieved with organocarbamate and organophosphate nematicides under field conditions.

If the eradication program fails or if it is not feasible then the pest is considered to be established. Failure to eradicate a newly introduced nematode in one field however will allow the nematode to spread and infest other fields. Cost of cultivation increases for other growers. Should an infestation of only one species of nematode become established the cost of production for many agricultural industries increases. Industry control costs would be greater. as infestation increase a larger proportion of production is affected. Eventually the proportion will be large to affect market prices qualities.

In case the nematodes are already spread to wide areas and eradication measures are not feasible we may go for either of the following depending upon need and prevailing situation:

- Rotation with alternative crops for 2-3 years
- Flooding for 8 weeks after having destroyed previous banana crop
- Fallow in absence of banana volunteers for 10-12 months
- Selection of disease-free suckers
- Use of in vitro-produced plants
- Paring diseased tissue from corm
- Paring and leaving large corms in sun for 14 days
- Immersing corms in hot water
- Coating corms with nematicide in mud
- Applying nematicide to planting hole and in-fill soil
- Regular spot applications with granular or liquid nematicide formulations.
- Planting resistant varieties
- Regular spot applications with granular or liquid nematicide formulations
- Heavy mulches with organic wastes may have beneficial root growth effects
- Propping fruiting stems with poles or with string guy ropes may prevent plants uprooting.
- Inoculation with mycorrhizal fungi

Containment of Spread

The banana burrowing nematode *R. similes* is widely distributed in most but not all tropical banana growing countries. It is of particular economic importance in commercial plantations in South and Central America, the Caribbean parts of Africa, Australia, Pacific Islands and some countries in South and Southeast Asia. The primary focus in containment is to minimize the potential spread of the nematode. Considerations include restricting movement of plant material, soil and drainage water from the infested area. In establishing perennials, preventing/disruption of root to root contact is important.

Pest Free Areas (PFA) for Production and Meeting Challenges of Bioterrorism

Significant production of many tropical crops occur in areas where their most damaging diseases do not occur. The continued production of these crops depends on the on-going exclusion of the causal agents. The long-range dissemination of plant pathogens can result from natural phenomena (e.g. trade winds) or human activities. Human-assisted dissemination is usually unintentional, but intentional movement also occurs. In response to 9-11, at World Trade Centre, USA there has been heightened interest in understanding and preventing bioterrorist threats to world agriculture. Clearly, devising tools for the intervention, and action plans for the arrival, of the thousands of diseases that threaten major and minor crops is not feasible. Even when these diseases affect a major crop,such as banana,limited resources make it impossible to interdict all pathogens.There is a need,then, for criteria with which the most serious pathogens could be identified and ranked.The banana specialists have demonstrated through models, representative threats from the major groups of pathogens which include a nematode *Pratylenchus coffeae* (root-lesion nematode) and *Banana bunchy top virus*.

Crop Loss and Income Losses

In West Africa surveys revealed that the two most damaging nematodes of banana and plantains are *Pratylenchus coffeae* and *Radopholus similes*. Both these nematodes have distinct regional distributions depending upon temperature and altitude (Bridge *et al.*, 1995). The *R. similes* is adapted to higher temperature and is consistently found below 900 m where most commercial production occurs. The lesion nematode has a lower temperature preference than *R. similes* is also a pest of plantations growing at higher altitudes where these crops are primarily grown by small landowners.

Most estimates of yield loss come from the use of nematicides that generally decrease populations of all species not only *R. similis* and can possibly cause other beneficial growth effects. Banana crop losses depend upon several factors including nematode species, banana cultivars, climate conditions and soil factors, as well as fungal, viral and biolochemical pathogens. Worldwide banana production losses of approximately 20% are caused by nematodes annually (Sasser and Freckman, 1987). The burrowing nematode causes brown lesions in the cortex and affected roots finally die. The damaged root system results in reduced water and nutrient uptake and poor anchorage of the plant. Yield is reduced and vegetative cycle is lengthened (Growen and Quénéhervé, 1990). Severely damaged plants may finally topple, leading to total loss of the bunch.

It is uncommon for bananas to be parasitized by monospecific nematode populations and the relative importance of the different species of other migratory endoparasites in mixed populations and their role in yield loss is not fully known. Where *R. similis* is the only major nematode present, yield improvement based on gross yield per hectare or weights of harvested bunches following nematicide application can be as high as 267%. Increase in banana and/or plantain yield has been recorded as 86% in Panama and 15% in Honduras, 207-275% in Puerto Rico (over a 3-year period) (Roman *et al.*, 1977), 5-30% in Australia, 71% in Ecuador and 267% in St Vincent (Gowen and Quénéhervé, 1990; Gowen, 1995) and 38% in South Africa (Jones and Milne, 1982). When *R. similis* is the main nematode species present in association with other nematodes, the yield improvement over untreated plants after nematicide application can range from 16 to 263% in Ivory Coast (Sarah, 1989; Gowen and Quénéhervé, 1990), 20-40% in Cameroon (Melin and Vilardebo, 1973;Vilardebo et al.,1988; Sarah, 1989), 35-40% in Madagascar (Beugnon and Vilardebo, 1974), 29-35% in Martinique (Gowen and Quénéhervé, 1990) and 46% in St Lucia (Gowen, 1975, Gowen and Quénéhervé, 1990).

Growers in Loss

Growers are affected through reduction in crop yields and increase in pest management costs needed to prevent crop damage or other degradation in quality. Growers are also affected by any changes in market prices for widespread establishment of exotic nematodes. With widespread infestation yield losses or increased costs of production may cause market prices to rise therefore increasing cost.

Marketing Sector in Loss

Restrictions of plant imports may be restricted by other countries. The burrowing nematode is a part of concern and has become an issue in the export commodities to at least Japan, Taiwan and European Union. Japan has published a list of plants reported to be hosts of the burrowing nematode and requires that the plants be accompanied by phytosanitary certificates that they are free from nematodes. Under the law citrus, strawberry nursery stock, carrot and other root crops could be prohibited for exports these countries or regulatory treatments could be required. As the reniform nematode is widespread, no impact to foreign exports has been identified.

Consumers in Loss

Consumers would be affected by higher food costs. Both increased product costs and yield losses put upward pressure on consumer prices. In addition people purchasing nursery plants for landscaping would face higher prices for those commodities.

Taxpayers Money Wasted

When public regulatory agencies are involved,issues related to taxes and budget allocation come into picture.Taxpayers may fund quarantine, border control measures to prevent entry and spread of nematodes. Public programs may be necessary to eradicate small infestations or undertake plant breeding programs to develop resistant varieties. Other important research areas include developing new nematicides, identifying natural enemies and decontaminating plants and export material.

National and International Quarantine Legislations and Policy Scenario

Today it is difficult to imagine that 100 years ago there were no government agencies anywhere in the world that specifically focused on regulating plant pests. Surprisingly it was the root knot nematode along with certain other pests that accompanied the cherry trees gifted from Japan to Washington DC, USA and efforts of "Father of Nematology Dr. N.A. Cobb" at that time on examination and retaining his argument, finally helped to convince the US Congress to pass the "Plant Quarantine Act of 1912".

After the World War–II, as a part of recovery plan the General Agreement on Tariffs and Trade (GATT) was negotiated in 1947 to open up markets for increased global trade. With rapidly changing conditions in trade and travel the need for effective international cooperation in preventing the spread of pests and diseases became urgent. In 1952, the International Plant Protection Convention (IPPC) was established as a subsidiary of the United Nation Food and Agricultural Organization, which has provided a framework for national,regional and global effort in plant protection.Member countries are required to establish a National Plant Protection Organization (NPPO) to survey crops for pests and diseases,report the outbreak and spread of pests, inspect plant and plant products moving in the international trade and issue Phytosanitary Certificate; publish its plant quarantine regulations and distribute them to others and strengthen advisory services and research capabilities.

The European Plant Protection Organisation (EPPO), a regional plant protection organization (RPPO) has provided a forum for scientifically based pest risk analysis (PRA) and currently there, ten nematodes listed among the pests recommended to be regulated among member countries. It has provided strong leadership in developing programs that certify the production of healthy plant propagative materials. Cooperative agreements have been reached in preventing the spread of nematodes to new areas.Nematodes that have been targeted in these certification programs are *Aphelenchoides* spp.,*Ditylenchus* spp., *Meloidogyne* spp.,*Pratylenchus* spp.and virus vector nematodes i.e. *Longidorus* spp.,*Paratrichodorus* spp.,*Trichodorus* spp.and *Xiphinema* spp. South Africa excludes 270 hosts to indirectly exclude nematodes such as *Aphelenchoides besseyi, Ditylenchus dipsaci* and *Radopholus similis.*

India has also published and circulated Plant Quarantine Order 2003 which includes requirements on requests for pre-import/import, pest risk analysis, transit and packaging requirements, processing procedures, observation to be taken on arrival and certification procedures after post entry growing and examination of *Musa* spp. The *Musa* spp. is listed under Schedule IV (List of plant/plant material and countries from where import is prohibited along with justification) and Schedule V (List of plants and plant materials with restricted import permissible only with the recommendation of authorized institutions with additional declaration and special conditions). All imports of *Musa* spp. Are subject to approval of Directorate of Plant Protection Quarantine and Storage, it needs recommendation, supervision, monitoring and testing by Director, National Research Centre-Banana at Trichi, Tamil Nadu and it is mandatory to grow imported consignment under post-entry quarantine for a period of 9-12 months. Plant health certificates are issued by exporting countries to confirm that the importing country's requirements have been complied with. Total prohibition of host plant material can be justified if risks of introducing dangerous new pests or diseases are judged to be too great to depend solely on inspection methods or quarantine treatments and such extreme measures are provided for in the International Plant Protection Convention. To avoid the contamination of burrowing nematode certain crops are prohibited in India, to enter from certain countries (Table 2).

Table 2 : Import of certain host crops, banned from entry from countries where burrowing nematode is quite prevalent.

Countries	Crops	Nematode
India, Indonesia, Oman, Singapore, Sri Lanka, Thailand, Pakistan, Philippines, Malaysia, Netherlands, Denmark, Germany, France, Belgium, United Kingdom, Uganda, Egypt, Ethiopia, Ghana, Gabon, Cameroon, Guinea, Kenya, Zaire, Zambia, Zimbabwe, Sudan, Senegal, Republic of Cote d'Ivoire, Somalia, Tanzania, Nigeria, Madagascar, Malawi, Republic of South Africa, Mozambique, Reunion, United States of America, El Salvador, Canada, Cuba, Guatemala, Guadeloupe Island, Costa Rica, Jamaica, Saint Vincent, Saint Lucia, Dominican Republic, Trinidad and Tobago, Nicaragua, Panama, Puerto Rico, Belize, Martinique Island, Mexico, Venezuela, Ecuador, Colombia, Suriname, Brazil, Peru, Australia, Tonga, Western Samoa, Papua New Guinea, Hawaiian Islands, Fiji	Underground portions of the live plant of avocado, turmeric, okra, cockscomb, coconut, *Colocasia esculentum*, sugarcane, ginger, *Canna edulis*, greater yam, tea, corn, potato, betel palm, peanut (excluding seeds without pod), and plants of the genera *Anthurium*, *Calathea*, *Maranta*, *Coffea*, *Piper*, *Musa*, *Philodendron* and *Beta* being capable of planting for cultivation. Banana burrowing nematode (*Radopholus similis*)	Banana Burrowing Nematode (*Radopholus similis*)

Source: Plant Quarantine Order 2003

Pest Risk Analysis

Quantification of crop losses attributable to nematodes is difficult because of close association between species, soil pests and pathogens and with environmental conditions. The nematode parasites of banana can be classified according to the damage caused. The most serious are those that destroy root cortex (*Radopholus, Pratylenchus, Helicotylenchus*).

Worldwide, banana production for export is often based on intensive monocropping systems that are generally not environmentally friendly. Banana crop is prone to attack by different pathogens like fungi, viruses, bacteria and nematode. Among the production constrains plant parasitic nematodes constitute one of the major limiting factor to production, causing extensive root damage, resulting in serious economic losses. Crop losses caused by nematodes which spread through suckers, are very high, with average annual losses estimated at about 20% worldwide (Sasser and Freckman, 1987).

India is the largest producer of banana in the world with a total production of 16.9 million tones per annum from 49,000 hectares having a share of 37% of the total fruit production (Singh, 2002). The major banana growing states are Tamil Nadu, Andhra Pradesh, Kerala, Karnataka,Gujarat,Maharashtra,Bihar,West Bengal and Assam. Among the states maximum area under production is with Tamil Nadu. A total of 140 species of nematodes belonging to 54 genera have been reported to be associated with rhizosphere of banana.Out of these 73 species of nematodes belonging to 33 genera have been recorded from banana from various parts of the country. The remaining 67 species, not reported from India, being of quarantine importance, have been assessed for risk analysis.

The important nematode problems assessed under climatic and environmental conditions of Tamil Nadu, in banana are the burrowing nematode, *R. similes*, followed by the root lesion nematode, *Pratylenchus coffeae*, root-knot nematode, *Meloidogyne incognita* and *M. javanica*; the cyst nematode *Heterodera oryzicola*, the reniform nematode, *Rotylenchulus reniformis* and the spiral nematode - *Helicotylenchus multicinctus* and *H. dihystera*.

The fact that a nematode species is a serious pest elsewhere does not mean that it would be equally damaging in importing country. The challenge is to speculate the probability that the nematode could arrive, become established and cause significant economic, ecological or societal damage. These risks are weighed against the benefits of introduction of germplasm, costs and probability of success of intervention. The question is not whether alien nematodes will be introduced into the state, but whether they can get chance to establish, can they be detected in time, can they be eradicated, contained or will become harmful after introduction.

Despite domestic, national, international quarantine and containment programs, plant feeding nematodes spread to new areas, new locations and cross country boundaries. Unless appropriate precautions are taken, many exotic nematodes may play a major role in Indian agriculture and urban landscapes. We need to estimate losses they may cause on inadvertent introduction, of each exotic nematode. Most are known to be distributed along with their primary agricultural hosts and are not prevalent or even present in natural habitats.

The nematode that countries most frequently regulate are ten. The PRA is a dynamic process. The number of countries regulating *Globodera rostochiensis* increased from 51 in 1982 to 106 countries in 2006. The ranking of nematodes in the top ten list has also been shifted in two and a half decades. For example, *Aphelenchoides besseyi* ranked 10th and regulated by nine countries but it ranked 2nd in 2008 with 70 countries regulating it, primarily to protect rice crops. The increase in the number of countries regulating *Bursaphelenchus xylophilus, Globodera pallida, Naccobus aberrans* and *Meloidogyne chitwoodi* is a result of research during the past two decades which has resulted in increased international awareness of these nematode pests.

The science of Nematology witnessed an exponential growth of knowledge in last two decades and this has been coupled with an opportunity for scientifically based regulatory nematode exclusion programs. In 1980, eleven countries regulated *Meloidogyne javanica* but in 2000 no country listed it as a prohibited nematode, reason being that by now all these countries have surveyed and found the nematode quite prevalent in the areas. The number of Agricultural Inspectors on United States borders has increased 40 folds during last 10 years and a large number of unfilled posts are being advertised month after month as qualified suitable applicants are not available.

Conclusions

The banana production is threatened by pest and disease pressure, which has been increasing during last few decades. A large number of plant parasitic nematodes are associated with banana and plantains throughout the world and several of these affect crop growth and yield (McSorley and Parrado, 1981). Among them the most alarming nematodes are the once spreading through the suckers/planting material, to almost all the areas wherever the land is under *Musa* cultivation. The virulent type of nematodes are burrowing nematode

(*Radopholus similis*), root lesion nematode (*Pratylenchus* sp.), root knot nematode (*Meloidogyne* sp.).

Low yields of banana are characteristic of small-scale holdings because of the combined effects of susceptible cultivars, increased pressure of pests and diseases and declining soil fertility. Although host resistance is the most sustainable management strategy, there are no domesticated or wild bananas with resistance to serious nematode pests. The degree of risk differs with the nematode involved, biology, life history, survival strategy of nematode species. Often lack of basic information impairs our ability to quantitatively assess the risk. In those instances we rely on qualitative assessment based on expenses with the pest in other areas of the world to determine whether there is cause for concern. Such assessments include a degree of uncertainty but provide direction for immediate action and future research.

Pest exclusion strategies such as regulatory certificate, sanitary procedures and quarantine are often overlooked as not being cost effective methods for nematode management. Nematodes are usually introduced into new areas with infested soil or plants. You can prevent nematodes from entering your farm by using only nematode-free suckers purchased from reliable nurseries. To prevent the spread of nematodes, avoid moving plants and soil from infested parts of the orchard. Do not allow irrigation water from around infested plants to run off, as this spreads nematodes. Nematodes may be present in soil attached to tools and equipment used elsewhere, so clean tools thoroughly before using them in your garden. Nematode attack can be prevented by using clean planting material in nematode free soil and growing the plants under strict quarantine conditions. As clean farming can make a difference: it can boost crop yields, and slow the spread of disease - crucially important to subsistence farmers, for who even cutting a percentage of loss can be lifesaving. And there have been considerable successes in some recent quarantine programs. For example the quarantine for *Globodera rostochiensis* and *Heterodera zeae* have been success stories in the United States. In contrast the quarantine measures for the soybean cyst nematode, *Heterodera glycines* failed in that country. This failure may have been as a result of the nematodes being more widely disseminated than initially realized. Internal quarantine measures developed and implemented by European and Mediterranean Plant Protection Organization have focused on *Aphelenchoides besseyi* for rice and strawberry, *Ditylenchus destructor*, *G. rostochiensis* and *G. pallida* on potato, *Radopholus citrophilus* on citrus and *Xiphinema americanum* on range of crops. More recently *Bursaphelenchus xylophilus* has been a focal point of concerning the importation of wood products in a number of countries.

To gain access to the Australian market, a Philippines exporter has to propose and then demonstrate to the satisfaction of the Australian Quarantine and Inspection Service (AQIS) the effectiveness of a multitude of quarantine measures under commercial conditions. In reducing the pest and disease risk on their export plantations to extremely low levels, these quarantine measures must be 100 percent effective, science based and verifiable. AQIS has estimated it would be at least two years and most likely much longer before a suite of risk management measures could be developed and appropriate monitoring and compliance systems could be put in place. It is worth noting that all costs associated with establishing the risk management regime are to be borne by the exporter, including the costs of stationing AQIS inspectors on the ground in the Philippines. This represents an enormous challenge for Philippine banana exporters who have chronic pest and disease issues, do not possess a quarantine culture and have yet to take up necessary to undertake the scientific work that will be required to develop and establish a reputable suite of quarantine measures.

The International Network for Improvement of Banana and Plantain (INIBAP) has been formed to co-ordinate the transfer and evaluation of *Musa* germplasm for disease resistance. The freer movement of germplasm has been made possible by the development of *in vitro* culture techniques thus overcoming the fear of further continental and intercontinental movement of some pests.

Eradication is difficult, if not impossible, unless the nematode is detected in a field at a very early stage (zum Felde *et al.* 2009). Significant costs may also accurse for early detection, sampling, processing etc. and ultimately impositions of quarantine restrictions on exports. Turkey has already imposed restrictions on import of unhulled rice due to a nematode *A. besseyi*. The challenge will continue to intercept nematodes early, identify accurately then using the best technology available to eradicate/contain or mange them as appropriate. It is well documented in the literature that eradication of nematode pests accidentally introduced into new environment are extremely difficult and expensive.

To strengthen phytosanitary measures, the future areas of work has four aspects:

a) *Prevention and legislation:* The legislation to quarantine the outbreaks needs to be upgraded to take account of advances in our knowledge of the disease and the field experience of current banana growers.

b) *Containment of new outbreaks*: Uprooting and burning the first individual plants as soon as they are detected appears to be the most effective method.

c) *Containment and restriction of spread from established infections:* The 'early intervention' approach where infected plants are detected early and killed has reduced the spread rate in existing infected plantings.

d) *Biological suppressants:* Promising results from overseas indicate this approach can be beneficial if combined with the three methods described above. Field trials need to be planned and applied on commercial orchards, to suggest effective recommendation.

References

Araya M, Vargas A, Cheves A, 1999. Nematode distribution in roots of banana (Musa AAA cv. Valery) in relation to plant height, distance from the pseudostem and soil depth. *Nematology*, 1(7/8): 711-716.

Beugnon M, Vilardebo A, 1974. Les nematodes du bananier a Madagascar:Aspect du problFme et son importance tconomique. *Fruits* 28: 607-612.

Binks RH, Gowen SR, 1996. Field evaluation of nematode infestations in *Musa* germplasm at FHIA (La Lima, Honduras). *Infomusa* 5(2):15-17.

Blake CD, 1966. The histological changes in banana roots caused by *Radopholus similis* and Helicotylenchus multicinctus. *Nematologica*, 12:129-137.

Bridge J, 1988a. Plant nematode pests of banana in East Africa with particular reference to Tanzania. In: INIBAP 1988. Nematodes and the borer weevil in bananas: present status of research and outlook. Proceedings of a workshop held in Bujumbura, Burundi, 7-11 December, 1987, 35-39.

Bridge J, 1988b. Plant-parasitic nematode problems in the Pacific Islands. *Journal of Nematology*, 20(2):173-183.

Bridge J, Price NS, Kofi P 1995. Plant parasitic nematodes of plantain and other crops in Cameroon, West Africa. *Fundamental and Applied Nematology*, 18(3):251-260.

Charles JSK and Venkitesan 1993. Pathogenicity of *Heterodera oryzicola* (Nemata: Tylenchida) toewards banana (Musa AAB cv. Nendran). *Fundamental and Applied Nematology* 16(4) : 359-365.

Cofcewicz ET, Carneiro RMDG, Randig O, Chabrier C and Quénéhervé P 2005. Diversity of *Meloidogyne* spp. On Musa in Martinique, Guadeloupe and French Guiana. *Journal of Nematology* 37: 313-322.

Colbran RC, Saunders GW, 1961. Nematode root-rot of bananas. Queensland *Agricultural Journal*, 87:22-24.

Collingborn FMB, Gowen SR, 1997. Screening of banana cultivars for resistance to *Radopholus similis* and *Pratylenchus coffeae*. *Infomusa*, 6(2):3.

Collingborn FMB, Gowen SR, Galan-Sauco V, 1998. Screening Indian cultivars of Musa for resistance or tolerance to *Radopholus similis* and *Pratylenchus coffeae*. Proceedings of the first international symposium on banana in the subtropics, Puerto de la Cruz, Tenerife, Spain, 10-14 November, 1997, *Acta Horticulturae*, No. 490:369-372.

Dochez Carine; Tenkouano, Abdou; Ortiz, Rodomiro, Whyte, Jim; De Waele, Dirk (2009) Host plant resistance to *Radopholus similis* in a diploid banana hybrid population. *Nematology* 11 (3) : 329-335

Dropkin VH, 1989. Introduction Plant Nematology. New York, USA: Academic Press.Dzhuraeva LM, 1976. Plant-hosts and distribution of gall nematodes in Tadzhikistan. Izvestiya Akademii Nauk Tadzhikskoi SSR (Ahboroti Akademijai Fanhoi RSS Tocikiston), *Biologicheskie Nauki*, No. 4(65): 45-51.

Duncan LW, Inserra RN, Thomas WK, Dunn D, Mustika I, Frisse LM, Mendes ML, Morris K, Kaplan DT, 1999. Molecular and morphological analysis of isolates of *Pratylenchus coffeae* and closely related species. *Nematropica*, 29(1):61-80.

Edmunds JE 1971. Banana nematodes and their control. Pflanzen Nach 24: 145-152.

Edmunds JE, 1969. Plant nematode problems of the Windward Islands.In: Peachey JE, ed. Nematodes of Tropical Crops. Technical Communication, No.40. St Albans, UK: *Commonwealth Bureau of Helminthology*, 142-148.

Esbenshade PR and Triantaphyllou 1985. Use of enzyme phenotypes for identification of *Meloidogyne* species. *Journal of Nematology* 17: 6-20.

Fernández Díaz-Silveira M, Ortega Herrera J, 1998. An overview of nematological problems in Cuba. *Nematropica*, 28(2):151-164.

Fortuner R, Merny G, 1979. Root-parasitic nematodes of rice. *Revue de Nematologie.*, 2(1):79-102.

Goodey JB,Franklin MT and Hooper DJ 1965.The nematode parasites of plants catalogued under their hosts. *Famham Royal, Bucks, UK, CAB.* 214pp.

Gowen SR, 1975. Improvement of banana yields with nematicides. British Insecticide and Fungicide Conference (8th), Brighton, 17-20 November, 1975. *Proceedings,* Volume 1, 121-125.

Gowen SR, 1995. Pests. In: Gowen S (ed.) Bananas and Plantains. London, UK: Chapman and Hall. pp 382-402.

Gowen, S.R., Quénéhervé, P., 1990. Nematode parasites of bananas, plantains ans abaca. In: Plant Parasitic Nematodes in Subtropical and Tropical Agriculture. M. Luc, R.A Sikora and J. Bridge (eds). CAB International pp. 431-460.

Hartman KM and Sasser JN 1985. Identification of *Meloidogyne* species on the basis of differential host test and perineal pattern morphology. Pp 69-77 in an Advanced Treatise on *Meloidogyne.* Vol II. *Methodology (Barker KR, Carter CC and Sasser JN, Eds.) North Carolina State University, Raleigh, USA.*

Hockland S, Inserra RN,Millar L, Lehman PS. 2006. International plant health: putting legislation into practice. [in Perry RN, Moens M, eds. 2006. *Plant Nematology.* Wallingford, UK:CABI Publ. 447 pp.] pp. 327–45.

Hooper, DJ, 1990. Extraction and Processing of Plant and Soil Nematodes. In: Luc M, Sikora RA, Bridge J, eds. *Plant Parasitic Nematodes in Subtropical and Tropical Agriculture.* Wallingford, UK: CAB International, 45-68.

Jonathan EI and Rajendran G 2000. Assessment of avoidable yield loss in banana due to root knot nematode, *Meloidogyne incognita. Indian J. Nematol.* 30 (2): 162-164.

Jones RK, Milne DL, 1982. Nematode pests of bananas. *Nematology in southern Africa.,* 30-37; [Sci.Bull.,Dep.Agric.Fish.,Repub.S.Afr. No.400].

Jones, D.R. 2007. "Diseases and pest constraints to banana production" ISHS/ ProMusa Symposium: *Recent advances in banana crop protection for sustainable production and improved livelihoods.* Greenway Woods Resort, White River, South Africa, 10th to 14th September 2007.

Kashaija, IN, Fogain R and Speijer PR 1999. Habitatmanagement for control of banana nematodes. p. 109–118. *In* E.A. Frison *et al.* (ed.) Mobilizing IPM for sustainable banana production in Africa. Proceedings of a workshop on banana IPM, South Africa. November 1998, INIBAP, Montpellier, France.

Koenning SR, Schmitt DP 1986. Integrated Pest Management of *Pratylenchus brachyurus* on soybean. *Nematropica,* 16(2):237-244.

Koshy PK,Sundararaju and Sosamma VK 1978.Occurrence and distribution of *Radopholus similis*(Cobb)Thorne in South India.*Indian J.Nematol.*8:49-58.

Krishnappa K, 1985. Nematology in developing countries: India - IMP Region VIII. In: Sasser JN, Carter CC, eds. An Advanced Treatise on *Meloidogyne.* Vol.I. Biology and Control. Raleigh, North Carolina State Graphics, 379-398.

Kumar RA, Kumar N, Poornima K and Soorianathasundaram K 2008. Screening of in-vitro derived mutants of banana against nematodes using bio-chemical parameters. *American-Eurasian Journal of Sustainable Agriculture* 2(3): 271-278.

Lehman PS 2004. Cost-benefits of nematode management through regulatory programs. In *Nematology Advances and Perspectives, Vol. 2*, ed. ZX Chen, SY Chen, DW Dickson, pp. 1133–77. CAB International Wallingford UK.

Luc M,Sikora RA and Bridge J (Eds.) 2005. Plant parasitic nematodes in sub-tropical and tropical agriculture.CAB International Wallingford UK, pp-871.

Luc M, Taylor DP, 1977. *Heterodera oryzae*. C.I.H. Descriptions of Plant-parasitic Nematodes Set 7, No. 91:3 pp.

Luc M, Vilardebó A, 1961. Les nématodes associés aux bananiers cultivés dand l'ouest Africains. *Fruits* 16:205-219.

Machon JE,Bridge J.1996.*Radopholus citrin*.sp(Tylenchida:Pratylenchidae) and its pathogenicity on citrus. *Fund. Appl. Nematol.* 19:127–33

Marin DH, Sutton TB and Barker KR, 1998. Burrowing nematode resistance of black sigatoka resistant banana hybrids. *Nematropica* 28: 241-246.

McSorley R and Parrado JL 1981. Population fluctuations of plant parasitic nematodes on banana in Florida. *Proc. Fla. State Hort. Soc.* 94: 321-323.

McSorley R, Parrado JL, 1986. *Helicotylenchus multicinctus* on bananas: an international problem. *Nematropica*, 16:73-91.

Melin P, Vilardebo A, 1973. Nematicide and hot water disinfection in the control of *Radopholus similis* in banana plantations. *Fruits*, 28(12):843-849, 909-912.

Merny G, Cadet P, 1978. Penetration of juveniles and development of adults of *Heterodera oryzae* on different plants. *Revue de Nematologie*, 1(2):251-255.

Minz G, Ziv D and Strich Harari 1960. Decline of banana plantations caused by spiral nematodes in the Jordan Valley and its control by DBCP. *Ktavin* 10: 147-157.

Nair KKR 1979 Studies on the chemical control of banana nematodes. *Agric. Res. J. Kerala* 17: 232-235.

Nair MRGK, Das NM, Menon MR, 1966. On the occurrence of the burrowing nematode, *Radopholus similis* (Cobb, 1893) Thorne, 1949, on banana in Kerala. *Indian Journal of Entomology*, 28:553-554.

Nguyen CN, Subbotin SA, Madani M, Trinh PQ, Moens M. 2003. *Radopholus duriophilus* sp. n. (Nematoda: Pratylenchidae) fromWestern Highland of Vietnam. *Nematology* 5:549–58.

Niere BI, Speijer PR and Sikora RA 1999. A novel approach to biological control of banana nematodes. ftp://ftp.gwdg.de/pub/tropentag/proceedings/1999/referate/STD_C5.pdf.

Philis J, 1971. Control of root-knot and spiral nematodes on bananas in Cyprus. *Plant Disease Reporter*, 55: 707-710.

Pocasangre, L.E. (2007) "Effect of combined inoculations od endophytic fungi on the bio control of *Radopholus similis*" ISHS/ProMusa Symposium: *Recent advances in banana crop protection for sustainable production and improved livelihoods.* Greenway Woods Resort, White River, South Africa, 10th to 14th September 2007.

Price NS, McLaren CG, Frison EA, Horry JP, de Waele D, 1996. Techniques for field screening of *Musa* germplasm. New frontiers in resistance breeding for nematode, Fusarium and Sigatoka. Proceedings of the workshop held in Kuala Lumpur, Malaysia, 2-5 October, 1995, pp. 87-107.

Quénéhervé P and Van Den Berg E 2005. Liste des nematodes phytoparasites (Nematoda : Tylenchida et Dorylaimida) des departements francais d Amerique : Guadeloupe, Matinique et Guyane. *Bulletin OEPP/EPPO Bulletin* 35: 519-530.

Quénéhervé, P and Cadet P, 1985. Etude de la dynamique de l'infestion en nématodes transmis par les rhizomes du bananier cv Poyo en Côte d'Ivoire. *Revue de Nématologie*, 8:257-263.

Rajendran G and Sivakumar CV 1996. Pathogenicity of the spiral nematode, *Helicotylenchus multicintus* and effect on yield and nutrient status of banana. *Pest Management in Horticultural Ecosystems* 2: 23-27.

Roman J, 1977. Observations on the association of *Pratylenchus brachyurus* with the dry rot of yam, *Dioscorea floribunda* in the tropical area of Mexico. *Nematropica*, 7(1):25-26.

Sarah JL 1989. Nematological review - banana nematodes and their control in Africa. *Nematropica* 19: 199-216.

Sasser JN and Feckman DW 1987. A world perspective on Nematology: the role of the society. In JA Veech and DW Dickson Eds. Vistas on Nematology. P 7-14. Society of Nematologists, Hyattsville,Maryland USA.

Siddiqi MR, 1973. *Helicotylenchus multicinctus.* C.I.H. Descriptions of Plant-parasitic Nematodes. Set 2, No. 23 Commonwealth Institute of Helminthology, St. Albans, 1-3.

Siddiqi MR, Hahn ML. 1995. *Radopholus bridgei* sp. n. (Tylenchida: Pratylenchidae) from Indonesia and its differentiation by morphological and molecular characters. *Afro-Asian J. Nematol.* 5:38–43

Sikora, R. A. *et al.* 2007. "The burrowing nematode of banana: strategies for controlling the uncontrollable" ISHS/ProMusa Symposium: *Recent advances in banana crop protection for sustainable production and improved livelihoods.* Greenway Woods Resort, White River, South Africa, 10th to 14th September 2007.

Singh HP 2002. Scenario of banana production, utilization and trade. Pp 1-17 in Global Conference on Banana and Plantain 28-31 October 2002.

Southey JF, 1986. Laboratory Methods for Work with Plant and Soil Nematodes. London, UK: HMSO, 202 pp.

Speijer PR, Kajumba C and Ssango F 1999 a. East African highland banana production as influenced by nematodes and crop management in Uganda. *International Journal of Pest Management* 45: 41-49.

Speijer PR, Kajumba C and Tushemereirwe 1999 b. Dissemination and adaptation of a banana clean planting material technology in Uganda. *InfoMusa* 8: 11-13.

Speijer PR, Nampala PM, Elsen A, Ekwamu A and De Waele D 2001. Reinfestation by nematodes and performance of hot water treated East African highland cooking bananas as perceived by farmers in Ikulwe, Iganga District, Uganda. *African Plant Protection* 7: 85-89.

Speijer, P.R., C.S. Gold, I.N. Kashaija, and E.B. Karamura. 1994. Assessment of nematode damage in East African highland banana systems. p. 191–193. *In* R.V. Valmayor *et al.* (ed.) Banana nematodes and weevil borers in Asia and the Pacific: Proceedings of a - conference on nematodes and weevil borers affecting bananas in - Asia and the Pacific. Malaysia, April 1994, INIBAP/ASPNET, Philippines.

Speijer,P.R.,J.Mudiope,F.S sango,and E.Adipala.1998.Nematode damage and densities at different plant growth stages of East African highland banana (Musa AAA), cv. Mbwazirume. *African Plant Protection* 4:19–25.

Stanton J, Mundo-Ocampo M, Baldwin JG, Kalplan DT. 2001. *Radopholus musicola* n. sp., a new pathogenic species from Australia (Nematoda: Pratylenchidae). *Nematology* 3:689–98

Stoffelen R, Verlinden R, Ngo Thi Xuyen, Swennen R, Waele Dde, 1999. Screening of Papua New Guinea bananas to root-lesion and root-knot nematodes. *InfoMusa*, 8(1):12-15.

Stover RH, 1972. Banana, plantain and abaca diseases. Kew, Commonwealth Mycological Institute., UK.

Stoyanov D, 1967. Nematodes parasitic on banana in Cuba and possibilities for their control. *Revista Agricultura, Cuba*, 1:9-47.

Strich-Harari D, Minz G, Peled A, 1966. The spread of spiral nematodes in banana roots and their control. *Israel Journal of Agricultural Research*, 16:89-94.

Sudha S and Sundararaju P 1996. Reaction of some banana cultivars to the burrowing nematode, *Radopholus similes*. Pp 341-343 in Proceedings of Banana - Improvement, Production and Utilization (HP Singh and KL Chadha, Eds.) National Research Centre for Banana, Trichy.

Sundaraju P. and Cannayane I, 2002. Management of root lesion nematode, *Pratylenchus coffeae* in banana cv. Nendran by using trap crop. National Symposium on Biodiversity and Management of Nematodes in Cropping System for Sustainable Agriculture, Nemalogical Society of India, Jaipur 11-13 November 2002, 91p.

Sundararaju P, Cannayane I and Ramesh R 2002. Penetration and development of *Pratylenchus coffeae* and *Meloidogyne incognita* in susceptible cultivars of banana. Pp 189 in Abstracts of 'National Seminar on Nematological Research in India at CS Azad University of Agri. & Tech., Kanpur, 17 December 1999.

Sundararaju P,Sasikala T and Cannayane 2003.Effect of *Verticillium chlamydosprium* on second stage juvenile mortality and egg hatching of *Meloidogyne incognita* under *in Vitro* conditions.*Current Nematology* 13:67

Tiwari SP, Dave GS, 1985. Burrowing nematode *Radopholus similis* (Cobb) Thorne, 1949 on banana in Madhya Pradesh. *Current Science India*, 54(24):1286.

Trinh PQ, Nguyen CN, Waeyenberge L, Subbotin SA, Karssen G, Moens M. 2004. *Radopholus arabocoffeae* sp. n. (Nematoda: Pratylenchidae), a nematode pathogenic to *Coffea arabica* in Vietnam, and additional data on *R. duriophilus*. *Nematology* 6:681–93

Vilardebo A, Boisseau M, Lassoudiere A, Melin P, Ternisien E, 1988. ExpTrimentation avec l'aldicarbe pour lutter contre Radopholus similis Cobb (Nematoda, Pratylenchidae) et Cosmopolites sordidus (Germar) (Coleoptera, Curculionidae) en bananeraie. 1: ExpTrimentation rTalisTe en Martinique at au Cameroun. *Fruits*, 43, 417-431.

Volkers EELA and Gamboa 1988. Nematodes on plantain. *Boletin informative Manejo integrado de plagas* No. 7, 3-4.

Vovlas, N., Avgelis, A., Goumas, D., Frisulo, S., 1994. A survey of banana diseases in sucker propagated plantations in Crete. *Nematologia mediterranea* 22: 101-107.

Zuckerman BM, Strich-Harari D, 1963. The life stages of *Helicotylenchus multicinctus* (Cobb) in banana roots. *Nematologica*, 9:347-353.

Zum Felde, A., Mendoza, A., Cabrera, J.A., Kurtz, A., Schouten, A., Pocasangre, L. and Sikora, R.A. 2009. The burrowing nematode of banana: strategies for controlling the uncontrollable. Acta Hort. (ISHS) 828:101-108. http://www.actahort.org/books/828/828_9.htm

❑❑❑

Phyto-nematodes in Crops: Their Identification, Treatment and Management, 2011
G.L. Sharma (editor), pp. 203-227
New India Publishing Agency, New Delhi (India)
E-mail : info@nipabooks.com; Website : www.nipabooks.com

Chapter 9

Nematodes Infestation in Plantation Crops and their Management

SUDERSHAN GANGULY AND PANKAJ

Introduction

Nematodes are cosmopolitan and ubiquitous in occurrence. These have been reported in all kinds of ecological niche. Accordingly, different kinds of nematode fauna had evolved with distinct morphological features and physiological variability. However, development and perception keep changing with the ambient environment. Agriculture has progressed with leap and bounce as human being has managed this to its maximum utility. The fast changes in agriculture definitely influenced the ecological niche and affected the flora and fauna. Consequently, specific and unspecific kind of relation evolved between host and parasite. With the thumb rule of ecological balance or homeostasis. In the dynamic Mother Nature, multispecies heterogeneous complex is sustained and nature do not perpetuate homogeneity.

The plant-parasitic nematodes are one of the major limiting factors in the production of plantation crops throughout the country. For centuries, man has been plagued by these microscopic organisms feeding on the roots of crop plants essential for his survival and well being. Even though nematodes are encountered in the rhizosphere of all the plantation crops, only a few are pathogenic and cause significant damage. There are several plant parasitic nematodes reported from the rhizosphere of different plantation crops among which, *Meloidogyne* spp, *Pratylenchus* spp., *Radopholus similis, Rotylenchulus reniformis, Helicotylenchus* spp.and *Tylenchorhynchus*spp.are economically important.

Plantation crops need to be well protected as they are the main foreign exchange earners of our country. An integrated nematode management programme involving nematode-free planting material, resistant/tolerant varieties of crops, organic manuring and other cultural methods, minimum quantity of nematicides and introduction of biocontrol agents viz., mycorrhizae, fungi and bacteria to the nursery potting mixture as well as to the planting pit is recommended. The plantation crops included in this chapter are coconut, areca nut, oil palm, banana, coffee, tea and rubber. Important nematode pathogens on these crops are listed in Table 1.

A. Nematodes of Coconut

The nematode pests of coconut have extensively been reviewed recently by Griffith and Koshy (22) and Koshy and Geetha (35). The important nematode pathogens on coconut are red ring nematode, *Rhadinaphelenchus cocophilus* (Cobb) Goodey and burrowing nematode, *Radopholus similis* (Cobb) Thorne.

Table 1 : Important nematode pests of plantation crops

Crops	Nematode	Disease/Symptoms
Coconut (*Cocos nucifera* L.)	*Rhadinaphelenchus cocophilus*	Red ring disease : Chlorosis at the tips of oldest leaves which later spread towards the bases, premature shedding of nuts, characteristic internal lesions which in a cross section of the stem appear as 2-4 cm wide orange to brick coloured ring continuous from the bole region to the top of the trunk.
	Radopholus similis	Stunting, yellowing, reduction in number and size of leaves and leaflets,

Contd...

		delay in flowering, button shedding and reduced yield, presence of small elongated orange coloured lesions on the tender creamy white roots which later enlarge and coalesce leading to severe root-rotting resulting in reduction of feeder roots.
		Generally large, oval, sunken, brown black lesions on the thick primary roots. Lesions on the main and lateral roots which later coalesce and cause root-rotting.
Arecanut (*Areca catechu* L.)	*Radopholus similis*	Red-ring disease: Drying of leaves which later turn brown, diseased internal tissues also appear brown.
Oil palm (*Elaeis guineensis* Jacq)	*Rhadinaphelenchus cocophilus**	Brownish discoloration with erect, short and often deformed leaves with suberized patches especially on the inner side of the leaf stalks known as little leaf diseases.
Coffee	*Meloidogyne exigua**	White to yellowish brown galls on the new roots, later turn dark brown with aging of roots, reduction in growth and defoliation of the infested seedlings are field symptoms.
	*M. incognita**	Peeling and cracking of the cortex of the roots, chlorosis, leaf fall, general decline, reduced growth and death of plants, localized swelling resembling galls on lateral roots and typical root galls on young plants.
	*M. coffeicola**	Peeling and cracking of roots without gall formation, above ground symptoms are yellowing, leaf gall, and decline in plant health leading to death.
Tea (*Camellia sinensis* (L) O. Ktze)	*Pratylenchus loosi*	Spindly growth of plant with reduced foliage, patchy growth in field, yellowish dull leaves, premature flowering and fruiting, less number of feeder roots.

Contd...

Banana (*Musa* spp.)	*R. similis*	Uprooting and toppling of plant especially those bearing fruits; appearance of dark red lesions on the outer part of the root, adjacent lesions coalesce, cortical root tissue atrophies and turn black; vegetative cycle lengthens and reduces bunch weight.
	Pratylenchus spp *Filipjev*	Stunting of plants, reduction in size and number of leaves, reduction in bunch weight, toppling of plant, extensive black or purple necrosis of epidermis and cortex (*P. coffeae*) or small brownish-red elongated flecks (*P. goodeyi*).
	Heterodera oryzicola Rao & Jayaprakash	Reduction in plant growth, bunch weight and loss of feeder roots.
	Helicotylenchus multicinctus (Cobb) Golden	Stunting of plants, lengthening of vegetative cycle, reduction in size of the plant and bunch weight, reduction in productive life of the plants, toppling on heavy infestation.

* Not reported in India.

Rhadinaphelenchus cocophilus

The"Red ring disease" of coconut caused by *R.cocophilus* and transmitted by *Rhynchophorus palmarum* L. has not yet been reported from India and other south east Asian countries. The infested trees may show "little-leaf" symptoms, and die within 3-4 months after initial symptoms appear. The disease was first reported from Trinidad by Hart (24) and later from Grenada (61). Red ring disease has been reported to occur in Latin America (Dominican Republic, Venezuela, Guyana, Surinam, French Guyana, Colombia, Ecuador, Peru, Mexico, Brazil, Panama, Nicaragua, Costa Rica, Honduras, Belize and EI Salvador) and West Indies (Trinidad, Tobago, Grenada and St.Vincent). It has also been reported to occur in Guatemala, but does not occur in the northern Caribbean Islands, Florida, Cuba or other parts of the world (11).

The disease is not very distinct in its early stages. The first external symptoms are visible only after full expression of the internal symptoms. Chlorosis first appears at the tips of the oldest leaves and spread towards their bases, younger leaves too get affected. The brown lower leaves may break across the petiole or the lower part of the rachis or may become partly dislodged at the base and hang down. Along with or prior to the development of the leaf symptoms, premature shedding of the nuts may also take place. As a result of severe internal damage, the crown often topples over about four to six weeks after the appearance of the first symptoms. The trunk then remains standing in the field for several months and finally decays (22).

The internal lesions are the most characteristic symptoms of red-ring disease. About 3-5 cm below the stem surface, scattered reddish dots of about 1 mm diameter appear first, which later coalesce and form a 2-4 cm wide orange–red ring. The ring extends to the whole length of the stem. In petiole, it takes a 'crescent' shape. In the diseased palm, soft white cortex of the root becomes orange to faint red colour and flaky and dry in texture. In dwarf cultivars, red colour gives way to shades of brown. Thus there is an internal brownish band instead of red ring. Nematodes are found in plenty in the infested parenchymatous cells of the stem, petiole and root. Although xylem vessels often get occluded by tyloses, nematodes have never been encountered in xylem vessels. Young coconut is very susceptible to the red-ring disease; trees of 3-10 years old are mostly affected, with highest incidence in trees of 4-7 years old. Occasionally younger or much older palms also get attacked. There is no report of any tree that has recovered after being affected.

The vector of the red-ring nematode is the palm weevil (*Rhynchophorus palmarum*). Larve of the palm weevil while feeding through the diseased tree gets infested with the nematode. Adult emerging out from the diseased tree acts as carrier. The female palm weevil makes punture in the softer tissue of the stem of the palm normally in the soft axils to a depth of 5-8 cm and deposits eggs together with the nematodes. The third stage juveniles can be found in the tracheal tracts of the insect from where they move directly to the ovipositor. The nematode infests only parenchymatous tissue in root, stem, leaf and artificially infested tender nut.

A diseased(red-ring)tree remains chemically attractive to all palm weevils including potential vectors. Red-ring disease in new groves generally begins when a vector weevil infects the palm. Fertile vectors developing from the heterozygous matings spread the disease in the grove.

Management

Effective management measures are employed in phases during development of the disease epidemic. The leaf axils of the diseased trees are sprayed thoroughly with 0.1% lannate (methomyl) insecticide solution which is followed by treatment with herbicides (weedicide"100" or silvisar "510") to kill the tree. Herbicide is applied by boring three holes around the the trunk of the tree slanting down downwards at an angle of 45^0 and at a level of about 15 cm above the soil. The holes are bored to depth of about 10cm using a 2 cm diameter pipe and are filled with 10ml of herbicide resulting death of the tree within 14 to 21 days

Trees in the advanced stages of the disease are cut down, and the remaining stumps are sprayed thoroughly with at least 4.5 liters of 0.1% lannate solution. After 14 days, the dried out remains of the trees are burnt using kerosene.

Traps or guard baskets are used to protect plantation from frequent outbreaks of the disease. Guard baskets are cylindrical, 1 m in height and 0.3 m in diameter, made of 2 cm mesh wire. The baskets are filled with chunks of fresh tissue from the diseased coconuts or grugru palms to attract beetle. The guard baskets are sprayed completely with about 4.5 liters of 0.1% lannate solution and placed on ground at the rate of one basket per acre. After two weeks, the tissue in the basket is burnt and replaced with fresh tissue.

Bio-control measures involving parasitization of vector weevils using several nematode species of Rhabditidae or Heteorhabditidae have been tried in Latin America (22).

The most important nematode pest of coconut in India is the burrowing nematode, *Radopholus similis*. The nematode infests a wide variety of inter-crops in coconut garden like banana, arecanut, black pepper, ginger, turmeric, sweet potato, groundnut, sugarcane, betel vine etc.

Radopholus similis

Burrowing nematode occurs in most tropical and sub-tropical areas of the world and has been reported from coconut palms in Florida, Jamaica, Sri Lanka and India (81, 82, 13, 51, 86, 40).

General decline symptoms like stunting, yellowing, reduction in number and size of the leaves and leaflets, delay in flowering, button shedding and reduced yield are exhibited by burrowing nematode infested palms. As these symptoms are often over looked and considered as due to nutrient deficiency, the definite method to identify burrowing nematode infestation is the presence of characteristic small, elongated, orange coloured lesions on the young creamy white roots during September to January. Consequent to nematode multiplication and parasitization, these lesions enlarge and coalesce to cause extensive rotting of the roots. Tender portion of the root where lesions and rotting are confined becomes spongy on heavy infestation. Lesions are not usually seen on the old hard dark brown root. Surface cracks develop on the semi-hard orange coloured root. One gram of the infested root yield as many as 4000 nematodes. Plumule, leaf base and haustorium's of seedlings are also susceptible to burrowing nematode infestation.

R. similis penetrates the absorbing region behind the root-cap covered with delicate epidermis by lysis of cells.Such entry points are 1-2 cells in diameter and surrounded by parenchymatous cells to a depth of 10-15 cells.Outer cortex harbors maximum number of cavities and nematodes.In the early stages of infestation,the root contains independent cavities separated by several cells.As a result of nematode multiplication and lysis of cell wall and cytoplasm, adjacent cavities merge with each other leading to almost complete destruction of cortex. Longitudinal section of the infested root shows eggs and all stages of the nematode with different orientation (34).Nematodes have not been observed in the stellar region or in the closely packed 4-6 layers of cortex cells just outside the endodermis i.e.inner cortex.The burrowing nematode is a migratory endoparasite and is capable of spending its entire life cycle within the root. The nematode takes 25 days at 25^0 - 28^0 C to complete one life cycle(J_1-J_5).Except the fourth stage male juvenile and adult male, all other juvenile stages and adult females are infective.

Under field conditions, in moist soil (27-36^0 C) the burrowing nematode survives for six months and in dry soil (29-39^0 C) for one month. In root of the stump of felled coconut palm, the nematode survives for upto six months (73). A mean soil temperature below 25^0 C and light rainfall coupled with availability of tender fleshy roots are ideal conditions for nematode multiplication.

Management

Cultural control : Application of oil cake, farm yard manure and green foliage specially *Gryricidia* to the basins, growing of intercrops like cacao that enriches the soil with sizeable quantities of shed-foliage which helps in the build up of beneficial organisms (22); growing of green manure crops like *Crotalaria juncea* in the basin and interspaces of coconut palms and its incorporation into soil at flowering as green manure, avoiding banana as shade crop in nurseries; use of less susceptible/tolerant cultivars or hybrids of coconut and other intercrops in infested areas have been recommended as cultural practices to reduce nematode incidence. Application of 50kg cow dung/farm yard manure, 2 kg neem/marotti oil cake and 25 kg green manure to the coconut basin has been tried and practiced.

Biological control: Introduction of bio-control agents like *Pasteuria penetrans, Paecilomyces lilacinus* and VAM to nursery potting mixture to reduce nematode population and for imparting resistance/tolerance to nematode infestation has been tried and suggested (39).

Chemical control : Soil application with phenamiphos or phorate @ 25 kg a.i/ha during September, December and May (33, 41) in nursery or phorate @ 10 g a.i/palm in June/July and September/ October in mature plantation or phorate @ 3 g a.i/vine/sucker at planting and again after three months in infested area, particularly in coconut based farming system involving black pepper and banana has been recommended.

B. Nematodes of Arecanut

A systematic survey carried out in Kerala, Karnataka and Tamil Nadu in India revealed the occurrence of 28 genera of plant parasitic nematodes in rhizospere of arecanut palms. *Radopholus similis* has been reported as the most pre-dominant species in the root samples (40, 59, and 74).

Burrowing nematode was first reported from soil around roots of arecanut palm in Mysore, India by Kumar *et al.* (49) and later by Koshy *et al.* (36, 37). The appearance of lesions and rotting is the characteristic of *R. similis* infestation on roots. The nematode produces small, elongated orange coloured lesions on the young, succulent, creamy-white to light-orange coloured portion of the main and lateral roots. On multiplication of nematodes these lesions coalesce and cause root-rotting. The tips of lateral and tertiary root on infestation become black. The thick primary roots arising from the bole of the palm exhibit large, oval, sunken lesions. The nematode infested areca palms show non-specific above ground symptoms like general yellowing and visible reduction in growth, vigour and yield. The nematode is found in inter and intracellular portions of the root-cortex but does not enter the stelar regions. The arecanut isolates of *R. similis* belongs to the banana race (31) with a haploid number of four chromosomes (n-4). Maximum population of *R. similis* in roots has been reported during October-November and minimum during March-June. The burrowing nematode takes 25-30 days to complete one life cycle (j_2-j_2) on arecanut seedlings at a temperature range of 21-31°C.

Management of Burrowing Nematode is Possible by Blending Following Measures

Cultural : Use only nematode free planting material of arecanut and other inter crops; avoid *R. similis* susceptible inter or mixed crops such as black pepper and banana, and protect them with nematicides in infested area as well.

Resistance: Cultivars Indonesia-6 (VTL-11), Mahuva-8 and Andaman-5 (VTL-29) are tolerant to *R. similis*. (38, 75). The hybrid VTL-11 x VTL I7 has been reported as highly resistant to *R. similis*.

Chemical: Control of *R. similis* infestation on arecanut in seedling stage can be achieved with application of aldicarb or fensulphothion @ 1 g a.i./seedling thrice a year for three consecutive years. In adult palms, ten fold increase in yield and reduction in nematode population are obtained with the application of fensulfothion @ 50 g a.i./palm and aldicarb @ 10 g a.i./palm in May/June, September/October and December/January for five years (76). Arecanut being masticator and chewed directly by the consumers, the frequency, dosage and time of application of nematicides have to be carefully regulated to avoid chances of accumulation of pesticide residues in the nut.

Management of Nematodes in Coconut and Aracanut

- Seedlings should be raised in nematode-free nursery
- Application of phorate @ 10 g a.i./palm in June-July and in September-October is beneficial.
- Growing green manure crops like *Crotalaria juncea* in the basins and interspaces of coconut palms and incorporating into soil at flowering are also helpful.
- Banana should be avoided as a shade crop in coconut nurseries.

C. Nematodes of Oil Palm

There are several plant parasitic nematodes reported from the rhizosphere of oil palm but only red-ring nematode, *Rhadinaphelenchus cocophilus* is known to cause major economic loss.It was identified in oil palm from Venezuela in 1953 (85) and later reported from Latin America (53)and Surinam (80). Oil palm was found to be susceptible to *R. similis* infestation on artificial inoculation under green house conditions (30). The leaves dry out and turn brown instead of the usual yellowing and then browning.Pathogenesis is longer in oil palm as compared to coconut.The young tree (ten years old palm) takes three to four years after infection to become virtually dead.The colonization of the nematode in the oil palm tissue is slow as compared to coconut. Single gram of infected tissue from red-ring zone of the coconut yields 5000-10000 nematodes as compared to less than 500 nematodes/g of the tissue from a similar region in oil palm. Another difference is that most nematodes are found outside necrotic zone, even in areas which show no necrosis such as the distal or the basal portion of the stem and occasionally in the rachis of the inflorescence. Eggs are present in the brownish spots which are found in the advancing area of the disease (22). Band of the necrotic tissue is very narrow. Necrotic areas indicate the evidence of plant reaction to the cellular damage caused by the nematode. In young groves the diseased palms are found clustered in 50 m radius which then expands. The diseased trees are found randomly distributed in older groves. A direct correlation between the levels of red-ring disease in coconut and oil palm exists in Latin America. Countries with high levels of red-ring disease in coconut groves also have high levels of red ring in oil palm groves.

Little leaf diseased palms can be identified by their erect, short and often deformed leaves with suberized patches especially on the inner side of the stalks. The diseased palms have a brownish discoloration. *R. cocophilus* is found in the discoloured tissues of the leaves, especially more in the petioles of the young folded leaves. It feeds ectoprasitically in the buds of the palms. Van Hoof and Seinhorst (80) in Surinam, observed that little leaf syndrome was associated with red-ring nematode infestation. The actual cause of the little leaf disease may be diverse and related to symptoms of other diseases.

A combination of different methods is advisable for the management of red-ring disease in oil palm.

(i) Destruction of the sources of infection (diseased trees and other hosts) helps to destroy a major portion of the inoculum

(ii) Weevil trapping is an important mode of pest control in oil palm estates. The traps are always sprayed with 0.1% lannate.

(iii) As the pathogenesis (rate of infection) is slow in oil palm, application of nematicides might prove effective in controlling the infestation.

D. Nematodes of Date Palm

Nematodes have not yet been turned out to be a limiting factor in date palm cultivation. Numerous nematodes like *Pratylenchus penetrans* (50) and *R. cocophilus* (22) though reported on date palm, but only root-knot nematodes *Meloidogyne* spp have been reported to cause economic damage. Though root-knot nematodes were found in the Coachella Valley of California,USA on date palm in 1925,the first report on their occurrence was made by Buhrer *et al.*(1933)(6).Jensen (26) reported *Meloidogyne incognita* (Kofoid & White) Chitwood infesting roots of date palms in nurseries. *M. javanica* (Treub)Chitwood is reported to severely damage or kill the date palm seedlings (8). *M. arenaria, M. hapla, M. incognita* and *M. javanica* have been reported to occur on date palms in Israel (55). More than 90% of the seedlings are killed prior to emergence on sowing in heavily infested soil and palms infested with nematodes become susceptible to secondary infection by fungi.

E. Nematodes of Banana

The most important and wide spread nematode pests associated with banana plantations throughout the world are *R. similis, Pratylenchus* spp and *Helicotylenchus multicinctus*. Sedentary endoparasites like *Meloidogyne* spp. and *R. reniformis* are also common in banana roots. In addition to the above, 146 species belonging to 43 other genera of nematodes have been found associated with *Musa* spp. In India, around 71 plant parasitic nematode species are known to be associated with banana plants (69) of which *R. similis* is the most serious. *R. reniformis, H. multicinctus, P. brachyurus* and *P. coffeae* have also been identi fied as pathogenic to banana where *R. similis* is absent. Multiplication and prevalence of these nematodes are high in older mono -culture plantations as compared to new ones. *M. incognita* occurs without any conspicuous damage to the roots of banana.

Radopholus similis

The disease caused by *R. similis* infestation on banana is known by different names in different parts of the world of which the most common are "black head toppling disease" and "toppling disease". The burrowing nematode *R. similis* was first observed in 1891 in the necrotic root tissues of *Musa* sp. in Fiji. Since then, the disease has been reported from all the tropical and subtropical regions except Israel, the Canary Islands, Cape Verde Islands, Cyprus, Crete, Mauritius and Taiwan. The most conspicuous symptom of *R. similis* infestation on banana is the toppling over or uprooting of the plants particularly those bearing heavy bunches. The nematode infestation reduces the anchorage of the plant and its ability to absorb water and nutrients. Depending on the intensity of the disease, the effect may vary from lenghthening of the life cycle to drastic reduction in bunch weight. The presence of dark red lesions on the outer part of the root is very clear.The nematode infestation is confined to the cortical tissue and not to the stele.Due to nematode multiplication, adjacent lesions coalesce and cortical tissue atrophies which later turns black. Migration and infestation of the nematodes from infected root to corm leads to the development of diffuse black lesions(52). Uprooting occurs commonly during windstorm as well as when heavy rains loosen the soil.

R. similis penetrates the root-tip and invades the entire length of the root. The females and all juvenile stages are infective. Inside root, the nematode occupies an intercellular position in the cortical parenchyma. Life cycle ranges from (egg-egg) 20 to 25 days at a temperature range of 24^{0}C to 32°C.

Pratylenchus spp.

Eight species of *Pratylenchus* have been reported as pests of *Musa* spp of which *P. coffeae* and *P. goodeyi* are the most important. *P. coffeae* earlier reported as *Tylenchus musicola* was first reported on the roots of the plantains. *P. goodeyi* was first observed by de Guiran and Vilardebo (23) on the banana roots in Canary Islands. Symptoms of *Pratylenchus* spp. infestation are similar to that of *R. similis* attack. Stunting of plants, lengthening of the vegetative cycle, reduction in size and number of leaves, reduction in bunch weight and productive life of the plant, and toppling of the plant occur. Roots on heavy infestation exhibit extensive black or purple necrosis of epidermal or cortical tissues often followed by root rotting and breakage. Outer parts of the corm also exhibit similar necrosis (5).

Helicotylenchus multicinctus

H. mulcinctus is wide spread on banana throughout the world.It is often encountered with *R. similis*. The symptoms with *H.multicinctus* infection are very similar to that produced by *R. similis*. Aerial symptoms like the reduction in size of the plant and bunch weight followed by overall stunting,reduction in productive life span and lengthening of vegetative cycle etc. are seen.On heavy infestation toppling of the plant may also occur. Small characteristic necrotic lesions develop on the outer root cortex. The development of the lesions is slow as compared to those produced by *R.similis*. On the primary roots, lesions often appear as superficial, shallow, reddish brown to black dashes which coalesce and cause root necrosis in case of heavy infestation.

Meloidogyne spp

Five species of root-knot nematodes have been reported on *Musa* spp. of which the most common are *M. incognita, M. arenaria, M. javanica* and *M. hapla*. Different species of the root-knot nematodes have been found to infest the same plant consequently the galls produced contain mixed species population (64, 60, 14). The most characteristic symptom of the root-knot nematode is the galling on primary and secondary roots often leading to bifurcation and distortion on roots.

Other Nematodes

Other nematodes reported on banana from different parts of the world are *R.reniformis* (Puerto Rico,India,Winward Islands), *Hoplolaimus pararobustus* (Ivory Coast), *H. mucronatus* (Papua New Guinea & Solomon Islands) and *H.microcephalus* (Papua New Guinea), *Cephalenchus emarginatus* (Ivory Coast).

Management

Cultural

- Use nematode-free planting material: Sets are disinfected by paring or chemical treatment, heat therapy or by a combination of all.
- Fallowing of field or crop rotation using non-hosts or intercropping using crop non-hosts : Intercropping with *Crotalaria* has been found effective in reducing *R. similis* population in root

and increasing plant growth (9) ; crops viz, sugarcane, sorghum, tobacco, cassava, grape fruit, pangola grass etc. are advised for crop rotation for reducing *R. similis* population.

- Application of neem cake @ 400 g per plant at planting and after 4 months reduces the population of *R. similis.*

Host Plant Resistance

None of the widely grown clone is known to be resistant to the important nematodes in India (54). However, banana cultivars viz, Yelakkabale, Kadali, Pedalimoongil, Kannan, Ayiramkapoovan, Pey Kannan, Vennattu Kannan and Anaikomban etc. have been reported to be tolerant to the burrowing nematode (63).

Biological

In India, VAM fungus, *G. fasciclatum* has been found to be effective in bringing down the population of *R. similis* and also their multiplication inside the banana roots (79).

Chemical

Paring of the diseased roots from the corm followed by dipping of pared sets in mud slurry and sprinkling with 1.2 g a.i. carbofuran or 1 g a.i. aldicarb per set. The sets are then shade dried before planting (66). Soil application of aldicarb, carbofuran or fenamiphos @ 2-4 g a.i./plant is effective in controlling *R. similis.* In new plantations, nematicides are applied in the planting pit. Post planting application is done in 50-100 cm radius around the plant and mixed with soil. Established plants are treated with nematicides every 4-5 months.

F. Nematodes of Coffee

Many species of nematodes are found associated with coffee in different parts of the world, of which the important pathogens are *Meloidogyne* spp. and *Pratylenchus* spp. On a worldwide basis the root-knot nematodes rank high on the list of pathogens affecting production of coffee. In India, although 50 species of plant parasitic nematodes have been reported from coffee, the important ones listed out are the coffe root-lesion nematode, *Pratylenchus coffeae* and *Hemicriconemoides* spp. *P. coffeae* has been reported highly destructive to *arabica* coffee (*Coffea arabica)* but least to robusta coffee *(C. canephora).*

Meloidogyne spp.

Root knot nematodes reported from different parts of the world can be separated into two groups.The most common and damaging species are *M. exigua, M. incognita and M. coffeicola.* The less prevalent species are *M. africana, M. areneria, M. decalineata, M. hapla, M. inornata, M. javanica, M. kiukuyensis, M. megadora, M. otteifai and M. thamesi* (47).

M. exigua causes typical round galls on the fresh roots emerged out after the first rain. The galls are initially white to yellowish brown and later turn dark brown on mature roots. Egg masses are produced beneath the root epidermis, in the cortex. Young coffee plants are more susceptible to *M. exigua* infestation than any other stage of the plants. Infested seedling in the field shows reduced growth and defoliation which on heavy infestation causes complete defoliation and death (7).

M. incognita causes peeling and cracking of cortical root tissue and produces galls on the roots. Darker dots are formed on the roots where the females are located. Above ground symptoms are often expressed as foliar chlorosis, leaf-fall, and general decline and reduced growth.

M. coffeicola causes peeling and cracking of roots but does not produce galls. The females are located in the older tissue especially on the tap root. The females generally lay egg outside the roots through the crack except a few that lay egg inside the roots. Above ground symptoms are general yellowing, leaf fall and decline leading to the death of the plant (7).

Pratylenchus spp.

P. branchyurus, P.coffeae, P. goodeyi, P. loosi and *P. pratensis* are the different lesion causing nematodes reported on coffee from different parts of the world (7).

P. coffeae has been reported as the most destructive nematode of coffee in South India (62). *P. coffeae* is known to cause "juvenile foot rot' of young coffee plants and spreading decline of mature coffee (44). The infested plants develop root lesions, the tap roots gets severed and feeder roots get reduced leading to complete destruction of the root system (56, 87). Production of large number of adventitious roots at the collar region gives a bearded appearance. During dry period, the plants become weak and get killed. The bearing plants exhibit "die-back" symptoms. The outer layer of the tap root and secondary root peel off and the feeder roots become extinct. The plants become prematurely old., requiring collar pruning and uprooting. The above ground symptoms include yellowing and shedding of old leaves, production of lankey stems, crinkling of leaves at the tip with a tufted appearance of the main stem. The disease symptoms appear in "patch" in the coffee fields which gradually spread to more areas.

Hemicriconemoides spp.

In Karnataka, Tamil Nadu, Kerala, *H. cococphilus, H. coffeae, H. gaddi,* and *Nothocriconema indicum* have been reported to cause considerable damage to both mature arabica and robusta coffee and are associated with "crinkled leaf disorder" of coffee (42). The disease is characterized by production of small, crinkled, cholorotic and leathery leaves on a loose and shaky stem with poor feeder roots. The internodes become shortened and new growth arising from these gives a bushy or "witches broom" appearance (45).

Other Nematodes of Coffee

Among the other nematode parasites of coffee, *R. reniformis* ranks first. It has been reported from Philippines, India, New Guinea, Fiji, Tonga and Western Samoa. *R. similis* is highly pathogenic to young arabica and robusta plants in Java. *R. similis* collected from black pepper (*Piper nigrum* L.) and banana (*Musa* sp.) fields does not infest and multiply on coffee in south India (12,42). *Trophotylenchulus obscurus* has been reported as pest of coffee in West Africa (84). In addition to these, many other nematodes belonging to various genera have been found associated with the coffee plants.

Management of Coffee Nematodes

Nematode control in coffee plantations is difficult. Effective control is possible only by blending different control measures.

Cultural

Maintenance of nursery away from the plantations is advised. After uprooting the dead plants, pits are left open throughout the summer prior to replanting for solar drying; nursery beds and jungle soil are also put on solar drying; replacement of affected *arabica* plants with tolerant varieties like *robusta* (*Coffea canephora*) and grafting of *arabica* on to *robusta* root stock has been recommended.

Host Plant resistance : Several lines of *C.canephora* and *C.congesis* and some progenies of *C.canephora,* Sarchimor (Vila Sarchi x Tinmor Hybrid), Icatu (*C.arabica x C.canephora*) are resistant to *M.incognita* (21). In India,*C.canephora* and *C.excelsa* (tree coffee) exhibit tolerence to *P.coffeae* while all *arabica* varieties have been found suscepitable (46).

Chemical

Application of nematicides like aldicarb, carbofuran and phenamiphos @ 1.6 to 6.0 g.a.i./plant once or twice during a year has been found effective. The first application is carried out in the beginning of the rainy season followed by second application three months later. Generally a furrow is dug along with both sides of the plant row close to the tree and nematicide is applied close to the plant and mixed well with the soil. The soil is kept wet at the time of nematicide application, Nemacur has been founbd to be very effective in the control of *P. coffeae* (43). Methyl bromide at the rate of 150 cm^3/m^3 soil is the most effective method for soil fumigation and disinfestations of nursery soil.

G. Nematodes of Tea

Although a large number of plant parasitic nematodes have been encountered from soil around the tea plants, the known and suspected pathogens are *Pratylenchus* spp., *R. similis, Meloidogyne* spp., *Helicotylenchus* spp., *R. reniformis, Hoplolaimus* sp., *Rotylenchus* sp., *Xiphinema* sp. and *Hemicriconemoides* sp.

Pratylenchus spp.

The lesion nematodes are known to infest tea in different parts of the world of which *P. loosi* and *P. brachyurus* have gained much attention. *P. loosi* is the most serious pest in Sri Lanka (15, 70) and Japan (27, 78) which also causes damage to tea in China and Bangaladesh. However, in India,though *P.loosi* was reported long back from Darjeeling (57),the pathogenicity is yet to be proved. Unlike *P. loosi,P.brachyurus* attcks only young plants (one-three year old), *P.brachyurus* and *Pratylenchus curvitatus* are reported as pathogenic to mature tea plants.

Typical symptoms of *P. loosi* infestation in the field are patches of unthrifty tea, infested plants showing spindly growth with a sparse foliage. The leaves turn dull and yellowish. Heavy infestation leads to premature flowering and fruit setting. Diseased roots appear brown and dry up; feeder root growth is reduced; large storage roots display dark brown necrotic lesions of different sizes. Heavily infested plants recover very poorly after pruning or fail to recover and die.

P. brachyurus attack leads to characteristic nutrient deficiency symptoms and the plants show stunted and unthrifty growth. Dark red lesions develop on the epidermal layer of the feeder root tissue on infestation.

Meloidogyne spp.

Barbar (3) made the first report of root knot nematode infestation of tea from Munnar in Kerala. Most of the root-knot species attack tea plants only at the nursery stage while *M. brevicauda* is a serious pest of mature tea. *M. brevicauda* has so far been recorded only from Sri Lanka (25), South India (83) and Darjeeling, W.B. (57). The root knot nematodes most commonly encountered in tea nurseries are *M. incognita, M. javanica, M. arenaria* and *M. hapla.* The species of root-knot nematodes attacking only young plants develop galls on the feeder roots as well as the tap root. Occasionally the collar region of the tea seedling also gets infested with *M. incognita* which leads to the development of characteristic galls on the affected region of such stem (4). Aerial symptoms of *M. brevicauda* infestation resemble those produced by root-lesion nematode of tea. Infested bush remains stunted with small, dull to yellowish coloured leaves. The roots often exhibited large galls with pinhole pits on their surface.

*Radopholus similis: R.*similis was first reported on tea from Indonesia,Java(88) and later from countries like Sri Lanka,China, Zimbabwe and S.Africa. Infested plants are stunted with pale oryellowish leaves and often exhibit premature flowering and setting of fruits.

Management of Tea Nematodes

As most of the nematode pathogens attack the young seedlings rather than mature tea plants, control has been largely confined to the treatment of nursery soil.

Physical

Heat treatment of nursery soil by spreading soil on a galvanised iron sheet heated up to 60-62^0 C with occasional moistening and stirring helps in reducing innoculum potential. Over heating of the soil may lead to "manganese toxicity" in nursery plants.

Cultural

Incorporation of organic matter including cow dung, decomposed plant residue etc. has been found to be effective in suppressing *P. loosi* population level (78). Incorporation of margosa seed cake, coconut copra residue and decomposed tea waste reduces the pathogenic effect of *P. loosi* (71, 17, 19). Regular forking of the soil helps to improve soil aeration and feeder root growth.

Cultivation of Cover Crops

African weeping love grass (*Eragrotis curvula*) has been reported to suppress population of *P. loosi* and *R. similis* (16,18) and *Meloidogyne* spp. in Malawi (1).

Chemical Control

Application of nematicides like carbofuran 3G or phorate 10 G @1 g/kg nursery soil for bed application or 1g/sleeve has been recommended which should be repeated every 3-4 months.

H. Nematodes of Rubber

A number of plant parasitic nematodes have been reported from the rhizosphere of rubber but none of these are known to be pathogenic. However, a root-knot nematode *M. incognita* is known to be pathogenic to the cover crop *Peuraria phaseoloides.*

I. Nematodes of Cocoa

Cocoa (*Theobroma cacao* L.) is one of the major cash crops in the tropical countries. West Africa grows about 60% of the world crop. In India, cocoa though cultivated since 200 years, was till recently, confined to a very small area. Today the area under cocoa is about 10,000 ha, of which Kerala state grows cocoa in 5,500 ha mainly as an intercrop in coconut and arecanut gardens. The paper presents the results of a preliminary survey on plant parasitic nematodes associated with cocoa in Kerala and Karnataka. Though many species of plant parasitic nematodes have been recorded from the rhizosphere of cocoa, precise information on the pathogenicity of any one of them except of *Meloidogyne incognita* is lacking. Results of the preliminary survey comprising 88 each of soil and root samples collected from Kerala and Karnataka showed that the most common root parasite is *Meloidogyne incognita* and the dominant nematode in the rhizosphere is

Helicotylenchus sp. Inoculation studies showed cocoa seedlings developed severe root galling with females with egg masses of *M. incognita* exposed to the root surface. Species belonging to *Meloidogyne, Tylenchorhynchus, Hoplolaimus, Xiphinema, Helicotylenchus, Rotylenchulus, Criconemoides, Longidorus, Radopholus* and *Trichodorus* were extracted from soil around roots of cocoa.

References

Anonymous 1960. *Tea research station, Tea Association of Nyasaland, Rhodesia Quarterly Newsletter* 18.

Bally, W. and Reydon, G,A 1931. De tegenwoordige stand Van het Vraagstuk van de Wortelaaltjes in de Koftiecultrn. *Archf voor de Koffiercultur, Indonesie* 2 : 23-216.

Barber, C.A 1901. A tea eel worm disease in S. India. *Tropical Agriculturist* 21 : 238.

Basu, S.D. 1976. Root-knot nematode, *Meloidogyne incognita* (Kofoid & White) Chitwood on stem of tea seedling. *Two & A Bud* 23 : 23.

Bridge, J. and Page, S.L.J. 1984. Plant nematode pests of crops in Papua New Guinea. *Journal of Plant Protection In the Tropics* I: 99-109.

Buhrer, E.M., Cooper, C. and Steiner, G. 1933. A list of plants attacked by the root-knot mematode *(Heterodera marioni). Plant Disease Reporter* 17 : 64-96.

Campos, V.P., Sivapalan, P. and Gnanapragasam, N.C. 1990. Nematode parasites of coffee, cocoa and tea. In: *Plant parasitic nematodes in subtropical and tropical agriculture* (eds. M.Luc, R.A. Sikora and J. Bridge) CAB International, U.K., pp. 387-430.

Carpenter, J.B. 1964. Root knot nematode damage to date palm seedlings in relation to germination and stage of development. *Date Growers Institute Report* 41 : 10-14.

Charles, J.S.K., Venkitesan, T.S., Thomas, Y. and Varkey, P.A 1985. *Indian Journal of Nematology* 15 : 186-190.

Dasgupta, M.K. and Rama, K. 1987. Plant parasitic nematodes associated with plantation crops. *Review of Tropical Plant Pathology* 4 : 289-304.

Dean, C.G. 1979. Red ring disease of *Cocos nucifera* L. caused by *Rhadinaphelenchus cocophilus* (Cobb, 1919) Goodey, 1960: An annotated bibliography and review. *Technical communication No 47, Commonwealth Institute of Helminthology,* St. Albans, U.K.

D'Souza, G.I. and Kumar, AC. 1969. Occurrence of *Radopholus similis* and *Pratylenchus vulnus* in India. In: *All India Nematology symposium,* New Delhi, Aug. 21-22, 1969 (Abstract).

Ekanayake, U.B.M. 1964. Annual report of the crop protection officer, coconut research institute for 1963. *Ceylon Coconut Quarterly* 15 : 50-57.

Fargette, M. 1987. Use of esterase phenotype in the taxonomy of the genus *Meloidogyne*. 2. Esterase phenotypes observed in West Africa populations and their characterization. *Revue de Nematologie* 10: 45-56.

Gadd, C.H. 1939 A destructive root disease of tea caused by the nematode, *Anguillulina pratensis. Tea Quarlerly* 12: 131-139.

Gnanapragasam,N.C. 1981. The influence of cultivating *Eragrostis curvula* in nematode infested soil, on the development on the subsequent build-up of populations in replanted tea. *Tea Quarterly* 50 : 160-162.

Gnanapragasam, N.C. 1984. *Report on the tea research institute of Sri Lanka,* 1984, pp. 67-75.

Gnanapragasam, N.C. 1986. *Report on the tea research institute of Sri Lanka,* 1986, pp. 69-78.

Gnanapragasam, N.C. 1987a. Effect of soil acidity on the population levels of *Pratylenchus loosi. International Nematology Network Newsletter* 4 : 23-24.

Gnanapragasam, N.C., Anpalagan, V.T., Dharrnasena, W.A.M., Ariyaratnam, V., Jayasinghe, P.R.R. and Navaratne, N. 1985. *Report of the tea research institute, Sri Lanka.*

Goncalves, W. and Ferraz, L.C.B. 1987. Resistencia do cafeeiro a nemtoides II. Testes de progenies e hibridos para *Meloidogyne incognita* race 3. *Nematologia Brasileira* 9 : 125-142.

Griftith, R. and Koshy, P.K. 1990. Nematode parasites of coconut and other palms. In: *Plant parasitic nematodes in subtropical and tropical agriculture* (eds. M. Luc, R.A. Sikora and J. Bridge) CAB International, U.K., pp. 363-386.

de Guiran, G. and Vilardebo, A 1962. Le bananier aux Itles Canaries. IV: Les nematodes parasites. *Fruits* 17 : 263-277.

Hart, .T.H. 1905. Coconut disease. *Bulletin of miscellaneous information, Royal Botamic Gardens, Trinidad* 6 : 241-243.

Hutchinson, M. T. and Vythilingam, M.K. 1963. Distribution of plant parasitic nematodes in the soils of tea estates in Ceylon. *Tea Quarterly* 34 : 119-126.

Jenson, H.J. 1961. The nematode menace. *American Nurseryman* 114: 7-8, 105-113.

Kane Ko, T. and Ichinohe, M. 1963. Notes on the nematode species and their bionomics associated with tea roots in Japan. *Japanese Journal of Applied Entomology and Zoology* 7 : 165-174.

Koshy, P.K. 1986a. *Research on plant parasitic nematodes.* Central Plantation Crops Research Institute, Kayangulam, Kerala, India, October 1972 to June 1986.

Koshy, P.K. and Nair, C.P.R. 1979. Control of *Radopholus similis* (Cobb, 1893) Thorne, 1949 in coconut nursery. *Indian Journal of Nematology* 9: 15-19.

Koshy, P.K. and Sosamma, V.K. 1975. Host range of the burrowing nematode *Radopholus similis* (Cobb, 1893) Thorne, 1949. *Indian Journal of Nematology* 5: 255-257.

Koshy, P.K. and Sosamma, V.K. 1977. Identification of physiological races of *Radopholus similis* populations intesting coconut, arecanut and banana in Kerala, south India. *Indian Phytopathology* 30 : 415-416.

Koshy, P.K. and Sosamma, V.K. 1978a. Studies on the population fluctuations of *Radopholus similis* in coconut and areca roots. *Indian Phytopathology* 31 : 180-183.

Koshy, P.K. and Sosamma, V.K. 1979. Control of the burrowing nematode *Radopholus similis*, on coconut seedlings with DBCP. *Indian Journal of Nematology* 7 : 32-33.

Koshy, P.K. and Sosamma, V.K. 1987. Pathogencity of *Radopholus similis* on coconut (*Cocos nucifera* L.) seedlings under greenhouse and field conditions. *Indian Journal of Nematology* 17: 108-118.

Koshy, P.K. and Geetha, S.M. 1994 Nematode pests of palms and cacao. In: *Nematode pest management in crops* (eds. D.S. Bhatti and R.K. Walia) CBS Publisher & Distributors, New Delhi.

Koshy, P.K., Sosamma, V.K. and Nair, C.P.R. 1975. Preliminary studies on *Radopholus similis* (Cobb, 1893) Thorne, 1949 infesting coconut and arecanut palms in south India. *Indian Journal of Nematology* 5 : 26-35.

Koshy, P.K., Sosamma, V.K. and Sundararaju, P. 1976. Yellow leaf disease: nematological studies. *Arecanut and Spices Bulletin* 8 : 37-41.

Koshy, P.K., Sosamma, V.K. and Sundararaju, P. 1979. Reactions of thirty one areca germplasm .collections to the burrowing nematode *Radopholus similis. Plant Disease Reporter* 63 : 433-435.

Koshy, P.K., Sosamma, V.K. and Sundararaju, P. 1991. Nematode management in coconut and coconut based cropping system. In: *International symposium on coconut research and development* II, *21-29* Nov. 1991, CPCRI, Kasargod, Kerala, pp. 4-5.

Koshy, P.K., Sundararaju, P. and Sosamma, V.K. 1978. Occurrence and distribution of *Radopholus similis* (Cobb, 1893) Thorne, 1949 in south India, *Indian Journal of Nematology* 8 : 49-58.

Koshy, P.K., Sundararaju, P., Sosamma, V.K.and Ravikumar, K. 1985. Efficacy of four systemic nematicides against *Radopholus similis* in coconut nursery.*Indian Journal of Nematology* 15:148-151.

Kumar, A.C. 1980. Studies on namatode in coffee soils of south India. 2. Occurrence of four species of *Hemicriconemoides* (Criconematidae) and studies on the life-history of *H. cocophilus. Journal of Coffee Research* 10: 4-11.

Kumar, A.C. 1982. Evaluation of nemacur 50 against the coffee root lesion nematode, *Pratylenchus coffeae. Journal of Coffee Research* 12 : 1-7.

Kumar AC. 1984. The symptoms and diagnosis of the disorder, 'spreading decline' (cannoncadoo die back) with a note on spread and control of the casual agent, *Pratylenchus coffeae. Journalof Coffee Research* 14 : 156-159.

Kumar, A.C. 1985. Investigations on the three species of *Hemicrionemoides* (Nematoda) associated with the "crinkle leaf' disorder in coffee. *Journal of Coffee Research* 15 : 90-98.

Kumar, A. C. 1992. Nematode pests of coffee. In : *Nematode pests of crops* (eds. D. S. Bhatti and R. K. Walia) CBS Publishers & Distributors, New Delhi, pp. 258-268.

Kumar, A.C. and Samuel, Stephen 1990. Nematodes attacking coffee and its management- A review. *Journal of Coffee Reserach* 20 : 1-27.

Kumar, A.C., Viswanathan, P.R.K. and D'souza, G.1. 1971. A study of plant parasitic nematodes of certain commercial crops in coffee tracts of south India. *Indian Coffee* 35 : 222-224.

Lamberti, F., Greco, N. and zaouchi, H. 1975. A nematological survey of date palms and other major crops in Algeria. *FAO Plant Protection Bulletin* 23 : 156-160.

Latta, R. 1966. Attempts to relate nematodes to lethal yellowing disease of coconut palms in Jamaica. *Tropical Agriulture Trinidad* 43 : 59-68.

Loos, C.A. and Loos, S.B. 1960b. The blackhead disease of bananas. *Proceedings of the Helminthological Society Washington* 27 : 189-193.

Malgutti, G. 1953. "Pudricion de Cogollo" de la palmera do aceite Africana *(Elaeis guineensis* Jacq) en Venezuela. *Agronomia Tropical* 3 : 13-21.

Menendez, T. and Shepherd, K. 1975. Breeding new bananas. *World Crops* 27: 104-112.

Minz. G. 1958. Two *Diplodia* species in date palms. *Ktavim* 8: 213-216.

Monteiro, AR. and Lordello, L.G.E. 1974. Encontro do nematoide *Pratylenchus coffeae* atacando cateeiro em Sao Paulo. *Revista de Agricultura,Piracicaba* 49 : 164.

Mukerjee, B. and Dasgupta, M.K. 1982. Community analyses of plant parasitic nematodes in tea plantation in W. Bengal. *Nematologia mediterranea* 10 : 1-7.

Mustika, 1. 1978. An observation on the relationship between namatode popualtion and yellow disease on balck pepper in Bangkak. *Pemberitaan Lembaga Penelitian Tanaman Industri. Indonesia* 30 : 11-19.

Nair, R.B. 1964. Nematodes from arecanut soils. *Arecanut Journal* 15 : 87-88.

Netsher, C. 1978. Morphological and physiological variability of species of *Meloidogyne* in West Africa and implications tor their control. *Mededelingen Landibouwhogeschool Wageningen* 46: 78-83.

Nowell, W. 1918. Red ring or root disease of coconut palms. *West Indian Bulletin* 17 : 189-192.

Palanichamy, L. 1973. Nematode problems of coffe in India. *Indian Coffee* 37 : 99-100.

Parvatha Reddy, P., Khan, R.M. and Agarwal. P.K. 1989. In: *Proceedings of third group discussions on nematological problems on plantation crops, Coimbatore,* pp. 30-31.

Pinochet, J. 1977. Occurrence and spatial distribition of root knot nematodes on bananas and plantations in Honduras. *Plant Disease Reporter* 61 : 518-520.

Pizarro, A.C. 1969. Nematodes associated with roots of coconut and other palms. *Phillipine Journal of Plant Industry* 34 : 155-158.

Rajagopalan, P. and Naganathan, T.G. 1977. *Tamil Nadu Agricultural University Annual Report* 6: 129.

Rama, K. 1987. *Ecology of plant parasitic nematodes affecting some major perennial crops in West Bengal.* Ph.D. Thesis, University of Visva-Bharathi, India.

Reddy, D.B. 1978. Insect pests, diseases and nematodes of areca palms *(Areca catechu* L.) and betel vine *(Piper betle)* recorded in Asia and the Pacific. *Technical Document* No.114, Plant Protection Committee for the South East Asia and Pacific Region, FAO.

Reddy, P.P. and Khan, R.M. 1994. Nematode pests of banana. In: *Nematode pest management in crops* (eds. D.S. Bhatti and R.K. Walia) CBS Publishers and Distributors, New Delhi.

Sivapalan,P.1972. Nematode pests of tea. In *Economic Nematology* (ed. J. M. Webster) Academic press, London, pp. 285-310.

Sivapalan, P. 1980. Report of tea research institute of Sri Lanka 1980, pp. 83-96.

Sosamma, V.K.and Koshy,P.K.1977.Additional hosts of the burrowing nematode *Radopholus similis* infesting coconut palm in south India. *Plant Disease Reporter* 61 : 760-761.

Sosamma, V.K. and Koshy,P.K. 1986. Survival of *Radopholus similis* In : *National conference -Plant parasitic nematodes of India -problems and progress,* 17-20 December 1986, IARI, New Delhi, India.

Sundararaju,P.and Koshy,P.K.1982a. Distribution of phytoparasitic nematodes on arecanut in south India. In: *Proceedings of the silver jubilee symposium on arecanut research and development,* 12-14, December, 1982, Vittal, south Karnataka, India, pp. 143-146.

Sundararaju,P.and Koshy,P.K. 1982b. Response of *Areca* collections against *Radopholus similis Indian Journal of Nematology* 12:404-405.

Sundararaju, P. and Koshy, P.K. 1986. Effect of different nematicides and neem oil cake in the control of *Radopholus similis* in yellow leaf disease affected arecanut palms. *Indian Journal of Nematology* 16 : 44-47.

Sundararaju, P., Sosamma, V.K. and Koshy, P.K. 1984. Additional host records of root-knot nematode *Meloidogyne incognita. Indian Journal of Nematology* 14 : 183-184.

Takaji, K. 1969. Nematodes confront tea plantations in Japan. *Japanese Agricultural Research Quarterly* 4 : 27

Umesh, K.C., Krishnappa, K. and Bagyaraj, D.J. 1988. Interaction of burrowing nematode, *Radopholus similis* (Cobb) Thorne and VA mycorrhizae. *Glomus fasciculatum* (Thaxt) Gerd. and Trappe in Banana *(Musa acuminata* Colla). *Indian Journal of Nematology* 18: 6-11.

Van Hoff, H.A. and Seinhorst, W. 1962. *Rhadinaphelenchus cocophilus* associated with little leaf of coconut and oil palm. *Netherland Journal of Plant Pathology* 68 : 251.

Van Weerdt,L.G.,Birchfield,W.R.and Esser,R.P. 1959a. Observations on some subtropical plant parasitic nematodes in F!orida. *Proceedings of the Soil Crop Science Society of F1orida* 19:443-451.

Van Weerdt, L.G., Martine, A.P. and Esser, R.P. 1959b. Results of a survey designed to determine the etiology of lethal yellowing of *Cocos nucifera. Proceedings of the F1orida State HortIcultural Society* 72 : 421-430.

Venkata Ram,C.S.1963.*Annual report of the united planters association of southern India.*Scientific Department,Appendix,pp. 31-36.

Vov!as, N. !987. Parasitism of *Trophotylenchulus obscurus* on coffee roots. *Revue de Nematologie* 10 : 337-342.

Webster, B.N. and Gonza!es, J. !959. Investigaciones sobre la efermedad "anillo rojo" en cocas y palma de aceite. In: *9th Convencion de la Association Venezolana el Avance de la Ciencia Sevicio shell para el Agricultor,* Estado Arajua, Venezuela.

Weischer, B. 1967. Report to the Government of lndia on plant parasitic nematodes. PL 24.52 Report No. TA 2332, FAO, Rome.

Whitehead, A.G. 1969b. Nematodes attacking coffee, tea and cocoa and their control. In : *Nematodes of tropical crops* (eds. J.E. Peachey) CAB, London, pp.238-250.

Zimmerman, A. !899. Het Voorkomen van nematoden. in de Wortels van sirihen thee I. De aaltjesziekte der sirih *(Piper betle).* II. *Tylenchus acutocaudatus* Zn in de thee. *Teysmannia* 10: 523-531.

Phyto-nematodes in Crops: Their Identification, Treatment and Management, 2011
G.L. Sharma (editor), pp. 229-240
New India Publishing Agency, New Delhi (India)
E-mail : info@nipabooks.com; Website : www.nipabooks.com

Chapter 10

Protected Cultivation: Nematode Problems and their Management

R. K. WALIA AND K. K. WALIA

Protected Cultivation

Protected cultivation (PC) may be defined as the "modification of the natural environment to achieve optimum plant growth" (Singh, 2005). Modifications can be made to both the aerial and root environments to increase crop yields, extend the growing season and permit plant growth during periods of the year not commonly used to grow open field crops. PC may also indicate comprehensive systems of 'Controlled Environmental Agriculture' in which all the aspects of natural environment are modified for maximum plant growth and economic return.

Advantages of PC: Excellent for cultivation of high value vegetable crops, ornamentals, medicinal plants, and horticultural nurseries etc. Vegetable crops can be grown under unfavorable conditions; certain vegetable crops can be grown round the year. Management of insect pests/diseases is easier. PC provides excellent opportunity to produce vegetables for export, and production of genetically engineered and micro-propagated vegetable varieties. Further, greenhouses also offer opportunities to grow high-quality products in large quantities on a small surface area. For example, in the Netherlands only 0.5% of the area in use for agriculture is covered with glasshouses. On this small area of 10,000 ha, about 20% of the total value of agricultural products is realized (van Lenteren, 2000).

Production Systems and Media for PC: Crops can be grown directly in natural soil under green-house cover (Geoponics), or cultivation may be done in containers having soil-less media comprising various blends of peat, vermiculite, sawdust etc. and supplemented with nutrient solution (Hydroponics). Growing of plants in a container in which roots are suspended and sprayed with a nutrient mist is called Aeroponics. Geoponics is the most widely practiced method in India.

Protected Structures: These are of various kinds, e.g., net houses (shade nets and insect-proof nets), plastic low tunnels/row covers (laid in open fields to cover rows of plants with transparent plastic film stretched over steel hoops), plastic mulch (covering the surface around the plant), trench cultivation (for growing vegetables in extreme winter, even at -15°C), floating plastic covers (plastic sheets to cover open field to protect crops from frost/snow), and green-houses (framed structures of variable shapes and sizes).

The underlying principle of green-houses is to create "green-house effect" whereby the major fraction of sunlight is absorbed by plants; plants emit long wave thermal radiations raising green-house temperature; during summer temperature is brought down by cooling devices. Based on climate control, green-houses can be classified into three types - High-tech or Climate controlled green-house; Semi-climate controlled green-house; and Naturally ventilated or Low cost green-house. High-tech green-houses have controlled temperature, relative humidity, CO_2, photoperiod, soil temperature, plant nutrients etc.

The total world area covered by green-house is about 300,000 ha, 50,000 ha of which are covered with glass, and 250,000 ha with plastic (Albajes *et al.*, 1999; Parrella *et al.*, 1999). Vegetables are produced on 195,000 ha and ornamentals on 105,000 ha (van Lenteren, 2000). According to another source, China leads the comity of nations with 380,000 ha under green-house cultivation, followed by Japan (51042 ha), based on 1999-2000 data. In India green-house cultivation is practiced only in about 1000 ha, but it is picking up fast (Table 1).

Table 1 : Area (ha) under green-house cultivation of horticultural crops in various countries

Country	1985-86	1990-91	1999-2000
China	13,130	200,000	380,000
Japan	27,300	45,033	51,042
South Korea	1,300	1,600	2,200
Spain	12,150	35,000	51,000
UK	2,600	2,650	2,500
USA	4,000	5,000	9,250
Morocco	660	5,000	10,000
Israel	400	1,500	5,200
USSR	6,900	4,850	3,250
India	50	100	1,000

Source: International Committee for Plastics in Agriculture, 2000

Pest Scenario in Green-houses - Certain Limitations : Growing vegetables and ornamentals in protected situation is very expensive and pest damage is not tolerated. Pest control is complicated by the virtually year-round culture of crops and by continuous heating during cold periods. These conditions provide excellent opportunities for the survival and development of a pest or disease once it has invaded the green-house. Some organisms that normally show diapauses when they occur in the field have adapted to the green-house climate by no longer reacting to diapause's inducing factors. As a result, rates of population growth are often much higher than in the field.

Green-house environment provides protection to pests from harsh environmental extremes. Specific chemicals suited for green-house conditions are lacking and there is no escape for toxic fumes of chemicals. Green-house grown crops require longer and frequent harvesting. It is difficult to use large sprayers due to limited space. Pesticide formulation is more phototoxic within the protected environment of the greenhouse. Pesticide residues are higher in green-house vegetables compared to field vegetable produce due to the absence of environment eroding forces such as UV light, rain etc.

Nematode Problems in Green-Houses

In the most widely practiced geoponic systems, the main sources of nematode infection are the infested soil and infected planting materials. Root-knot nematodes (*Meloidogyne* spp.) have emerged as the major problems in green-house vegetables in many countries - Netherlands (Cools & Stolk, 1984), USSR (Lashkova & Danilov, 1982), Italy (Calabretta & Privitera, 1985), Jordan (Muhammad *et al.*, 1991), Germany (Arndt, 1994), Italy (Fiume & Parisi, 1995), Canada (Belair & Tremblay, 1995), Japan (Tanaka *et al.*, 2000), Egypt (El-Rab, 2000), Korea (Kim, 2001), Spain (Verdejo-Lucas *et al.*, 2003), China (Huang *et al.*, 2005), Brazil (Charchar & Aragao, 2005), including India (Sharma et al., 2007). In case of ornamental crops, lesion nematodes - *Pratylenchus* spp. (Gabriel & Lucke, 1979; Schreiner, 1986; Lung *et al.*, 1997), root-knot nematodes - *Meloidogyne* spp. (Nagesh & Reddy, 2000; D'Errico *et al.*, 2003), and foliar nematode - *Aphelenchoides ritzemabosi* (Vovlas *et al.*, 2005) have been reported.

Ideal environment (soil temperature and moisture), availability of favourable host and intensive cultivation in green-house cultivation are most suited for unimpeded build-up and multiplication of nematodes (Fig. 1). Once introduced, the nematode populations shoot up exponentially and maintain damaging (pathogenic) levels continuously, resulting in huge crop losses. Further, unlike insect pest and other disease problems, nematode problems are relatively difficult to manage and perhaps impossible to eradicate.

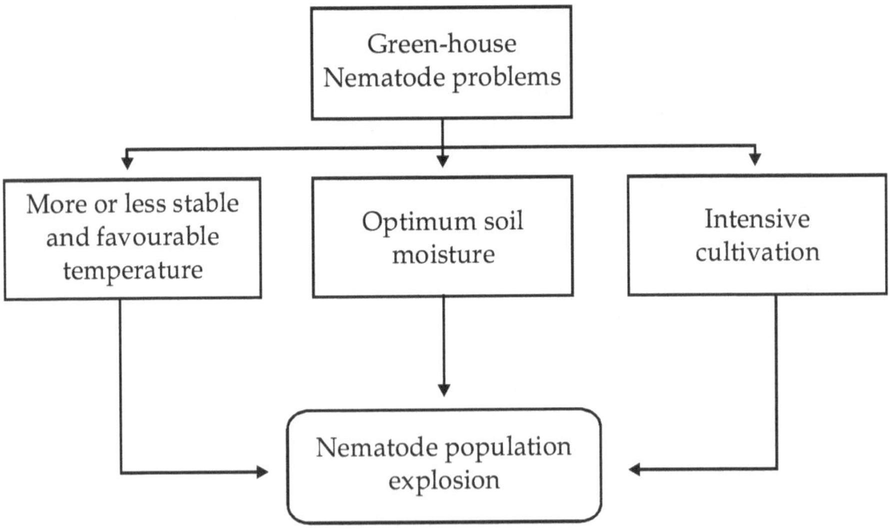

Fig. 1 : Ecological factors favouring nematode population development in green-houses.

Constant monitoring of pests and diseases in general, and nematodes in particular is the first and foremost requirement in protected cultivation. Lack of diagnostic above-ground symptoms and ignorance about nematode problems is chiefly responsible for their non-detection in the initial stages.

Work done in India: In India, poly-house cultivation of horticultural crops has recently been started. Root-knot nematodes, predominantly *Meloidogyne incognita* has emerged as the main problem. Studies were conducted on Hi-tech cultivation of hybrid tomato, sweet pepper and hybrid cucumber under Indo-Israel project at Indian Agricultural Research Institute (IARI), New Delhi. Crop management practices such as fumigation with formaldehyde @ 250 l/ha of poly-house every year, removal/changing of top soil and regular supplement of carbofuran 3G @ 1 kg a.i./ha were followed regularly. Despite all these measures, *M. incognita* problem persisted in tomato and cucumber. The intensity of infestation and crop damage increased exponentially even in plots treated with formaldehyde before planting. The rate of growth of population density and crop damage was much faster in the poly-house conditions than usually seen in open fields. Sweet pepper was least damaged due to the nematode infestation while, tomato and cucumber were the most severely affected. About 20 and 45% damage in tomato and cucumber crop was observed, respectively due to *M. incognita* (Sharma *et al.*, 2007) (Fig. 2).

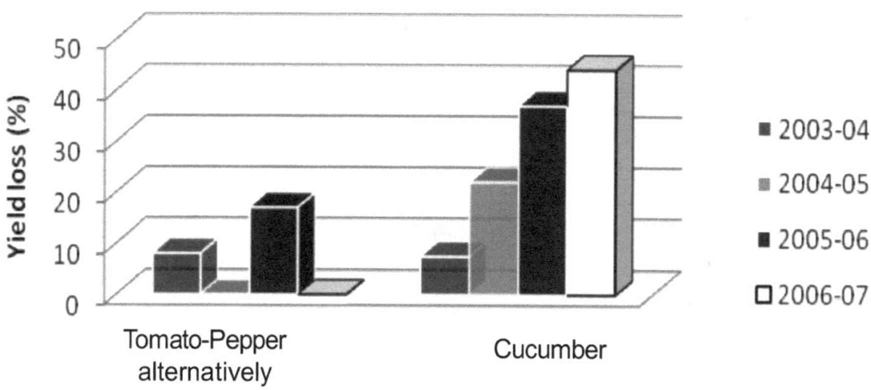

Fig. 2 : Per cent yield loss in poly-house I (Tomato-Pepper) and poly-house II (Cucumber) caused by *Meloidogyne incognita*
(Source: Sharma *et al.*, 2007)

Further studies conducted on nematode population dynamics in the Indo-Israel poly-house cultivation at IARI, New Delhi indicated *M. incognita* as a major problem. The nematode problem was undetected in the first two years, but subsequently (2003 onwards), infection was severe on tomato. Tomato crop suffered root-knot damage in alternate years (2003-04 and 2005-06), whenever it was grown (Fig. 3). The nematode was able to survive for one year without producing tangible symptoms on pepper during 2004-05. Root-knot nematode symptoms did not appear in the initial phase of tomato crop. Plant length was not affected as all the plants attained 5-7 m length. Symptom of root-knot damage appeared in the lower part of the plant. Yellowing and wilting appeared in tomato in the middle phase of crop due to enhanced reproduction of nematode and the population density rose from 1 to 30 J2 per cc soil during 2005-06 (Fig. 3). The nematode multiplied by 10 folds during 2003-04 while during 2005-06 it multiplied by 30 folds on tomato. Sweet pepper rotated with hybrid tomato in poly-house I was not attacked by root-knot nematode (Sharma *et al.,* 2007).

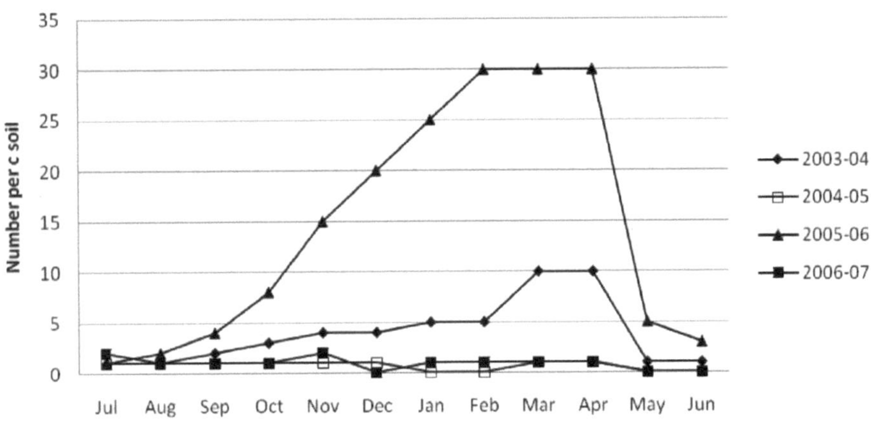

Fig. 3 : Population dynamics of *Meloidogyne incognita* in poly-house I (Tomato-Pepper)
(Source: Sharma *et al.*, 2007)

In cucumber, however, two crops are taken in a year; therefore, infestation of *M. incognita* spread very fast from few patches to whole area. Seeding mortality occurred in the initial phase of the crop. The progress of symptoms was fast as infected seedlings showed reduced number and size of fruits with time. The nematode population exhibited two peaks, commensurate with the cropping seasons (Fig. 4) (Sharma *et al.*, 2007).

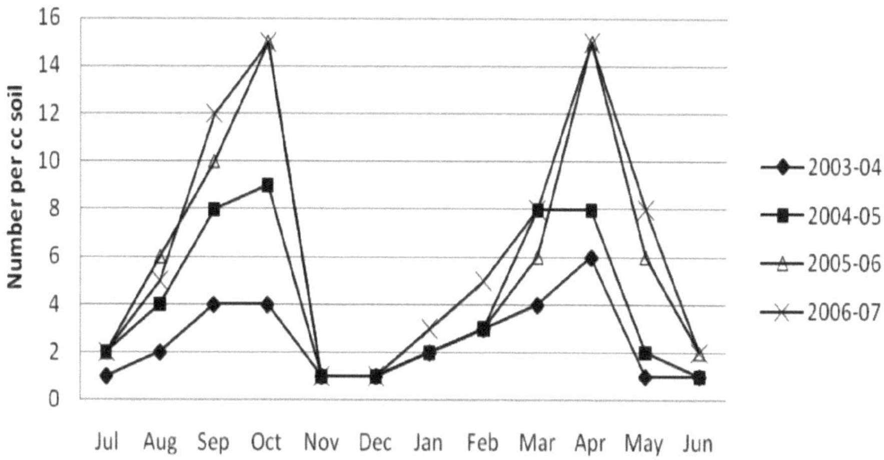

Fig. 4 : Population dynamics of *Meloidogyne incognita* in poly-house II (Cucumber) (*Source*: Sharma *et al.*, 2007)

Another major nematode, the reniform nematode - *Rotylenchulus reniformis* was observed in IARI poly-houses, however, build-up of this nematode was not recorded over the years. Perhaps, the fast multiplication of *M. incognita* resulted in the suppression of reniform nematode. Prevalence of *Tylenchorhynchus brevilineatus* was also recorded in all the poly-houses, before and after treatment of formaldehyde although population density was observed to be below 1 J2 per cc soil. The population of free-living nematodes was also meager (< 1 nematode per cc soil) than in open cultivation (Sharma *et al.*, 2007).

Effect of chemical nematicides (carbofuran 3G, phorate 10G, triazophos 40EC) and neem seed powder (NSP) was tested at IARI, New Delhi. The chemicals were applied @ 2 kg or 2 l a.i./ha in two split doses one month apart; first dose was applied 30 days after transplanting. NSP was used @ 100 kg /ha. Root-knot population increased 8-fold in carbofuran treated plots; but two months after treatment it declined to 3-fold. In phorate-treated plots, nematode population increased 15 times, but second application reduced it to 4 times, which increased again to 10 times after 2 months. Triazophos-treated plots recorded minimum population. NSP also showed a gradual increase of 1-20 times (Fig. 5). The high temperature and humidity also had adverse effect on the efficacy of the applied chemicals. It was concluded that the management of root-knot nematodes on tomato under poly-house conditions could be achieved with triazophos when applied at 2 l a.i./ha in two equal split doses (Sharma *et al.,* 2008).

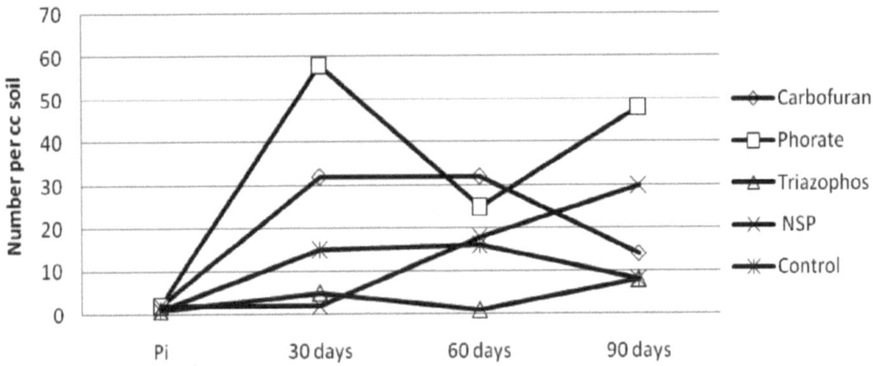

Fig. 5 : Effect of chemical nematicides and neem seed powder on root-knot nematode population in poly-house
(*Source*: Sharma *et al.*, 2008)

In another study conducted on ornamental crops at Indian Institute of Horticultural research(IIHR),Bangalore,yield loss of 26.6% in carnation and 31.1% in gerbera was assessed due to *M.incognita* by the application of phenamiphos @ 2.5 kg a.i./ha (Nagesh & Reddy, 2000). Other root-knot nematode management options in the poly-houses have been documented by Nagesh & Reddy (2005),who observed pre-plant application of dazomet (40 g/sq. m.) alone and in combination with neem cake 1 kg and *Paecilomyces lilacinus* significantly reduced root-knot count on gerbera.

Nematode Management Options and their Feasibility in PC

All the management options available for open-field cultivation may not find application in green-houses. The uniqueness of green-house cultivation permits cultivation of only selected vegetables crops; hence crop rotations involving non-hosts, antagonistic/trap crops become redundant. Similarly, practices like time of sowing, flooding, fallowing are also irrelevant for this system. Among the cultural practices, use of healthy planting material and general sanitation are of paramount importance. Soil solarization can also be very effective but warrants keeping the space without cultivation for a few weeks. Physical methods like hot water treatment offer excellent options in bulbous ornamentals. Replacement of top soil helps in the reduction of nematode inoculum but the endoparasitic nematodes are carried deep into the soil along with roots. Use of steam or hot water (Hayashi, 1979) through network of underground pipes is practiced in certain high-tech green-houses but involves high energy consumptions.

Use of nematode resistant cultivars unequivocally is the best remedy for nematode management, particularly in green-house cultivation, but unfortunately, nematode resistant material is not available in most cases, save tomato. In cherry tomato which is commonly cultivated in green-houses, no resistant cultivars have been developed so far. Resistant root-stocks grafted with desired germplasm scions (Graf *et al.*, 2001) may provide long-term solutions of protection against persistent root-knot nematode problem.

Treatment of soil with a general biocide (formaldehyde) will alleviate most of the resident fauna and flora, including nematodes, at the start of cultivation. This is a general practice which is good for nematode management. Studies have revealed that, by and large, the impact of nematicides is ephemeral and the residual populations shoot up due to favourable environmental conditions. For green-house applications, if at all required, it is best to use liquid formulations which can be applied with drip irrigations. Application of soil pesticides through drip irrigation systems offers many benefits in that it - (i) increases a grower's flexibility, (ii) reduces the potential for worker exposure, and (iii) allows growers to continue using the same drip lines, plastic mulch, and residual nutrients for successive crops (Ajwa *et al.*, 2002). Disadvantages are the dependence on soil water movement and soil characteristics to achieve uniform treatment of the bed, especially when dealing with fumigants.

The potential of bio-control agents perhaps can best be harnessed in green-house cultivation. The favourable and stable environmental conditions both for nematode pests and their fungal (*Paecilomyces lilacinus*, *Pochonia chlamydosporia* etc.) and bacterial (*Pasteuria penetrans*) parasites, help in the establishment and amplification of introduced bio-agents. Initial supplements with compatible nematicides may help in the gradual build-up of suppressive soils; consequently, nematicides can be withdrawn from the system completely.

Van Lenteren (2000) has discussed specific advantages of biological pest control in green-houses. Besides the general advantages, the specific reasons that make growers working in green-houses prefer biological control (in relation to nematodes) are -

a. there are no phytotoxic effects on young plants, and premature abortion of flowers and fruit does not occur,

b. release of natural enemies takes less time and is more pleasant than applying chemicals in humid and warm greenhouses,

c. with biological control there is no safety period between application and harvesting fruit; with chemical control one has to wait several days before harvesting is allowed again;

d. biological control is permanent; once a good natural enemy - always a good natural enemy, and

e. biological control is appreciated by the general public.

Specific integration of nematode management options such as - initial and periodic monitoring for nematode infestations during growth season, selection of nematode resistant varieties, planting nematode-free and healthy seedlings, general sanitation practices, application of liquid formulations of nematicides through drip irrigation system, and use of biological control agents,can be judiciously framed in the IPM schedules designed for particular cropping and pest/disease green-house systems.

In the light of little information available on the subject, the views expressed herein are only empirical and the nematode problems of PC systems need further detailed investigations to formulate tangible solutions.

References

Ajwa, H. A., Trout, T., Mueller, J., Wilhelm, S., Nelson, S. D., Soppe, D. and Shatley, D. 2002. Application of alternative fumigants through drip irrigation systems. *Phytopathology* 92: 1349-1355.

Albajes, R., Gullino, M. L., van Lenteren, J. C. and Elad, Y. (Eds.) 1999. *Integrated Pest and Disease Management in Greenhouse Crops*. Kluwer Publishers, Dordrecht.

Arndt, M. 1994. (Which nematodes are injurious in vegetables grown under glass?). *Gartenbau Magazin* 3(5): 14-15.

Belair, G. and Tremblay, N. 1995. The influence of chitin-urea amendments applied to an organic soil on a *Meloidogyne hapla* population and on the growth of greenhouse tomato. *Phytoprotection* 76 (2): 75-80.

Calabretta, C. and Privitera, S. 1985. (Considerations on the present position in controlling root-knot nematodes of horticultural crops under protected cultivation and real prospects for other methods). *Tecnica Agricola* 37 (3-4): 243-257.

Charchar, J. M. and Aragao, F. A. S. de 2005. (Reproduction of *Meloidogyne* spp. on tomato and cucumber varieties under plastic greenhouse and field). *Nematologia Brasiliera* 29 (2): 243-249.

Cools, M.H.and Stolk,J.H. 1984. (At last, root-knot nematode resistance and quality combined in one cultivar).*Groenten en Fruit* 39 (47): 38-39.

D'Errico, F.P.,Caprio,E. and Russo, G. 2003. (Meloidogyne hapla on rose in the province of Naples). *Informatore Fitopatologico* 53 (7/8): 55-56.

El-Rab, S. M. G. 2000. Studies on the effect of *Damsis* sp. dry matter (*Ambrosia martimala* L.) as natural material to control nematodes of plastic-houses cucumber in Egypt. *Egyptian Journal of Horticulture* 27 (3): 373-383.

Fiume, F. and Parisi, B. 1995. (Control of root-knot nematode and corky root on tomato grown in protected cultivation). *Informatore Fitopatologico* 45 (7/8): 44-50.

Gabriel, G. and Lucke 1979. Nematodes, danger for roses under glass. Gb- + -Gw: 79 (14): 330-332, 334.

Graf, V., Augustin, B. and Laun, N. 2001. (Grafting of tomato plants - protection against root gall nematodes and cork root disease). *Gemuse Munchen* 37 (3): 8-12.

Hayashi, I. 1979. (Sterilization by buried hot-water pipes, against soil-borne pests and diseases. I. Adaptations of the technique for green-house roses). Bulletin of the Kanagawa Horticultural Experiment Station 26: 60-72.

Huang-TingTing, Liu-BingLu, Wang-ChangYi and Liu-ShuQin 2005. (A new tomato hybrid for protected cultivation - 'Meilisha'. *Acta Horticultrae Sinica* 32 (1): 179.

Kim, D. G. 2001. Distribution and population dynamics of *Meloidogyne arenaria* on oriental melon (*Cucumis melo* L.) under greenhouse conditions in Korea. *Journal of Nematology* 9 (2): 61-68.

Lashkova, I. A. and Danilov, L. G. 1982. (Results of trials of different means of control against gall nematodes). In: Sbornik Nauchnykh Trudov Vsesoyuznogo Nauchno Issledovatel skogo Instituta Zashchity Rasteneii Biologicheskie osnovy bor by s nematodami (Guskova, L. A. & Chumakova, A. E., eds.), pp. 50-52.

Lung,G.,Fried,A. and Schmidt, U. 1997. (Biological control of nematodes with the enemy plant *Tagetes* spp.). *Gesunde Pflanzen* 49 (4): 111-118.

Muhammad, A. G., Suwwan, M. A. and Abu-Gharbieh, W. I. 1991. Silver thiosulfate effects on yield and growth of plastic house tomato in root-knot nematode infested soil. *Emirates Journal of Agricultural Sciences* 3: 41-65.

Nagesh, M.and Reddy,P.P.2000.Crop loss estimation incarnationand gerbera due to the root-knot nematode,*Meloidogyne incognita* (Kofoid & White) Chitwood. *Pest Management in Horticultural Ecosystems* 6 (2): 158-159.

Nagesh, M. and Reddy, P. P. 2005. Management of carnation and gerbera to control root-knot nematode, *Meloidogyne incognita* in commercial polyhouses. *Nematologia Mediterranea* 33: 157-162.

Parrella, M. P., Stengard Hansen, L. and van Lenteren, J. C. 1999. Glasshosue environments. In: Bellows, T. S., Fisher, T. W. (Eds.) *Handbook of Biological Control*. Academic Press, New York, pp. 819-839.

Schreiner, U. 1986. (Nematode damage to sweet peas under glass). *Deutscher Gartenbau* 40 (23): 1076-1077.

Sharma, H. K., Pankaj and Jagan Lal 2008. Effect of chemical on root-knot nematode, *Meloidogyne incognita* infecting tomato under polyhouse cultivation. *Indian Journal of Nematology* 38: 124-125.

Sharma, H. K., Pankaj, Gaur, H. S. and Balraj Singh 2007. Nemic population dynamics on hybrid tomato, sweet pepper and hybrid cucumber under polyhouse cultivation. *Indian Journal of Nematology* 37: 161-164.

Singh, B. 2005. *Protected cultivation of vegetable crop*. Kalyani Publishers, 168 pp.

Tanaka, T., Ichikawa, K., Matsusaki, M. and Sakai, K. 2000. (Effect of soil disinfection using solar heating with soil warming system to *Meloidogyne* sp. and *Ralstonia solanacearum*). *Research Bulletin of the Aichi ken Agricultural Research Center* 32: 105-110.

van Lenteren J. C. 2000. A greenhouse without pesticides: fact or fantasy? *Crop Protection* 19: 375-384.

Verdejo-Lucas, S., Sorribas, F. J., Ornat, C. and Galeano, M. 2003. Evaluating *Pochonia chlamydosporia* in a double-cropping system of lettuce and tomato in plastic houses infested with *Meloidogyne javanica*. *Plant Pathology* 52 (4): 521-528.

Vovlas, N., Minuto, A., Garibaldi, A., Troccoli, A. and Lamberti, F. 2005. Identification and histopathology of the foliar nematode, *Aphelenchoides ritzemabosi* (Nematoda:Aphelenchoididae) on basil in Italy. *Nematology* 7(2) :301-308.

Phyto-nematodes in Crops: Their Identification, Treatment and Management, 2011
G.L. Sharma (editor), pp. 241-265
New India Publishing Agency, New Delhi (India)
E-mail : info@nipabooks.com; Website : www.nipabooks.com

Chapter 11

Interactions between Nematode Phytopathogens and *VA*-Mycorrhizae

R. K. CHUGH AND HARI CHAND

Introduction

Biological control of plant pathogens is currently accepted as a key practice in sustainable agriculture. Plant parasitic nematodes and mycorrhizal fungi are commonly found inhabiting the same soil and colonizing roots of their host plants. These two groups of microorganisms exert a characteristics, but opposite effect on plant health. The potential role of mycorrhizal fungi as biocontrol agent for the control of nematode plant diseases has recently received considerable attention. The literature on the various aspects of VAM research such as interaction of VAM- mycorrhizal fungi with nematode pathogens (Sikora, 1995, Pinochet *et al.* 1996), biotechnology and future of VAM commercialization (Wood and Cumming, 1992) and trends in mycorrhizal research (Sahay *et al.* 1998) has been reviewed. The damage due to nematodes diseases is generally decreased in mycorrhizal fungi has shown an antagonistic influence on the population of plant parasitic nematodes (Nagesh *et al.* 1999, Kantharaju *et al.* 2002, Aparajita-Borah and Phukan, 2003, Sharma *et al.* 2003). Establishment of VAM in plant generally confers resistance to nematode parasitism or adversely affects nematode reproduction (Bhagwati *et al.* 2000, Kantharaju *et al.* 2002, Nagesh *et al.* 1999). Useful data generated on VAM in recent years has opened up the prospects for their practical applications. In recent years, therefore, major interest has centered on its relevance in biological control of soil borne diseases caused by fungi and nematodes which are otherwise difficult to control by conventional methods. In our earlier article (Jalali and Hari chand 1988), the work done on the subject during eighties was reviewed. Sikora (1995) has reviewed earlier literature on the subject. Pinochet *et al.* (1996) have reviewed interactions between migratory endoparasitic nematodes and arbuscular mycorrhizal fungi in perennial crops. Interactions between arbuscular mycorrhizas and plant parasitic nematodes in agro-ecosystems have also been reviewed (Rosen Dahl and Rosen Dahl 1990). In the present article interactions between VAM and nematode pathogens reported after eighties are reviewed. The factors affecting disease control and mechanisms involved are also described.

1) Interaction of VAM with nematode pathogens of legume crops

Soybean

Carling *et al.* (1989) recorded the effects of two VAM fungi (*G. margarita* and *G. etunicatum*), *M. incognita* and P fertilization (0, 25, 50 and 150 g/g of soil) on soybean growth and yield in greenhouse. The eggs production on VAM plants (eggs/root system and egg/g root) was suppressed at the lowest P rate as well as by increased P fertilization, indicating that induced resistance was possibly due to improved P nutrition in the host. *M. incognita* had no effect on root colonization by VAM fungi, but at the lowest P rates, the nematode stimulated sporulation by *G. margarita* and suppressed sporulation by *G. etunicatum*. Oyekanmi *et al.* (2007) studied the interaction of two soybean genotypes through root knot nematode and application of *Bradyrhizobium japonicum*, *Trichoderma pseudokoningii* and *Glomus mosseae*. They reported that these organisms provide equal or better protection against root knot nematode damage, than the synthetic nematicide carbofuran.

Blackgram

Interaction of *G. fasciculatum* with *M. incognita* on black gram (*Vigna mungo*) was recorded by Sankaranarayanan and Sundarababu

(1997b). Inoculation of black gram with *G. fasciculatum* 15 and 20 days earlier than inoculation with *M. incognita* controlled the nematode population and also increased biomass production. When *G. fasciculatum* was applied 5 days earlier than the nematode, VA mycorrhizal colonization and spore production were reduced. Sankaranarayanan and Sundarababu (1997a) observed the effect of oil cakes and nematicides on the growth of blackgram inoculated with VAM fungus (*G. fasciculatum*) and root-knot nematode (*M. incognita*) in pot studies. Pot soil was either treated with *Azadirachta indica, Madhuca longifolia, Ricinus communis* or groundnut oilcakes, 15 days prior to sowing of blackgram or inoculated with mycorrhizal fungi at 10g inoculum/plant, 22 days prior to sowing. Thirty seven days later, nematodes were inoculated at one juvenile /cc soil. One week following nematode application phorate 10G was applied. All treatments were effective but the greatest reduction in nematode population was found in oil cake treatments. In a further study (Sankaranarayanan and Sundarababu, 1997b) recorded the effect of leaf extracts on growth of black gram inoculated with *G. fasciculatum* and *M. incognita* in pot experiments. There was significant increase in all the growth parameters of black gram due to the addition of leaf extracts of *Calotropis procera, Targetes erecta, Catharanthus rosea* and *Bougainvillea spectabilis*. Leaf extracts recorded highest spore population and mycorrhizal colonization than the nematodes but the latter recorded less gall index and nematode populations.

Sankaranaryanan and Sundarababu (1998) observed the effect of *Rhizobium* on the interaction of vesicular arbuscular mycorrhizae and root knot nematode on black gram. The combination of *Rhizobium* and VAM resulted in the least nematode population density and maximum spore, mycorrhizal colonization, total nitrogen and phosphorus content of black gram.

Sankaranarayanan and Sunderbabu (2001) studied the inoculation of Rhizobium sp.,phosphobacteria or VAM either alone or in combination increased the growth of the plants or reduced the nematode population. Among the treatments,combined inoculation Rhizobiumsp.,phosphobacterium on VAM and nematode, increased all the growth pa rameters, phosphorus content, spore population, mycorrhizal colonization and reduced gall index and nematode population significantly. Nematode inoculation to mycorrhizal plants reduced the nitrogen and phosphorus content of the plants. But inoculation of either Rhizobium or phosphor bacterium or both, increased nitrogen and phosphorus content of black gram. Bornali-Mahanta *et al.* (2002) noted the effect of *G. fasiculatum* and *M. incognita* on black gram. The results showed that seedling inoculation with VAM before nematode infestation increased shoot length and reduced number of galls and egg masses on black gram. Bornali-Mahanta and Phukan (2004) recorded effect of *G. fasciculatum* on penetration, development and survival of *M. incognita* on blackgram. The plants inoculated with VAM + nematode after 10 days and VAM and nematode after 15 days showed significantly reduced second stage juvenile penetration and ultimately delayed the development of different stages of *M. incognita* compared with the plants inoculated with VAM and nematode simultaneously and nematode alone. Significant decreases in nematode penetration, delayed development, number of galls, egg masses, eggs per mass and final nematode population in soil were recorded in plants inoculated with VAM + nematode after 15 days followed by the plants inoculated with VAM+ nematode after 10 days.

Sorghum

Association of vesicular arbuscular mycorrhizal fungi and plant parasitic nematodes in forage sorghum (*Sorgum bicolar* L.) was observed by Jain and Hasan (1986). In their study soil/root samples collected from the rhizosphere of sorghum revealed two VAM fungi, *G. fasciculatum* and *G. mosseae* and three nematodes, *Helicotylenchus dihystera, P. ratylenchus* zeae and *Tylenchorhynchus vulgaris*. Where roots had >50% VAM infection the nematode numbers were lower. Presence of nematode did not adversely affect VAM sporulation. It was concluded that VAM fungi could be used to manage plant parasitic nematode populations.

Beans

Oliveira and Zambolim (1988) studied the interaction of endomycorrhizal fungus *G. etunicatum* and *Meloidogyne javanica* in green house using different inoculum levels on bean (*Phaseolus vulgaris*). Inoculation was performed at sowing, with *G. etunicatum* inoculum levels of 0, 500, 1000, 2000 chlamydospores/plant and *M. javanica* levels of 0, 100, 1000, 5000 and 10000 eggs/plant. Root infection rates did not differ 15 days after inoculation of *G. etunicatum*. With increasing inoculum density of nematodes, plant growth was reduced, but had no effect on mycorrhizal colonization or chlamydospore production by the fungus. There was no significant increase in damage to the plant when inoculum levels of *M. javanica* exceeded 500 eggs/plant. Decreased nematode egg production was observed in the presence of *G. etunicatum*. Levels above 1000 chlamydospores of *G. etunicatum* did not differ in stimulating plant growth or in having antagonistic effect on the nematode. Osman *et al.* (1990) inoculated bean seedlings in pots with *Glomus* spp. and *M. incognita* either alone or simultaneously. Gall index and final nematode populations were significantly increased with simultaneous inoculations. There was a significant decrease when nematodes were inoculated 15 and 39 days after the fungus. In a similar study (Osman *et al.* 1991), when the bean plants were infected with *M. incognita* 15 or 30 days after mycorrhizal infection, the development of nematode was inhibited. Application at the same time reduced fungal development.

Chickpea

Pandey *et al.* (2005) studied the management of root knot nematode (*M. incognita)* and *Fusarium oxysporium* f.sp. *ciceri* using ecofriendly methods. The highest shoot length, root weight, chlorophyll content, number of pods/plant, 100 seed weight, per cent mycorrhizal colonization and chlamydospore population was obtained when combination of all the treatments were used. Similarly number of galls/ plant, egg mass per plant, nematode soil population and wilt per cent was lowest with *G. fasciculatum* + *Trichoderma viride* + *Paecilomyces lilacinus* + neem oilseed cake.

Cowpea

In a study (Jain and Sethi, 1988) influence of time of inoculation was recorded on the interaction between vesicular arbuscular mycorrhiza, *M. incognita* and *Heterodera cajani* on cowpea. Inoculation of cowpea seedlings with *G. fasciculatum* or *G. epigaeum (G. versiforme*) one to 2 week prior to inoculation with *M. incognita* or *H. cajani,* alleviated the deleterious effect of both nematodes to a considerable extent. Early establishment of *G. fasciculatum* reduced gall formation by *M. incognita* and cyst formation of *H. cajani*. Prior establishment of *G. versiforme* resulted in increased numbers of cyst of *H. cajani*. The presence of nematodes reduced mycorrhizal infection and spore production. Ahmed and Alsayed (1991) evaluated influence of *G. macrocarpum* on the penetration and development of *M. incognita* in cowpea in glass house tests. *G. macrocarpum* significantly reduced nematode numbers in cowpea roots and soils. Similarly a reduction in number of arbuscules was observed in the presence of *M. incognita*. It was suggested that VAM fungi-infected roots stimulated plant growth by improving host nutrition and minimizing the damage caused by nematode perhaps by change in root exudates causing fewer nematodes to be attracted to and penetrate the plant roots. Santhi and Sundarababu (1995a) recorded effect of phosphorus on the interaction of vesicular arbuscular mycorrhizal fungi with *M. incognita* on cowpea in a pot experiment. Plants with VAM were more resistant to *M. incognita* than those without. Positive correlation was observed between the phosphorus levels and the nematode population, whereas, negative correlation existed between phosphorus levels and the VAM spore population and colonization. The effect of soil type on mycorrhizal fungi (*G. fasciculatum, G. versiforme, G. epigaeum*) and *M. incognita* interactions on cowpea was studied by Santhi and Sundarababu (1995b). The effect of different soil types on *G. fasciculatum* and *G. etunicatum* was the same and both species preferred the pot soil mixture followed by sandy soil, clay loam, black loam, whereas *G. versiforme* preferred the pot soil mixture followed by red loam, clay soil, sandy soil and black loam. *G. fasciculatum* was superior in reducing the nematode population in soil. The pot mixture (red soil/sand/FYM in a ratio 2:1:1) increased growth as well as spore production and colonization, reducing the nematode population.

Pigeonpea

Siddiqui and Mahmood (1995) made some observations on the management of wilt disease complex of pigeon pea by treatment with a vesicular arbuscular fungus and biocontrol agents for nematodes. A vesicular arbuscular fungus, *Gigaspora margarita* and two biocontrol agents (*Paecilomyces lilacinus* and *Verticillium chlamydosporium* were used for the management of the nematode *Heterodera cajani* and the fungus *Fusarium udum*. All the three management agents, alone or in combination, applied to plants inoculated with pathogens increased shoot dry weight, number of nodules, phosphorus content and reduced nematode multiplication and wilting index. Simultaneous use of the biocontrol agents and the vesicular arbuscular fungus gave better control of the disease complex than did their individual application. In a further study (Siddiqiue and Mahmood, 1996), *Glomus mosseae, Trichoderma harzianum* and *Verticillium chlamydosporium* were used alone and in combination for the management of wilt disease complex of pigeon pea. Treatment of plants inoculated with pathogens increased plant length, shoot dry weight, number of nodules and reduced nematode multiplication and wilting index. Simultaneous use of biocontrol agents against pathogens gave better control than their individual application. *T. harzianum* had an adverse effect on root colonization by *G. mosseae*. The highest reduction in nematode multiplication was observed when all the three biocontrol agents were used together. Nageswari and Mishra (2005) evaluated the efficacy of neem and powder, neem-based pesticidal formulations, soil solarization and VAM against *H. Cajani* on pigeon pea (cv. Pusa 33). The effective integrated integrated nematode management schedules were: a) soil application of neem seed powder at 50kg/ha + soil solarization (covering of plots with transparent 400-gauge thick polythene sheet for 4 weeks) + VAM at 100 kg/ha, and b) seed treatment with neem seed powder at 10% w/w + soil solarization + VAM at 100 kg/ha. The performance of these schedules was almost similar as they resulted in the growth and grain yield, and greatest reduction in nematode population.

Goswami et al. (2007) found that treatment of FYM, karanj oilseed cake and VA mycorrhiza reduced the disease incidence caused by root knot nematode, *Meloidogyne incognita* and root wilt fungus, *Fusarium udum* to a great extent with the most promising improvement in the growth parameters.

2) Interaction of VAM with nematode pathogens of vegetable crops

Tomato

Al- Momany and Al-Raddad (1988) used seven isolates of *Glomus* spp. I ncluding *G.mosseae* and *G. fasciculatum* to control tomato wilt caused by *F. oxysporum* f.sp. *lycopersici*. Mycorrhizal plants inoculated with *F. oxysporum* had significanty higher root and shoot weights than plant inoculated with only pathogen. Root colonization by mycorrhizae in tomato plants was not reduced by *Fusarium*. The *Glomus* isolates reduced the percentage of *Fusarium* infection in roots at different rates. M66 minimized *Fusarium* infection of tomato by up to 65%. Biological control of Fusarium crown and root rot of tomato using *T. harzianum* and *G. intraradices* was attempted by Datnoff *et al.* (1995) in field trials. Significant decreases in disease incidence were obtained with treatments of *T. harzianum* in 1993, *G. intraradices* in 1991 and combination of *T. harzianum* and *G. intraradices* in both the years. Bochow and Abou-shaar (1990) reported the influence of VA-mycorrhizas (*Glomus caledonicum*) in tomatoes on the intensity of corky root disease caused by *Pyrenochaeta lycopersici* in a pot experiments.

Combinations of VAM-application with additional nitrogen and potassium fertilization of plants and soil inoculation by *Bacillus subtilis* str.T99 resulted in lowering the disease index and the highest plant growth and yield. It was suggested that the phytosanitary role of the VAM is especially effective when integrated with other plant protection measures. Management of *M. incognita* on tomato by integrating endomycorrhiza (*G. mosseae*) with oil cakes has been reported (Rao *et al.* 1995, Reddy *et al.*1998). Mycorrhizal seedlings of tomato, when transplanted in neem cake-amended soil, were the least infected by *M. incognita.* An increase in plant growth parameters and a decrease in root-knot index and final population of nematode were observed in treatments where mycorrhizal seedlings were transplanted in neem cake-amended soil. Amendment of neem cake affected VAM fungus favourably. Similar were the observations when *G. fasciculatum* was integrated with caster cake for the management of root knot nematode in tomato (Rao *et al.*1997). Interaction between *G. fasciculatum,* M. *incognita* and fungicide in tomato was noted by Abha Mishra and Shukla (1996). Pre-establishment of *G. fasciculatum* to nemadote increased plant growth, decreased the size and number of galls and improved NPK uptake compared with plants inoculated with nematode alone or pre-inoculated with nematode to VAM. Fungicidal combinations (Carbendazim, aluminium-tris and chlorothalonil) with VAM and nematode predicted an overall decrease in plant growth. However, simultaneous applications of Carbendazim + *G. fasciculatum* improved plant growth. Reduced uptake of NPK was recorded in all the treatment combinations of VAM and nematode when combined with fungicides. In interaction between *G. fasciuculatum* and *M. incognita* on tomato, the time of inoculation also mattered (Abha Mishra and Shukla, 1997). Simultaneous inoculations of *G. fasciculatum* with *M. incognita* caused the greatest reduction in number and size of the nematode-induced root galls. Neelima Mittal *et al.* (1991) concluded that presence of *G. fasciculatum* alone in the roots was inhibitory to the formation of nematode galls. Sundarababu *et al.* (1993a) also reported that VAM enhanced plant growth, reduced nematode population and increased yields. Rao and Gowen (1998) attempted bio-management of *M. incognita* on tomato by integrating *Glomus deserticola* and *Pasteuria penetrans.* The combination significantly reduced the number of egg masses in root systems, and increased the parasitization of females by *P. penetrans.* Root colonization by *G. deserticola* was not affected by *P. penetrans.* It was suggested that *G. deserticola* and *P. penetrans* can prove useful for sustainable management of *M. incognita* on tomatoes.

Sundarbabu *et al.* (1993b) studied influence of time of inoculation on interaction between vesicular arbusular mycorrhiza and *M. javanica* on tomato. In their studies prior establishment of mycorrhiza could offset the damaging effect of the nematode and improve plant growth. Interaction of *G. mosseae* and *P. lilacinus* on *M. javanica* in tomato was evaluated by Al-Raddad (1995). Inoculation of plants with VAM suppressed gall index and average number of galls/root system by 52% and 66% respectively, compared with seedings inoculated with *M. javanica* alone. The presence of *G. mosseae* and *P. lilacinus*, together or separately in the layer manure completely inhibited root infection by nematode. Mycorrhizal colonization was not affected by the layer manure treatment or by root inoculation with *P. lilacinus*. Rivas-Platero *et al.* (1998) observed interaction between *Glomus* spp and *Meloidogyne arabicida* in tomato. *M. arabicida* multiplication rate was reduced by 50.5% by *Glomus* spp. and the gall index in tomato roots by 13%.*Glomus* spore production was stimulated by *M.arabicida*.The phosphorus content was higher in the *M. arabicida* + *Glomus* pots than *Glomus* alone ones. Nagesh *et al.* (1999) studied correlation between *G. fasciculatum* spore density, root colonization and *M. incognita* infection on tomato. The nematode populations, egg masses/plant and root gall index were negatively correlated with spore density. Further, at least 2spores/g soils were required for 50% root colonization under pot conditions. However, under nursery conditions, 4spores/g soils were required to obtain 50% root colonization which reduced nematode populations in root (in terms of egg masses/plant) by 59 and 61%, respectively. In studies of Wang-Yan Ling and Hu-Zheng Jia (2000) *G. mosseae* had an inhibitory effect on *M. incognita* infection in tomatoes in Hubaei, China. Bhagawati *et al.* (2000) reported that mustard cake and *G. etunicatum* were equally effective in reducing the damage caused by *M. incognita* and/or *F. oxysporum* f.sp. *lycopersici* in tomato. When mustard cake and VAM were applied together, the combined efficacy was much greater than single treatments. The population of pathogens was reduced. Kantharaju *et al.* (2002) evaluted 8 isolateds of *G. fasciculatum* against root knot nematode (*M. incognita*) race-1 on tomato. Nematode population in soil, number of galls and egg masses per root system were lowest in Shimoga Banana isolate followed by Shimoga Areca isolate.

Nasr *et al.* (2005) inoculated tomato seedlings with 125, 250 and 500 chlamydospores of the vesicular arbuscular mycorrhiza, *Glomus* sp., in the nursery. Nematode (*Meloidogyne incognita*) inoculum, i.e.

350 juveniles per plant, was added to the pots as 2cm deep trench around the stem base. P was added as rock phosphate at concentrations of 75 or 100% of recommended dose (75kg/ha). VAM sporulation decreased in the presence or absence of M. incognita. On the other hand, root weight increased in the presence of nematode infection regardless of the P rate and VAM inoculum density. Rubose et al. (2006) studied the interactions of *Paecilomyces lilacinus* strain 251 with the arbuscular mycorrhizal fungus *Glomus intraradices* and their significance for the control of *Meloidogyne incognita* on tomato in greenhouse experiments. They reported that application of *P. lilacinus* had no effect on the frequency and intensity of tomato root colonization by *G. intraradices.* Likewise, decline of the nematophagus fungus densities after single application in soil was not affected the presence of the mycorrhizal fungus. Single application of *P. lilacinus,* as pre-planting soil treatment, resulted in significant reduction of nematode damage. In contrast, mycorrhizal inoculation did not provide sufficient biocontrol. Combined application of the two agents did not enhance root protection compared to single treatments. Double treatment of mycorrhized seedlings with *P. lilacinus,* as drench and pre- planting soil treatment, 4 and 1 week before transplanting respectively, resulted in the highest reduction of the nematode damage. Goswami and Tewari (2007) studied the synergistic effect of *M. incognita and F. oxysporium f.sp. lycopersici* in causing disease complex on tomato. Soil amendment of vesicular arbuscular mycorrhiza *(Glomus fasciculatum)* @100 chlamydospores per kg soil and *T. harzianum* @ 7g/kg (2x10(8) spore load) as talc based formulation along with leaf power (0.05%w/w) of *Tinospora longifolia* managed the disease complex by reducing its severity and making the host tolerant.

Reimann et al.(2008)found that plant promoting rhizobacterium *Rhizobium etli* G12 seems to be responsible for accelerated arbuscular establish -ment in tomato roots in concomitant inoculation. Therefore, R. etli is regarded to be a mycorrhiza helper bacterium. Combined inoculation of *R.etli with G.intraradices* in biocontrol of *Meloidogyne incognita* on tomato led to additive effects. Single inoculation of the microorganisms reduced galling by 24 to 39% while dual inoculation resulted in a 60% reduction.

Sweet Potato

Kassab (1990) studied relationship between initial population densities of nematodes (*Criconemella* spp., *Rotylenchus reniformis* or

Tylenchorhychus spp.) and yield of mycorrhizal and non-mycorrhizal sweet potato plants. Increasing population density from 100-1000 nematodes/plant caused an increase in final population densities of all nematodes. However, population density of *R. reniformis* and *Criconemella* spp. was greater on mycorrhizal plants than non-mycorrhizals. Conversely, population density of *Tylenchorhynchus* spp. was higher on non-mycorrhizal plants than on mycorrhizals. Growth of fibrous roots and yield of storage roots of mycorrhizal plants were significantly higher than those non-mycorrhizals,except that of fibrous roots infected with *Criconemella* spp. where the converse was true.

Okra

Sharma and Trivedi(1997) studied the concomitant effect of *Paecilomyces lilacinus* and vesicular arbuscular mycorrhizal fungi on root-knot nematode infected okra. Combined application of *Glomus fasciculatum* and *G.mosseae*,two weeks prior to inoculation with *Meloidogyne incognita* resulted in increased fresh and dry weight of root,shoot and okra fruit as compared to only nematode treated plants in pot experiments.The combined effect of the mycorrhizal fungi and *P. lilacinus* gave maximum reduction in nematode galls.Jothi *et al.*(2000) attempted management of *M.incognita* on okra by integrating non-host and endomycorrhiza.In their studies pre inoculation with VAM enhanced growth and yield of okra and significantly reduced the population of *M. incognita* (from 994 to 240 nematodes/200g soil). Sharma *et al.* (2003) studied the effects of plant growth promoting microbes and vesicular arbuscular mycorrhiza on root knot nematodes (*M.incognita*) infecting okra. Microbes were applied in the top soil in earthen pots alone or in combination with phorate.The maximum reduction in number of galls was obtained with VAM followed by *Azotobater* and *Azosprillum* alone. Combined application of phorate and microbes increased the root-knot galling.Similar trend was noticed with egg mass and eggs/egg mass count.

Egg Plant

Rao *et al.* (1998a) conducted screen house experiments to integrate the use of the vesicular arbuscular mycorrhiza, *Glomus mosseae* with a nematophagous fungus *Paecilomyces lilacinus*, for the management of *Meloidogyne incognita* on egg plant. Both biological control agents did not affect each others colonization on the roots resulting in an additive effect of both on the management of *M. incognita*. The final nematode population was significantly less in the treatment where *G. mosseae* and *P. lilacinus* were integrated. Rao *et al.* (1998b) conducted experiments to study the feasibility of integration of *Glomus fasiculatum* and *R. communis* cake for the management of *Meloidogyne incognita* on egg plant. Seedlings colonized with mycorrhiza were least infected by *M. incognita* when transplanted in soil which was amended with caster cake.

Aparajita-Borah and Phukan (2003) observed interaction of *Glomus fasciculatum* and *Meloidogyne incognita* in brinjal. In their studies mycorrhizal plants showed relatively higher growth than non-mycorrhizal plants. The treatment with VAM at 600 spores/pot + *M. incognita* was best for reducing galls, egg masses and nematode population.

Aprajita- Borah and Phulkan (2006) found that plant inoculated with vesicular arbuscular mycorrhiza alone had significantly increased plant growth parameters, nitrogen, crude protein, phosphorus, potassium, total sugar, total protein and total free amino acids contents compared to uninoculated plants and nematode inoculated plants. The plants inoculated with VAM + nematode after 10 days (V+N) significantly decreased the number of galls, egg masses, and final nematode population in soil, increased nitrogen, crude protein, phosphorus, potassium, total sugar, total phenol free amino acids in the roots of brinjal as compared to nematode alone, nematode prior to VAM (N+V) and VAM +nematode simultaneously.

3) Interaction of VAM with nematode pathogens of other crops

Cotton

Saleh and Sikora (1989) studied the effect of fungicides (Quintozene, carbendazim and benomyl) on the interaction between the endomycorrhizal fungus *G. fasciculatum* and root- knot nematode *M. incognita* on cotton in the greenhouse. *G. fasciculatum* did not affect plant growth or nematode population density 30days after simultaneous inoculation at planting, but caused significant increases in plant growth and a significant reduction in *M. incognita* population density and reproduction 50 days after inoculation. The fungicides did not alter nematode population on non-mycorrhizal plants. Benomyl and carbendazim, however, significantly reduced *G. fasciculatum* antagonistic acitivity towards *M. incognita* reproduction when applied at planting. The results demonstrated that fungicides concentration and time of application are important factors influencing the antagonistic interaction between endomycorrhizal fungi and root-knot nematodes on cotton.

Sreenivasan and Sundarababu (2007) studied the efficacy of commercial formulations of VAM (*Glomus mosseae*), plant growth promoting rhizobacterium (Pseudomonas *fluorescens*) and the antagonistic fungus (*Trichodernma viride*) against reniform nematode, Rotylenchulus reniformis, in cotton under greenhouse conditions. Among the three biological control agents tried, *G. mosseae* as soil application and *P. fluorescens* as seed treatment performed well in improving the plant growth and yield of cotton. *G. mosseae* as soil application recorded the maximum reduction of nematode population in roots (72.1%) and soil (81.1%), which was at par with the seed treatment with *P. flurescens*. These treatments also significantly reduced the number of eggs/egg mass. *T. viride,* both as seed treatment and soil application, was less effective in the management of *R. reniformis*.

Coffee

Vaast *et al.* (1998) recorded influence of root-lesion nematode, *Pratylenchus coffeae* and two arbuscular mycorrhizal fungi, *Acaulospora mellea* and *Glomus clarum* on coffee (*Coffea arabica* L.) in green house experiment. The late mycorrhizal inoculation (added simultaneously with nematode) did not enhance coffee tolerance to *P. coffeae*. In the presence of *P. coffeae,* late mycorrhizal plants were P deficient during the entire experiment and their foliar P concentration remained as low as that of non mycorrhizal plants. After 7.5 months, nematodes decreased mycorrhizal colonization of late plants and their biomass. In contrast, early mycorrhizal inoculation (4 months before inoculation) with either of the mycorrhizal species improved the tolerance of coffee to *P. coffeae*. Root colonization by mycoorhizas was not significantly reduced by *P. coffeae*. These studies confirmed earlier observations (Vaast, 1997) that an early VAM inoculation gives better results than simultaneous inoculation with nematodes in coffee. The interactions between mycoorhizal fungi (*Entrophospora colombiana* and *Gigaspora margarita*) and *M. exigua* in coffee were observed by Rivas Platero and Andrade (1998). The mycorrhizal fungi reduced the multiplication rate of *M. exigua*. Both the fungi reduced the gall index by 20% compared to nematode alone. The nematode stimulated sporulation of the fungi.

Cardamom

Thomas *et al.*(1989)noted the effect of inoculation with 6 species of VAM fungi,individually and different combinations with *M. incognita*, simultaneously and sequentially,on cardamom (*Elettaria cardamomum*) at 6 and 12 months after inoculation. Compared with the control, the growth of cardamom plants was improved by inoculation with VAM fungi, whereas nematode inoculation reduced it. The growth response induced by VAM fungi was similar when inoculation was made simultaneously and after nematode inoculation.VAM fungi significantly reduced the nematode population in roots.*Gigaspora margarita* and *G.fasciculatum* were the most effective VAM fungi, able to promote maximum growth response in the absence as well as in the presence of root-knot nematodes.

Maize

Babu *et al.* (1998) reported that mycorrhizal fungi *Glomus fasciculatum* and *Azosprillium* and phosphobacterium when applied alone or in different combination in soil field reduced the population of *Pretylenchus zeae* and recorded the highest cob yield of maize.

Tobacco

Subhashini (1990) studied the role of VA mycorrhiza in controlling certain root diseases of tobacco. Inoculation with *G. fasciculatum* reduced the intensity of infection by root knot nematode but had no effect on orobanche infestation. Influence of a vesicular arbuscular mycorrhiza on the development and reproduction of root knot nematode in tobacco was recorded by Krishna Prasad (1991). The nematode infestation of seedlings, number of galls and egg masses per infested seeding were significantly reduced by VAM inoculation.

Berseem

Jain *et al.* (1998a) investigated the influence of the *Glomus fasciculatum* on *Meloidogyne incognita* and *Tylenchorhynchus vulgaris* infecting berseem (*Trifolium alexandrium*). Gall formation by *M. incognita*, and multiplication of both the nematodes was reduced by 28-50 per cent. Conversely, the prior presence of nematodes reduced the per cent root colonization by VAM fungi.

Lucerne

Jain *et al.* (1998b) reported that pre- inoculation of lucerne plants with *Glomus fasciculatum* two weeks before introduction of either

Meloidogyne incognita or *Tylenchorhynchus vulgaris* significantly reduced the adverse effects of nematodes on total biomadss of the plant. Simillarly, gall formation and also the multiplication of both nematode species were considerably reduced. In contrast, prior presence of the nematodes reduced mycorrhizal colonization.

Future Prospects

A detailed analysis of MYCOLIT, a comprehensive bibliographic database of mycorrhizal research showed that much research on VAM has focused on nutrient dynamics, inoculum production, mycorrhiza formation, and morphology and physiology of both vesicular-arbuscular mycorrhizas and ectomycorrhizas. There has been a shift of attention from ectomycorrhiza to vesicular-arbuscular mycorrhiza since 1970. Very little work has been done on the kinds of mycorrhizas. Barriers to progress in some areas, such as our inability to grow the Glomales in asexinic culture, have stimulated large number of methodological papers over the past 15 years, while the genetics and molecular biology, especially of VAM, and have been neglected. Ecological studies were more numerous than any other kind.

Several studies have certainly revealed that mycorrhiza induced resistance in plants, while in some fewer cases either no effect or an increase in the disease severity has been observed. The critical evaluation of VAM as a biocontrol agent is therefore one of the challenging areas, and it holds a great promise in our future research endeavors. The effect of vesicular-arbuscular mycorrhizal fungi on plant diseases has not been mostly demonstrated under field conditions, although there are some promising results that support the conclusions of the greenhouse studies. Study on parasitic ability of VAM fungi on plant parasitic nematodes is required so that proof of successful parasitism of VAM would explore the possibility of managing the nematode population directly acting as biocontrol agent. Few available reports on integration of VAM with other biocontrol agents have given promising results. This aspect and utilization of VA-mycorrhiza as a factor in integrated plant protection remains to be studied in detail. VA-myrrohizae has to be considered a part of the environment and of the plant itself, which is able to influence its disease resistance. Since interactions between the host plants, VA-fungus, soil and climate will determine the mycorrhizal effect on plant growth, research efforts need to be directed towards finding appropriate host-fungus combinations adapted to well define soil types and climatic conditions.

Table 1 : Effect of VAM on soil borne nematode pathogens

Pathogen	Host Plant	Effects in mycorrhizal plants*			Reference
		Disease damage	Infection	Reproduction	
Meloidogyne javanica	Bean	-	...	-	Oliveira and Zambolim (1988)
Meloidogyne incognita	Bean	-	...	...	Osman *et al.* (1994)
Meloidogyne incognita	Bean	-	...	-	Osman *et al.* (1991)
Meloidogyne incognita	Black gram	-	...	-	Sankaranarayanan and Sundarababu (1990)
Meloidogyne incognita	Black gram	-	...	-	Sankaranarayanan and Sundarababu (1997a)
Meloidogyne incognita	Black gram	-	...	-	Sankaranarayanan and Sundarababu (1997b)
Meloidogyne incognita	Black gram	-	...	-	Sankaranarayanan and Sundarababu (1998)
Meloidogyne incognita	Black gram	-	...	-	Bornali Mohanta *et al.* (2002)
Meloidogyne incognita	Black gram	-	-	-	Bornali Mohanta and Phukan (2004)
Meloidogyne incognita	Cardamom	-	...	-	Thomas *et al.* (1989)
Pratylenchus coffeae	Coffee	-	...	...	Vaast *et al.* (1998)
Pratylenchus coffeae *Meloidogyne konaensis*	Coffee	-	...	...	Vaast (1997)
Meloidogyne exigua	Coffee	-	...	-	Rivas Platero and Andrade (1998)

Meloidogyne incognita	Cotton	...	-	-	Saleh and Sikora (1989)
Meloidogyne incognita *Heterodera cajani*	Cowpea	-	...	-	Jain and Sethi (1988)
Meloidogyne incognita	Cowpea	-	-	-	Ahmed and Alsayed (1991)
Meloidogyne incognita	Cowpea	-	...	...	Santhi and Sundarababu (1995)
Meloidogyne incognita	Cowpea	...	...	-	Santhi and Sundarababu (1995)
Meloidogyne incognita *Tylenchorhynchus vulgaris*	Lucerne	-	...	-	Jain *et al.* (1998)
Heterodera cajani	Pigeon pea	...	...	-	Siddiqui and Mahmood (1995)
Heterodera cajani	Pigeon pea	...	...	-	Siddiqui and Mahmood (1996)
Meloidogyne incognita	*Piper nigrum*	-	-	-	Sivaprasad *et al.* (1990)
Meloidogyne incognita	*Piper nigrum*	-	-	-	Sivaprasad *et al.* (1992)
Helicotylenchus dihystera *Pratylenchus zeae* *Tylenchorhynchus vulgaris*	Sorghum	....	...	-	Jain asnd Hasan (1986)
Meloidogyne incognita	Soybean	...	...	-	Carling *et al.* (1989)
Criconemella spp. *Rotylenchulus reniformis* *Tylenchorhynchus spp.*	Sweet potato	+ - -	... ...	+ + -	Kassab (1990)
Meloidogyne incognita	Tobacco	...	-	...	Subhashini (1990)
Meloidogyne incognita	Tobacco	...	-	-	Krishna Prasad (1991)
Meloidogyne incognita	Tomato	-	...	...	Neelima *et al.* (1991)

Meloidogyne javanica	Tomato	-	...	...	Sundarababu *et al.* (1993)
Meloidogyne incognita	Tomato	-	...	-	Sundarababu *et al.* (1993)
Meloidogyne javanica	Tomato	-	-	...	Al-Raddad (1995)
Meloidogyne incognita	Tomato	-	...	...	Abha Mishra and Sukula (1996)
Meloidogyne incognita	Tomato	-	...	-	Rao *et al.* (1997)
Meloidogyne incognita	Tomato	-	-	...	Abha Mishra and Sukula (1997)
Meloidogyne arabicida	Tomato	-	...	-	Rivas- Platero *et al.* (1998)
Meloidogyne incognita	Tomato	...	-	-	Rao and Gowen (1998)
Meloidogyne incognita	Tomato	-	...	-	Reddy *et al.* (1998)
Meloidogyne incognita	Tomato	-	...	-	Nagesh *et al.* (1999)
Meloidogyne incognita	Tomato	...	-	...	Wang-Yan Ling and Hu- Zheng Jia (2000)
Meloidogyne incognita	Tomato	...	...	-	Bhagawati *et al.* (2000)
Meloidogyne incognita	Tomato	-	...	-	Kantharaju *et al.* (2003)
Pratylenchus Zeae	Maize	...	...	-	Babu *et al.*(19980
Meloidogyne incognita *Tylenchorhynchus vulgaris*	Berseem	-	...	-	Jain *et al.* (1998)
Meloidogyne incognita	Okra	...	...	-	Jothi *et al.* (2000)
Meloidogyne incognita	Okra	-	...	-	Sharma *et al.* (2003)
Meloidogyne incognita	Egg plant	...	...	-	Rao *et al.* (1998a)
Meloidogyne incognita	Egg plant	-	...	...	Rao *et al.* (1998b)
Meloidogyne incognita	Brinjal	-	...	-	Aparajita Borah and Phukan (2003)
Meloidogyne incognita	Brinjal	-	...	-	Aparajita Borah *et al.* (2003)

The management practices enhancing the combined effect of surviving and introduced fungi need to be developed. In order to establish biocontrol efficacy of mycorrhizal fungi there are many inherent obstacles to overcome. These include: strain selection, mass production of mycorrhizal inoculants, storage, the relationship between inoculum potential and infectivity of VAM fungi, methods of application and the effects of nutrients on mycorrhizal responses. For mass production of VAM fungi, protoplast fusion with soil culture microorganisms which are harmless to plant growth may be attempted in future. Taxonomic consideration at molecular level warrants immediate attention.

Investigation of new technologies is needed, especially genetic manipulation of fungal symbionts to induce new biotypes for improved performance, adaptation of native microbial agents, new disease control requirements and improved compatibility with chemical fungicides and nematicides. Side effects of pesticides used in normal farming systems on non-target mycorrhizal endophytes also need to be evaluated more critically for developing viable plant protection strategies. Only fragmentary information is available on the mechanisms involved in the suppression of the severity of soil borne diseases by VAM, endophytes. Although the improvement of plant nutrition, compensation for pathogen damage and competition for photosynthesis or colonization/infection sites have been claimed to play a protective role in VAM symbiosis, information is scarce, fragmentary or even controversial, particularly concerning other mechanisms. Such mechanisms include: anatomical or morphological VAM induced changes in the root system, microbial changes in rhizosphere populations of mycorrhizal plants and local elicitation of plant defense mechanisms by VAM fungi. Information needs to be gathered on these mechanisms, if these fungi are to be fully utilized for meaningful integrated crop disease management. Future research should be directed towards correcting the various imbalances revealed by this analysis.

References

A1-Momany and A1-Raddad, A., 1988. Effect of vesicular-arbuscular mycorrhiza on Fusarium wilt of tomato and pepper. *Alexiandria Journal of Agricutural Research,* 33(1): 249-261.

A1-Raddad, A.M., 1995. Interaction of *Glomus mosseae* and *Paecilomyces lilacinus* on *Meloidogyne incognita* of tomato.*Mycorrhiza*: 5 (3):233-236.

Abha Mishra and Shukla, B.N., 1996. Interaction between *Glomus fasciculatum, Meloidogyne incognita* and fungicide in tomato. *Indian Journal of Mycology and Plant Pathology*, 26 (1): 38-44.

Abha Mishra and Shukla, B.N.,1997. Interaction between *Glomus fasciculatum* and *Meloidogyne incognita* on tomato. *Journal of Mycology and Plant Pathology*, 27 (2): 199-202.

Ahemed, S.S. and Alsayed, A.A., 1991. Interaction between the vesicular arbuscular mycorrhiza *Glomus macrocarpus* and *Meloidogyne incognita* infecting cowpea. *Annals of Agricultural Science*, 29 (4): 1765-1772.

Aparajita-Borah and Phukan, P.N., 2003. Effect of interaction of *Glomus fasciculatum* and *Meloidogyne incognita* on growth of brinijal. *Annals of Plant Protection Sciences*, 11(2): 352-354.

Babu, R.S., Sankaranarayan, C. and Joshi, G., 1998. Management of *Pratylenchus zeae* on maize by biofertilizers and VAM. *Indian Journal of Nematology*, 28(1):77-80.

Bhagwati, B., Goswami, B.K. and Singh, C.S. 2000. Management of disease complex of tomato caused by *Meloidogyne incognita* and *Fusarium oxysporum* f.sp. *lycopersici* through bioagent. *Indian Journal of Nematology*, 30 (1):16-21.

Bochow, H. and Abou-Shaar, M. 1990. On the phytosanitary effect of VA-mycorrhiza in tomatoes to the corky-root disease. *Mikrobiologie*, 145(3): 171-176.

Bornali, Mahanta and Phukan, P.N., 2004. Comparative efficacy of *Glomus fasciculatum*, neem cake and carbofuran for the management of *Meloidgyne incognita* on *Vigna mungo*. *Annals of Plant Protection Sciences*, 12(2): 377-379.

Bornali, Mahanta., Phukan, P.N. and Mahanta, B., 2002. Effect of *Glomus fasciculatum* and *Meloidogyne incognita* on blackgram. *Annals of Plant Protection Sciences*, 10(2): 407-408.

Carling, D.E., Roncadori, R.W. and Hussey, R.S., 1989. Interactions of vesicular-arbuscular mycorrhizal fungi, root-knot nematode and phosphorus fertilization on soyabean. *Plant Disease*, 73(9): 730-733.

Datnoff, L.E., Nemec, S. and Pernezny, K., 1995. Biological control of Fusarium crown and root rot of tomato in Florida using *Trichoderma harzianum* and *Glomus intraradices*. *Biological control*, 5(3): 427-431.

Jain, R.K. and Hasan, N., 1986. Association of vesicular-arbuscular mycorrhizal (VAM) fungi and plant parasitic nematodes with forage sorghum (*Sorghum bicolar* L.) *Sorghum Newsletter*, 29: 84.

Jain, R.K. and Sethi, C.L., 1988. Interaction between vesicular arbuscular mycorrhiza, *Meloidogyne incognita* and *Heterodera cajani* on cowpea as infuenced by time of inoculation. *Indian Journal of Nematology*, 18(2): 263-268.

Jain, R.K., Hasan, N., Singh, R.K. and Pandey, P.N., 1998a. Influence of the endomycorrhizal fungus, *Glomus fasciculatum* on *Meloidogyne incognita* and *Tylenchorhynchus vulgaris* infecting berseem. *Indian Journal of Nematology,* 28(1): 48-51.

Jain, R.K., Hasan, N., Singh, R.K. and Pandey, P.N., 1998b. Interaction between plant parasitic nematodes, *Meloidagyne incognita, Tylenchorhynchus vulgaris* and VAM fungi, *Glomus fasciculatum* on lucerne. *Annals of Plant Protection Sciences,* 6(1): 37-40.

Jalali, B.L. and Hari Chand., 1988. Role of VAM in biological control of plant diseases. *In*: Mycorrhizae for Green Asia (Eds. A. Mahadevan, N. Raman and K. Natarajan). Centre for Advanced Studies in Botany, University of Madras, Guindy campus, Madras. pp. 209-215.

Jothi, G., Mani M.P., Rajeshwari, Sundarababu, R., 2000. Management of *Meloidogyne incognita* on okra by integrating non-host and endomycorrhiza.*Current Nematology,* 11(1-2): 25-28.

Kantharaju, V., Babu, N.P., Krishnappa, K. and Reddy, B.M.R., 2002. Biodiversity in indigenous isolates of VAM, *Glomus fasciculatum* against *Meloidogyne incognita* race-1 on tomato. *Proceedings of National Symposium on Biodiversity and and Management of Nematodes in Cropping Systems for Sustainable Agriculture, Jaipur,India,*11-13 Nov.2002: 9-14.

Kassab, A.S., 1990. Aspects of the host-parasite relationships of nematodes and sweet potato. II. Relationships between initial population densities of *Criconemella* spp, *Rotylenchus reniformis* or *Tylenchorhynchus* spp. and yield of mycorrhizal and non-mycorrhizal plants. *Annals of Agricultural Science* (Carico), 35(2): 1107-1116.

Krishna Prasad, K.S. 1991. Influence of vesicular arbuscular mycorrhiza on the development and reproduction of root-knot nematode affecting flue cured tobacco. *Agro-Asian Journal of Nematology,* 1(2): 130-134.

Nagesh, M. Reddy, P.P., Kumar, M.V.V., Nagaraju, B.M., 1999. Studies on correlation between *Glomus fasciculatum* spore density, root colonization and *Meloidogyne incognita* infection on *Lycopersicon esculentum. Zeitschrift fur Pflanzenkrankheiten und Pflanzenschutz.,* 106: 82-87.

Neelima Mittal., Mamta Sharma., Geeta Saxsena and Mukerji, K.G., 1991. Effect of VA mycorrhiza on gall formation in tomato roots. *Plant Cell Incompatibitity Newsletter,* 23: 39-43.

Oyekanmi, E.O., Coyne, D.L., Fagade, O.E. and Osonubi, O., 2007. Improving root knot nematode on two soyabean genotypes through the application of *Bradyrhizobium japonicum, Trichoderma psedokoningii* and *Glomus mosseae* in full factorial combinations.*Crop Protection,* 26(7): 1006-1012.

Oliveira, A.A.R. and Zambolim, L., 1988.Influence of different inoculum levels on the interaction between *Glomus etunicatum* and *Meloidogyne Javanica* on bean. *Pesquisa Agropecuaria Brasileira,* 23(4): 371-378.

Osman, H.A. Korayem, A.M., Ameen., H.H. and Badr-Eldin, S.M.S., 1990. Interaciton of root-knot nematode and mycorrhizal fungi on common bean (*Phaseolus vulgaris* L.) *Anzeiger fur Schadlingskunde Pflanzenschutz Umweltschutz,* 63(1): 129-131.

Osman, H.A., Korayem, A.M., Ameen., H.H. and Badr-Eldin, S.M.S., 1991.Interaction of root-knot nematode and mycorrhizal fungi on common bean (*Phaseolus vulgaris* L.) *B, eitrage Zur Tropischen Landwirtchhaft und Veterinarmedizin,* 29 (3): 341-346.

Pandey, R.K., Goswami, B.K. Satyendra, S., 2005. Management of root knot nematode and fusarium wilt disease complex by fungal bioagents, neem oilseed cake and/or VA mycorrhiza on chickpea. *International Chickpea and Pigeonpea Newsletter,* 12: 32-34.

Pinochet, J., Calvet,C.,Camrubi,A.and Fernandez, C., 1996. Interactions between migratory endoparasitic nematodes and arbuscular mycorrhizal fungi in perennial crops : a review. *Plant and Soil,* 185(2): 183-190.

Rao, M.S. and Gowen,S.R.,1998.Bio-management of *Meloidogyne incognita* on tomato by intergrating *Glomus deserticola* and *Pasteuria penetrans. Zeitschrift Fur Pfulnzenkrans kheiten und Pflanzenschutz,*105(1): 49-52.

Rao, M.S., Reddy, P.P. and Mohandas, S., 1995. Effect of integration of endomycorrhiza (*Glomus mosseae*) and neem cake on the control of root-knot nematode on tomato. *Zeitschrift fur Pfulanzenkrantkheiten und Pflanzenschtuz,* 102(5): 526-529.

Rao, M.S., Reddy, P.P. and Mohandas,M.S.,1998a.Bio-intensive management of *Meloidogyne incognita* on egg plant by integrating *Paecilomyces liiacinus* and *Glomus mosseae. Nematologia Mediterranea,* 26(2): 213-216.

Rao, M.S., Reddy, P.P., Sukhada, M., Nagesh, M., Pankaj., 1998b. Management of root-knot nematode on egg plant by integrating endomycorrhiza. (*Glomus fasciculatum*) and castor (*Ricinus communis*) cake. *Nematologia Mediterranea,* 26(2): 217-219.

Rao, M.S., Somasekhar, N. and Nageshi, M., 1997. Management of root-knot nematodes, *Meloidogyne incognita* in tomato nursery by integration of endomycorrhiza, *Glomus fasciculatum* with caster cake. *Pest Management in Horticultural Ecosystems,* 3(1): 31-35.

Reddy, P.P.,Nagesh, M., Divappa, V. and Kumar, M.V.V., 1998. Management of *Meloidogyne incognita* on tomato by integrating endomycorrhiza, *Glomus mosseae* with oil cakes under nursevery and field condition. *Zeitschrifit fur Pfaunzenkrantkeiten und Pflanzenschutz,* 105(1): 53-57.

Rivas Platero,G.G.and Andrade,J.C.,1998.Interaction of mycorrhizal fungi and *Meloidogyne exigua* in coffee.*Manejo Integrado de Plagas*,49:68-72.

Rivas Platero, G.G., Miranda, T.R. and Andrade, J.C., 1998. Vesicular arbuscular fungus, *Glomus* sp. integration with *Meloidogyne arabicida* in tomato. *Manejo Integrado de Plagas*, 47: 41-43.

Rosendahl, C.N. and Rosendahl, S., 1990. The role of vesicular arbuscular mycorrhiza in controlling damping-off and growth reduction in cucumber caused by *Pythium ultimum*. *Symbiosis*, 9(1-3): 363-366.

Sahay,N.S.,Sudha.,Arehana Singh,and Verma,A.,1998.Trends in endomycorrhizal research. *Indian Journal of Experimental Biology*, 36(11): 1069-1086.

Saleh, H.M. and Sikora, R.A., 1989. Effect of quintozen, benomyl and Carbendazim on the interaction between the endomycorrhizal fungus *Glomus fasciculatum* and the root-knot nematode *Meloidogyne incognita* on cotton. *Nematologica*, 34(4):432-442.

Sankaranarayanan, C. and Sundarababu, R., 1994. Interaction of *Glomus fasciculatum* with *Meloidogyne incognita* inoculated at different timings on blackgram (*Vigna mungo*). *Nematologia Mediterranea*, 22(1): 35-36.

Sankaranaraynan, C. and Sundarababu, R., 1997a. Effect of oil cakes and nematicides on the growth of blackgram (*Vigna mungo*) inoculated with VAM fungus (*Glomus fasciculatum*) and root-knot nematode (*Meloidogyne incognita*).*Indian Journal of Nematology*, 26(2): 144-147.

Sankaranaraynan, C. and Sundarababu, R., 1997b. Effect of leaf extracts on growth of blackgram inoculated with vesicular arbuscular mycorrhiza (*Glomus fasciculatum*) and root knot nematode (*Meloidogyne incognita*). *Indian Journal of Nematology*, 27(1): 128-130.

Sankaranarayanan, C. and Sundarababu,R.,1998.Effect of *Rhizobium* on the interaction of vesicular-arbuscular mycorrhizae and root-knot nematode on blackgram (*Vigna mungo*). *Nematologia Mediterranea*, 26(2): 195-198.

Santhi, A. and Sundarababu, R., 1995a. Effect of phosphorus on the interaction of vesicular-arbuscular mycorrhizal fungi with *Meloidogyne incognita* on cowpea. *Nematologia Mediterranea*, 23(2): 263-265.

Santhi, A. and Sundarababu, R., 1995b. Effect of three species of VAM viz *Glomus fasciculatum, G. versiforme, G. etunicatum* and root knot nematode *Meloidogyne incognita* on cowpea growing in different types of soil. *International Journal of Tropical Plant Diseases*, 13(1): 63-68.

Sharma, W. and Trivedi, P.C., 1997. Cocomitant effect of *Paecilomyces lilacinus* and vesicular-arbuscular mycorrhizal fungi on root-knot nematode infected okra. *Annals of Plant Protection Sciences*, 5: 70-74.

Sharma, H.K., Pankaj and Mishra, S.D. 2003. Effect of plant growth promoter microbes on the root-knot nematode *Meloidogyne incognita* on okra. *Current Nematology,* 14(1/2): 57-60.

Siddiqui, Z.A. and Mahmood, I., 1995. Some observations on the management of wilt disease complex of pigeonpea by treatment with a vesicular arbuscular fungus and biocontrol agents for nematodes. *Bioresource Technology,* 54(3): 227-230.

Siddiqui,Z.A.and Mahmood,I.,1996.Biological control of *Heterodera cajani* and *Fusarium udum* on pigeonpea by *Glomus mosseae, Trichoderma harzianum* and *Vericillium chamydosporium.Israel Journal of Plant Sciences,* 44(1): 49-56.

Sikora, R.A.,1995.Vesicular arbucular mycorrhizae: their significance for biological control of plant parasitic nematodes.*Biocontrol,*1(3). 29-33.

Subhashini, D.V., 1990. The role of VA mycorrhiza in controlling certain root disease of tobacco. *In: Trends in Mycorrhizal Research. Proceedings of the National Conference on Mycorrhiza* (Eds. B.L. Jalali and Hari Chand). Haryana Agricultural University, Hisar. pp.102.

Sundarababu, R., Sankaranarayanan, C. and Vadivelu, S., 1993a. Interaction of mycorrhiza species with *Meloidogyne incognita* on tomato. *Indian Journal of Nematology,* 23(1): 121-123.

Sundarababu, R., Sankaranayanan, C. and Santhi, A.,1993b. Interaction between vesicular-arbucular mycrrhiza and *Meloidogyne incognita* on tomato as influenced by time of inoculation. *Indian Journal of Nematology,* 23(1): 125-126.

Thomas, G.V., Sundararaju, P., Ali, S.S. and Ghai,S.K.,1989. Individual and interactive effects of VA mycorrhizal fungi and root-knot nematode, *Meloidogyne incognita* on Cardamom. *Tropical Agriculture,* 66(1): 21-24.

Vaast, P., Caswell-Chen, E.P. and Zasoski, R.J., 1998. Influences of a root-lesion nematode, *Pratylenchus coffeae,* and two arbuscular mycorrhizal fungi, *Acaulospora mellea* and *Glomus clarum* on coffee (*Coffea arabica* L.). *Biology and Fertility of Soils,* 26(2): 130-135.

Vaast, P., 1997. Interaction of endomycorrhiza of *Arabica coffee* and the chemical control of nematodes *Pratylenchus caffeae* and *Meloidogyne konaensis. In* : Dix-Septieme colleque Scientifique International sur le Café, Naiobi, Paris, France., Association Scientifique Internationale du cafe. pp. 564-571.

Wang-Yan-Ling and Hu-Zheng Jia., 2000. Effect of VA mycorrhiza on nematode infection in tomato. *Journal –of- Hauzhong-Agricultural- University,* 19(1): 25-28.

Wood, T. and Cummings, B. 1992. Biotechnology and the future of VAM commercialization.*In:Mycorrhizal Functioning:an integrative plant fungal process*(Ed. M.F. Allen), Chapman and Hall,London,U.K.pp: 468-487.

Wyss, P.,Boller,T.and Wiemken,A.,1990.Effects of high phosphorus supply on the interaction of soybean roots with *Glomus mosseae* and *Rhizoctonia solani*: degree of infection and accumulation of the Phytoalexin glyceollin. *Symbiosis*, 9(1-3): 383-387.

Rumbos, C., Reimann., Kiewnick,S and Sikora, R.A., 2006. Interactions of Paecilomyces lilacinus strain 251 with the arbuscular mycorrhizal fungus *Glomus intraradices* 99.

Phyto-nematodes in Crops: Their Identification, Treatment and Management, 2011
G.L. Sharma (editor), pp. 267-292
New India Publishing Agency, New Delhi (India)
E-mail : info@nipabooks.com; Website : www.nipabooks.com

Chapter 12

Nematodes of Plantation Crops and their Management

M.S. SHEELA

Introduction

Plantation crops are high value commercial crops which are broadly classified as estate crops consisting of crops like tea, coffee and rubber and small holders plantation crops like coconut, cashew, arecanut, cocoa etc. Spices especially black pepper, cardamom, vanilla, tree spices like clove, nutmeg, cinnamon and allspice are generally grown as mixed crops in perennial cropping systems. In India, plantation crops including spices are grown in an area of three million hectares and generate an annual income of more than Rs, 5,53,630 million. They are valued for multifarious uses to meet the human needs such as food, oil, industrial raw materials, beverages and confectionary items. Pests and diseases are the major limiting factors in increasing the productivity of these crops. Among the pests, nematodes have got a major role in devastating the pepper plantations by the involvement in slow wilt disease, nematode diseases of cardamom nurseries, root-knot nematode infestation in ginger, turmeric, clove, nutmeg, allspice, lesion nematode infestation in tea, coffee, etc.

1. Coconut

Many different nematodes have been found in diverse forms of association with the living coconut palm (Govindankutty & Koshy 1979; Koshy & Banu 2002). But the major nematode disease affecting the crop is red ring disease caused by Bursaphelenchus cocophilus. Fortunately it was not found in India. The only nematode known to cause severe damage leading to malfunction in the coconut is *R.similis* (Koshy *et al.*,1975). The burrowing nematode,*R.similis*,occurs in most tropical and subtropical areas of the world and has been reported from coconut palms in Florida,Jamaica,Sri Lanka and India (Van Weerdt *et al.*, 1959a, 1959b; Ekanayake, 1964; Latta, 1966; Weischer, 1967; Koshy *et al.*, 1975).

Coconut seedlings are raised by sowing seednuts in the interspaces of coconut plantations in Kerala, India. Most of the nurseries in Kerala and Tamil Nadu (South India) are infested by *R. similis.* One year old coconut saplings raised in these infested nurseries harbour large populations of the nematode in roots internal and external to the husk. Such seedlings when distributed for planting help in the dissemination of the nematode over long distances.

Barett (1906) reported that in Trinidad 95% of crop losses were due to root disease/red ring disease caused by the nematode, Rhadinaphelenchus cocophilus. Koshi (1994) reported that Burrowing nematode, *R. similis* causes upto 30% reduction in yield of coconut on heavy infestation.

Management of the burrowing nematode on a perennial palm such as coconut with a massive root system is difficult, especially under the high density multispecies cropping system that exists along the West Coast of South India involving susceptible crops like arecanut, banana, black pepper, betel vine, ginger, turmeric, etc. Unlimited use of nematicides for the control of the burrowing nematode may cause problems of residual toxicity in coconut water and copra (Habeebullah *et al.*, 1983). Apart from this, it may also lead to residual toxicity in the products of the intercrops. Therefore, control of nematodes by field application of nematicides alone is not a practical proposition.

Cultural Methods

Organic Amendments

The cultural practices existing in Kerala and Karnataka (India) are the application of neem/marotti (*Hydrocarpus*) oil cakes at 2-4 kg/palm/year, farm-yard manure @50kg/palm/year,and green foliage and tender stem of *G. maculata* to the basins @50kg/palm/year.

Intercropping

The growing of green manure crops like cowpea, *Crotolaria* or *Sesbania* in the basins and interfaces during June to August and ploughing in of the entire crop at flowering help in reducing the burrowing nematode population and enriching the nutritional status of the soil. In addition, the growing of intercrops like cacao that enrich the soil with sizeable quantities of shed foliage which helps in the build-up of beneficial organisms and antagonistic micro organisms that may inhibit nematode multiplication (Koshy *et al.,* 1991 a & b; Koshy *et al.,* 2002).

Biological Methods

Significant increase in width and leaf area has been recorded in coconut seedling that received AMF alone. Increase in shoot weight, root weight and decrease in lesion indices occurs in seedlings inoculated with mycorrhizae prior to *R. similis.* Prior inoculation of mycorrhizae is more effective against *R. similis* than simultaneous inoculation of mycorrhizae and nematode even though root infection and spore production is slightly reduced in nematode inoculated plants, the lesion index and root nematode population can be successfully lowered with mycorrhizae infection. A mixture of mycorrhizae consisting of multiple endophytes, viz. *Acaulospora bireticulata, G. fasciculatum, G. macrocarpum, G. mosseae, G. versiforme, Sclerocystis rubiformis* and *Scutellospora nigra* was found effective in improving the plant growth and reducing *R. similis* infestation of coconut seedlings (Sosamma, 1994).

Introduction of *P. lilacinus, P. penetrans and Arbuscular Mycorrhizae* in to potting mixture contained in poly bags of coconut nurseries and again in planting pit at the time of transplantation of coconut seedlings in the field helps in better establishment of plants and imparts better growth by offering protection against *R. similis* (Koshy, 1998; Sosamma *et al.,* 1998).

A new isolate of *Pasteuria* parasitising *R. similis* in Kerala, India has great potential for use in IPM. The infective propagules of *Pasteuria* adhered to the cuticle of adult males, females and juveniles of *R.similis* (Sosamma, 1999, 2000, 2002) Introduction of *P. ilacinus, P.penetrans* and mycorrhizae into potting mixture contained in plastic bags of coconut nurseries and again in the planting pit at the time of transplantation of coconut seedlings in the field helps in better establishment of plants and imparts better growth by offering protection against *R.similis* (Koshy, 1998; Koshy *et al.*, 1998; Sosamma *et al.*, 1998). *Catenaria vermicola* was also found parasitsing *R. similis* in Kerala (Sosamma, 2000).

Plant Resistance and Tolerance

All the coconut cultivars (29 exotic, 15 indigenous and 15 hybrids) screened for resistance to *R. similis* in India were found susceptible in varying intensities. The dwarf cultivars; Kenthali and Klappawangi, recorded the least nematode multiplication and lesion indices. Similar reactions were noticed in hybrids such as Java Giant x Kulasekharam Dwarf Yellow, Kulasekharam Dwarf Yellow x Java Giant, Java Tall x Malayan Yellow Dwarf and San Ramon x Gangabondam (Sosamma *et al.*, 1980, 1988; Sosamma, 1984).

Chemical Methods

Dipping the seedlings in 1000 ppm DBCP for fifteen min was effective in controlling nematodes in seedlings in coconut nurseries (Koshy & Sosamma, 1979). Effective control of *R. similis* can be obtained with soil application of phenamiphos or phorate at 2.5 kg a.i./ha during September, December and May in infested coconut nurseries (Koshy & Nair, 1979; Koshy *et al.*, 1985).

The following measures are suggested towards developing an integrated management schedule for *R. similis* infestation on coconut palms (Koshy *et al.*, 2002).

1. Application of cow dung (50 kg), oil cakes (2-4 kg) and green manuring with *Glyricidia maculata* (50kg)/palm/year to the basins.
2. Growing *Crotolaria juncea*, cow pea or Sesbania in the basins and interspaces and incorporating into the soil by ploughing in at flowering stage.
3. Application of phorate at 10 g a.i./palm twice yearly (in June-July and in October- November in India).
4. Avoid growing of bananas as a shade crop in coconut nurseries.

5. Use of nematode free planting material of cocont & other intercrops
6. Use of tolerant or less susceptible cultivars or their hybrids in infested areas.
7. Cut and remove all roots external to the husk of seedlings raised in the field before planting.
8. Raise coconut seedlings in potting mixture enriched with bioagents such as *P. lilacinus, Pasteuria penetrans* and mycorrhyzae in plastic bags.
9. Introduce bio agents in to the planting pits while planting in the main field.
10. Apply phorate@ 3g a.i./plant to inter crops such as banana, black pepper and arecanut in June-July and October- November.

2. Arecanut

A number of nematodes have been reported from the rhizosphere of arecanut (Nair,1964;Weischer,1967;Pizarro,1969; Koshy *et al.*, 1976, 1978; Reddy, 1978; Koshy *et al.*, 1981; Sundararaju & Koshy, 1982a; Sundararaju *et al.*, 1984; Rama, 1987; Dasgupta & Rama, 1987; Subramaniyan *et al.*, 1988) but only *R. similis* is known to be an important parasite of the palm. A number of other palms have been reported as hosts of *R. similis* and it would not be unexpected if nematode problems with some of these other palms became apparent in the years ahead.

Biological Method

Sudha et al. (2000) reported that *P. lilacinus* significantly reduced the population of *R. similis* in soil and root of arecanut.

Plant Resistance/Tolerance

The cultivars Indonesia 6 (VTL-11) Mahuva 8 and Andaman-5 (VTL-29e) are tolerant to *R. similis* (Koshy *et al.*, 1979; Sundararaju & Koshy, 1982b). The cultivar Indonesia-6 (VTL-11) and Singapore (VTL-17) are known to yield 15% more nuts over local South Canara cultivar. Thus, these cultivars could profitably be recommended for *R. similis* infested areas. The hybrid VTL-11 x VTL-17 is highly resistant to *R. similis.*

Chemical Methods

As arecanut is chewed directly by many consumers, dosage, frequency and time of application of nematicides and their persistence on arecanut have to be done carefully to avoid residues in the nut.

Experiment carried out under field conditions revealed that fensulfothion and aldicarb at 1 g a.i./seedling applied thrice a year for three consecutive years gave control of *R. similis* both in soil and roots. Increase in plant growth with regard to shoot length, shoot weight, root length, root weight, number of leaves and collar girth with fensulfothion were 46, 168, 33, 173, 25 and 41% respectively over control plants after three years (Sundararaju & Koshy, 1986).

Field experiments carried out in arecanut monocrop, arecanut + banana and arecanut + banana + pepper to evaluate the efficacy of neem oil cake and phorate singly and in combination for control of *R.similis* in the cropping system. Even though, all the treatments were significantly superior over the untreated control, the best treatment in these experiments was 1.5 g of phorate in combination with 1 kg neem oil cake which controlled the *R. similis* population in arecanut and subsidiary crops very well (Sudha and Sundraraju, 1998).

Integrated Methods

Combination of Organic Amendments and Biocontrol Agents

The combined effect of organic amendments and biocontrol agents viz. *P. lilacinus, P. penetrans* and Arbuscular Mycorrhizal Fungi (AMF) against *R. similis* infecting arecanut (Areca catechu var. Mangala) was conducted using sandy loam soil amended with various organic matters. The organic amendments used were neem and marotti oil cakes, leaves and tender shoots of sun hemp and Glyricidia, vermicompost, cow dung and coir pith. Maximum nematode control (95%) was recorded in soil amended with Glyricidia leaves and bio agents. Significant reduction in root lesion index and maximum leaf area was recorded in these plants. Increase in height and root growth was maximum in plants grown in coir pith amended soil with bio agents which were at par with plants grown in soil enriched with Glyricidia leaves and bio agents.

Control of *R. similis* on arecanut is difficult under the high density, multispecies, subsistence farming systems involving perennial crops such as coconut, banana, black pepper, betel vine, cardamom and cacao. Use of nematicides for the control of burrowing nematode on coconut or arecanut may cause problems of residual toxicity. The following control measures are suggested: 1) use of nematode-free planting material of arecanut and other intercrops; 2) avoiding *R. similis* susceptible intercrops such as black pepper and banana in infested areas; 3) use of resistant/tolerant cultivars of arecanut, when available, and other crops in farming systems; and 4) apply 5-10 kg green manure preferably Glyricidia or Crotolaria ; 5) apply 1 kg of neem oil cake/ palm/year, and 6) apply phorate @1.5g a.i./plant to the root zone of arecanut, banana and black pepper in June-July and October-November in arecanut based farming systems (Koshy, 1994; Sudha & Sundaraju, 1998).

Oil palm is also affected by *R. similis*, but due to limited scale of cultivation it was not yet raised as a limiting factor in production.

3. Cashew

To date, there has been no conclusive report of any damage caused by plant parasitic nematodes to cashew plants. Nematodes do survive in this plant rhizosphere but without causing any apparent damage. In Brazil, nematodes such as *Criconemoides* sp., *Scutellonema sp. and Xiphinema* sp. have been found around roots of cashew trees (Lima *et al.,* 1975). In India, the nematode *Basirotyleptus caudatus* was also detected in the rhizosphere of adult plants. However, pathogenic association was proven (Jairajpuri, 1966).

Plant parasitic root nematodes of cashew were surveyed in Southeastern Nigeria. Five genera and eleven species of plant parasitic root nematodes showed association with cashew trees. Based on degree of virulence, frequency of occurrence and population density, the genera *Xiphinema sp., Scutellonema sp. and Criconemella* sp. were most important. *Triphorus sp. and Rotylenchulus* sp. caused no appreciable damage (Agu, 2007).

4. Tea

Root-knot nematode are the most frequently encountered in the tea growing regions of India and most of the species of *Meloidogyne* except *M.brevicanda* are associated only with nursery plants.Twenty one species of nematode belonging to 14 genera have been reported in association with tea roots from India.

Detailed assessment on crop losses in tea caused by plant parasitic nematode have been carried out almost entirely in Sri lanka. Although

Pratylenchus lose has been recorded from several locations. Significant damage to tea has been observed mostly at elevations of 900-1800 m,.The decline in yield in such areas has been estimated to be around 225-350 kg made tea/ha/annum (Vissen,1959).

Chemical Method

Disinfestation of nursery soil and pre planting application of chemicals (fumigents) for the control of root-knot nematode are recommented in the infected area.

Cultural Method

Incorporation of organic matter into nematode infested soil to suppress the incidence of parasitic nematodes. Cultivation of marigold (*T.erecta and T.patula*)as a pre plant crop considerably reduces the population of *Meloidogyne* spp .Intercropping of marigold also recommended.

5. Coffee

Quite a few nematodes attack coffee (Coffea spp.) throughout the world.The important ones are *Meloidogyne spp., Pratylenchus spp., Hemicriconemoides spp.,R. similis and R. reniformis*. Of these, only the species of *Pratylenchus and Hemicriconemoides* infect Arabica coffee (*C.arabica*) in India (Kumar,1984 and 1985). Robuta coffee (*C.canephora*) exhibit tolerance against the nematode attack.

Cultural Methods

Attempts to check nematodes through the means such as addition of soil ameliorants like oil cakes, coffee cherry husk or organic matter or using the chemical nematicides have a limited value.

Plant Resistance

Robusta is fairly tolerant to the nematode attack and hence the planters have been advised to replace the nematode infected arabica with either robusta or arabica-robusta grafted plants. Wedge-cleft grafting of Arabica scion on robust root-stock at topee stage has been adopted with a fair degree of success (Kumar, 1988).

6. Black Pepper

Plant parasitic nematodes are an extremely important limiting factor in black pepper cultivation. The damage due to nematode is more severe in extensive production systems like black pepper nurseries. Although a sizeable number of plant parasitic nematodes have been recorded on black pepper, only burrowing nematodes (*Radopholus similis)* and root knot nematodes *(Meloidogyne spp.*) can be regarded as economically important.

Krishna Menon (1949) reported mortality upto 10% of paper vines due to slow wilt disease caused by Radopholus similes.

Being a perennial crop trailed on live standards, any management measures need to be repeated every year under Indian conditions. The perennial multi-cropping systems are ideal situations where the burrowing nematode multiplies and causes heavy damage to all the susceptible crops. The high density multi- species cropping pattern does not permit use of nematicides, as most of the crops are export - oriented and tender berries are consumed without any processing or cooking or as pickles. The situation is further complicated where arecanut and coconut are used as live standards as they are very good hosts of *R.similis.*

Cultural Methods

Mulching and bio-fumigation ameliorate symptoms of slow-decline/pepper yellows. Addition of chopped leaves of *Glyricidia maculata* (10g per kg of soil) as green manure reduced populations of *R.similis* and increased the plant growth (Jasy and Koshy, 1992). Application of botanicals like neem oil cake reduced root-knot nematodes (Ramana *et al.,* 1982).

Among the large number of live standards used in black pepper gardens, coconut and arecanut are good hosts of *R.similis* while several of these are highly susceptible to root knot nematodes too. However, *Garuga pinnata* Roxb. and *Macaranga indica* Wight are not susceptible to root knot nematodes and the popular live standards, *Erythrina indica* Lank. and *G. sepium* Walp are less susceptible. Several intercrops like banana, ginger and turmeric and a large number of weeds are also susceptible to these nematodes (Koshy *et al.,* 1977; Ramana, 1986). Large numbers of weeds that are found in pepper gardens have been recorded as hosts of the root-knot nematodes.

Plant Resistance and Tolerance

A number of black pepper germplasm accessions, including wild types, were screened against *M.incognita and R. similis* by several workers (Venkitesan and Sethi, 1978; Koshy and Sundararaju 1979; Jacob and Kuriyan 1979; Mohandas and Ramana, 1983; Ramana and Mohandas, 1986, 1987b; Koshy, 1987b). Wild collection, Vittal No.430, *Piper hymenophyllum and P.attenuatum* recorded least (less than 30 per cent) root reduction and minimum reproduction of *R.similis*. No resistance or tolerance was found to the nematode on screening a number of cultivated and wild germ plasm, intercultivar hybrids and open pollinated seedlings, except for *P. kolubrinum,* which is now widely used as a root-stock to graft cultivated pepper plants (Ramana et al., 1987b; Ramana, 1992). A total of 101 cultivars, 74 accessions of wild Piper sp. and 140 intercultivar hybrids, were screened against *M.incognita* of which one cultivar, CLT-P-812 was found resistant (Ramana and Mohandas, 1986, 1987b). This accession was later released as 'Pournami' for cultivation in root knot infested areas (Ravindran *et al.,* 1992). Some of the wild related species of Piper are also resistant to root knot nematodes (Ramana, 1992). The biochemical changes in this variety consequent to nematode infection have been studied (Eapen *et al.,* 1999).

Chemical Methods

Root-knot infestation in black pepper nurseries has been a serious problem in several nurseries. Fumigation of nursery potting mixture was effective in checking the infestation (CPCRI, 1974, Koshy, 1986; Mohandas and Ramana, 1987). A number of pesticides have been found effective in reducing nematode populations on black pepper. Aldicarb/carbofuran/phorate at 3g a.i /vine applied in May/ June and again in September / October resulted in the remission of foliar yellowing and reduction in nematode populations. Among the above three nematicides, phorate was superior (Ramana, 1986; Mohandas and Ramana, 1987; Lokesh and Gangadharappa, 1995; Sundararaju and Sudha, 1998).

Biological Methods

By fortifying potting mixtures with biological control agents, nematode- free rooted cuttings can be raised. A number of organisms have been tested and found effective in suppressing root knot nematodes. Promising among these are *P. lilacinus* (Ramana, 1994; Sosamma and Koshy, 1997), *Pochonia chlamydosporia* (*Verticillium chlamydosporium*) (Sreeja *et al.,* 1996), *Pasteuria penetrans* (Sosamma and Koshy, 1997), Bacillus spp. (Sheela, 1991, Sheela and Venkitesan, 1992, 1994 and 1995, Sheela *et al.,* 1993) and *Pseudomonas fluorescens*

(Eaten *et al.*, 1997). A number of rhizobacteria that are antagonistic to root knot nematodes have been isolated recently (Beena *et al.*, 2001). Black pepper plants pre inoculated with vesicular arbuscular mycorrhizal fungi like *G. fasciculatum, G. etunicatum, G. mossae; Gigaspora margarita* recorded a significant increase in growth even in the presence of root knot nematodes (Anandaraj *et al.*, 1991; Sivaprasad *et al.*, 1990, 1992). Endophytic bacteria that suppress root knot nematodes also have been isolated from black pepper (Aravind *et al.*, 2004).

However, there are only very few successful attempts to control *R. similis* by use of any fungal bioagents, probably due to the migratory endoparasitic nature of this nematode (Geetha, 1991; Ramana, 1994). The mycorrhizal fungus, *G. fasciculatum* suppressed burrowing nematode infestation (Anandaraj *et al.*, 1996c). Recently, rhizobacteria that suppressed *R. similis* infesting pepper were identified in green house studies (Beena *et al.*, 2003). Endophytic bacteria that suppress root knot nematodes also have been isolated from black pepper (Aravind *et al.*, 2004).

The following integrated method can be suggested for the management of nematodes in pepper

a) Use nematode free rooted cuttings for raising new plantations

b) For the nematode infested vines, apply phorate/carbofuran @ 1 g a.i/vine twice in a year. The first dose is to be given in May-June at the *onset* of monsoon and the second dose in the last week of October or early November. The chemicals are to be applied around the vines in shallow basins and incorporated into the soil.

c) Root knot nematode can be effectively managed by application of talc based formulation of *Bacillus macerans* @ 10 g/vine in basins (106 cfu /g) at the time of planting of vines or just before the monsoon period in established plantations.

7. Ginger

Plant parasitic nematodes belonging to 17 genera were reported on ginger (Sundararaju *et al.*,1979b, Rama & Dasgupta, 1985; Kaur, 1987; Ramana and Eapen, 1998) the most important parasites are *Meloidogyne* spp., *R.similis and P.coffeae*. In Kerala, *M.incognita, R.similis andPratylenchus* spp. were the major nematode species found in the rhizhosphere of ginger (Mammen, 1973; Charles, 1978; Charles and Kurian,1982; Sheela *et.al,* 1995). Several species of root knot nematodes viz. *M. arenaria, M.hapla, M.incognita and M. javanica* have been reported as parasites of ginger in various countries.Other nematodes observed include *R.reniformis, H.multicinctus, Tylenchorynchus spp., Hoplolaimus indicus, Criconemoides* spp. and *Xiphenema* sp. (Sheela *et al.,* 1995).

In Qeensland, Australia, sever infestation of ginger rhizomes by root-knot nematode,Meloidogyne spp. reduces yield by 57% as determined by fumigation (Pegg *et al.,*1974). In Fiji, burrowing nematode, Radopholus similes has been reported from more than 50% of the total area with a rate of infection ranging from 10-50% resulting in the yield reduction of about 40% (Sunder raju *et al.,*1979).

Integrated Method

Neem cake followed by carbofuran treated plots showed minimum gall index in ginger (Mohanty et al., 1995). Sheela et al. (1995) reported that application of neem cake @ 2.50 t ha^{-1} at the time of planting and carbofuran 1 kg a.i. ha^{-1} forty five days after planting were effective in reducing *M. incognita* population in soil and root, root-knot count and increasing the yield of ginger. *G. fasciculatum* was very effective in controlling root-knot nematode in spices like ginger, cardamom, pepper (Sivaprasad and Sheela, 1998).

8. Turmeric

A number of species of plant parasitic nematodes have been reported in association with turmeric in India (Nirula and Kumar, 1963; Sundararaju *et al.,* 1979b; Dasgupta and Rama, 1987; Gunasekharan *et al.,* 1987; Haider *et al.,* 1998a; Rama, 1987; Mani & Prakash, 1992; Routaray *et al.,* 1987b; Bai *et al.,* 1995) of which *Meloidogyne spp., R.similis and P.coffeae* are of economic importance. Ayyar (1926) first reported root knot nematode infestation in turmeric. Two species of root knot nematodes, *M.incognita and M.javanica* have been reported on turmeric, but most investigations have been concerned with *M.incognita.*

Cultural Method

Immersing turmeric rhizomes in hot water at 55° C for 10 min. or 45° C for 50 min. can kill *M. incognita* inside the rhizomes (Chen *et al.,* 1986) and this could be used for establishing nematode-free multiplication plots but is unlikely to be economic for large scale field use.

Plant Resistance

The cultivars and breeding lines 5379-1-2, 5363-6-3, Kodur, Cheyapuspa 5335-1-7, 5335-27, Ca-17/1, Cli-124/6, Cli-339, Armoor, Duggirala, Guntur-1, Guntur-9, Rajampet, Sugandham and Appalapadu have been reported as resistance to M. incognita (Mani *et al.*, 1987; Gunasekharan *et al.*, 1987). The species *C.zedoaria* is more resistant to *M.incognita* than *C.domestica* in China (Chen *et al.*, 1986). In Andra Pradesh, India, the high yielding varieties like PCT8, PCT 10, Suguna, and Sudharshana were free from root knot nematode infestation (Rao *et al.*, 1994). Recently eight turmeric accessions (Acc. Nos. 31, 82, 84, 142, 178, 182, 198 and 200) were identified as resistance to root knot nematode (Eapen *et al.*, 1999).

Chemical Methods : Application of DBCP at 15 l a.i./ha 15 days prior to planting results in yield increase of 253-270 per cent compared with 59-187 per cent increase in yield with application of phenamiphos at 2.5 kg a.i./ha one day before planting (Patel *et al.*, 1982). Aldicarb and carbofuran applied at 1kg a.i./ha increase yield by 71 per cent and 68 per cent respectively over control, with a cost benefit ratio of 1:6 in aldicarb and 1:2 in carbofuran treatments (Gunasekharan *et al.*, 1987). Carbofuran at 4 kg a.i./ha applied in rows to 4-months-old turmeric crop has resulted in a 81.6 per cent reduction in root knot nematode population against 45 per cent increase in untreated plots (Mani *et al.*, 1987). Similarly, application of carbofuran or phorate (@1kg a.i./ha) reduced root knot nematodes (Haidar *et al.*, 1998b).

Biological Methods

Biocontrol agents like *V.chlamydosporium, P.lilacinus, Fusarium* sp., *Aspergillus* sp., and *Scopuloriopsis* sp. controlled root knot nematodes in field trials (Ramana *et al.*, 2002).

9. Cardamom

The major nematode problem is caused by the root-knot nematodes, *Meloidogyne* spp., although the lesion nematode,*P.coffeae* and the burrowing nematode,*R.similis* are known to cause root rotting. (D'Souza *et al.* 1970; Kumar *et al.*, 1971; Khan and Nanjappa, 1972; Viswanathan *et al.*, 1974; Sundararaju *et al.*, 1979b).Reniform nematode, *R. reniformis*,is also recorded on cardamom.(Eapen, 1995a).

Ali (1984) reported a yield loss of 32-47% in cardemom plants due to root-knot nematode infestation from the result of a nematicidal experiment.

Even though there is a general awareness about nematodes infesting cardamom, the recommended control measures are not being followed by many growers.

Cultural Methods

It is advisable to change nursery sites every year, but this is not always practicable in view of the difficulties involved in getting suitable sites having facilities for irrigation. Hence, disinfestations of the nursery beds need to be carried out every year. Disinfestations of nursery beds with methyl bromide at 500 g/10m^2 is effective in controlling root-knot infestation in both primary and secondary nurseries (Ali & Koshy, 1982).

A large number of annual weeds present in the cardamom plantations and the common shade trees, *Erythrina indica and E.lithosperma,* are hosts of root-knot and help in the build up of nematode populations (Muniappan, 1993). No resistance to root-knot nematodes has been found and the popular cardamom cultivars, Malabar, Mysore and Vazhukal are all susceptible (Hegde *et al.*, 1993; Eapen, 1995).

Chemical Methods

Drenching of nursery beds with fenamiphos significantly reduced root-knot nematodes (Ali, 1986). Aldicarb, carbofuran, phorate at 5, 10 or 15 kg a.i. /ha respectively, have been applied in primary nurseries of cardamom for control of M. incognita. None of the nematicides treatment totally prevented nematode infestation but there was significant reduction in root-knot densities. Aldicarb at the very high level of 15 kg a.i. /ha reduced nematode numbers by 90 per cent (Ali, 1987). Application of aldicarb/ carbofuran/phorate at 5g and 10ga.i./ plantand neem oil cake at 500g and 1000g / plant twice a year increases yield of cardamom plants infested with M. incognita from 47 to 88 per cent. Maximum yield was obtained by the addition of neem oil cake at a rate of 1000g / plant followed by 500 g / plant (Ali, 1984). However, in another study, application of phorate @ 2.5-5.0 g a.i. / plant reduced the nematode population and increased the yield by more than 40 per cent. (Eapen, 1995).

Coconut

Barett (1906) reported that in Trinidad 95% of crop losses were due to root disease/red ring disease caused by the nematode *Rhadinaphelenchus cocophilus.*

Koshy (1994) reported that the burrowing nematode *Radopholus similis* causes upto 30 per cent reduction in yield of coconut on heavy infestation.

Coffee

Goncalves *et al.* (1978) reported that a very important source of loss due to root-knot nematode is the total destruction of the coffee seedling enforced by law when root-knot is found in nurseries. In Sao Paulo State in Brazil, 3231952 seedlings were destroyed from 1976 to 1977.

The different types of losses caused by root-knot nematodes (*Meloidogyne sp.*) can be summarized:

1. Yield decreases.
2. Destruction of seedlings in nursery.
3. Unemployment in traditional coffee producing areas.
4. Decrease of the farmer's income by cultivation of a less profitable crop.
5. Losses of investment on equipment or machines specific for this crop.
6. Increases in the cost of coffee production due to nematicide application. From the research stand point, yield loss has tended to preoccupy scientists, to exclusion of the other causes of loss (Campos *et al.,* 1990).

Bridge (1984) reported that in Tanzania, it is estimated that yield losses of trees severely infested with the African coffee root-knot nematodes will be in the region of 20% in optimum conditions, extending to the point of non-productivity.

Barett (1906) reported that in Trinidad 95% of crop losses were due to root disease/red ring disease caused by the nematode *Rhadinaphelenchus cocophilus*.

Koshy (1994) reported that the burrowing nematode *Radopholus similis* causes upto 30 per cent reduction in yield of coconut on heavy infestation.

Tea

Detailed assessments on crop losses in tea caused by plant parasitic nematodes have been carried out almost entirely in Sri Lanka. Although

Pratylenchus lose has been recovered from several locations, significant damage to tea has been observed mostly at elevations of 900 to 1800 m. The decline in yield in such areas has been estimated to be around 225 to 350 kg made tea/ha/annum (Visser, 1959).

Pepper

Krishna Menon (1949) reported mortality up to 10% of pepper vines due to slow wilt disease caused by *Radopholus similis.*

Cardamom

Ali (1984) reported a yield loss of 32-47% in cardamom plants due to root-knot nematode infestation from the results of a nematicide experiment.

Ginger

In Queensland, Australia severe infestation of ginger rhizomes by root-knot nematode *Meloidogyne* sp. reduces yields by 57% as determined by fumigation (Pegg *et al.,* 1974).

In Fiji, burrowing nematode, *Radopholus similis* has been reported from more than 50% of the total area with a rate of infection ranging from 10-50% resulting in yield reduction of about 40% (Sundararaju *et al.,* 1979).

References

Agu, C.M. 2007. Survey and effects of plant parasitic root nematodes of cashew (*Anacardium occidentale*) in South-Eastern-Nigeria. *Journal of American Science* 3(2):50-54.

Ali, S.S. (1984). Preliminary observations on the effect of some systemic nematicides and neem oil cakes in a cardamom field infested with root-knot nematodes. *Proceedings of the Sixth Symposium on Plantation Crops,* Kottayam, India, 16-20 December: 215-223.

Ali, S.S. 1984. Occurrence of root-knot nematodes in cardamom plantations of Tamil Nadu. In: *Proc. PLACROSYM - VI 1982.* (eds) Bavappa, K.V.A., Iyer, R.D., Das, P.K., Sethuraj, M.R.,Ranganathan, V., Nambiar, K.K.N., Chacko, M.J. Indian Society for Plantation Crops, Kasaragod, Kerala, India, pp. 615-620.

Ali, S.S. 1986. Occurrence of root-knot nematodes in cardamom plantations of Karnataka. *Indian J. Nematol,* 16: 269-270.

Ali, S.S. 1987. Effect of three systemic nematicides against root-knot nematodes in cardamom nursery. *Nematologia mediterranea,* 15:155-158.

Ali, S.S. and Koshy, P.K. 1982. Occurrence and distribution of root-knot nematodes in cardamom plantations of Kerala.

Ananadaraj, M., Ramana, K.V. and Sarma, Y.R. 1991. Interaction between vesicular arbuscular mycorrhizal fungi and *Meloidogyne incognita* in black pepper. In: *Mycorrhizal Symbiosis and Plant Growth.* (eds.) Bagyaraj, D.J. and Manjunath, A. University of Agricultural Sciences, Bangalore, India, pp.110-112.

Ananadaraj, M.,Ramana, K.V. and Sarma, Y.R. 1996. Suppressive effect of VAM on root damage caused by *Phytophthora capsici, Radopholus similis* and *Meloidogyne incognita* in black pepper. In: *Impact of Diseases and Pests in Tropical Forests,* (eds.) Nair, K.S.S., Sharma, J.K. and Varma, R.V., Kerala Forest Research Institute, Peechi, India, pp. 232-238.

Aravind, R., Antony, D., Santhosh,J.E., Kumar, A. and Ramana, R.V. 2004. Isolation and *in vitro* evaluation of endophytic bacteria against root-knot nematodes infesting black pepper (*Piper nigrum* L.). *National Nematology Symposium, 17-19 November 2004, University of Agricultural Sciences, Bangalore.*

Bai,H.,Sheela,M.S.and Jiji,T.1995.Nemic association and avoidable yield loss in turmeric,*Cucurma longa* L.*Pest Mgmt.hort.Ecosystems* 1: 105-110.

Barrett, O.W. (1906). Notes upon miscellaneous crops. *Proceedings of Agricultural Society of Trinidad,* 7: 303.

Beena, B., Eapen, S.J. and Ramana, K.V. 2001. Antogonistic rhizobacteria of root knot nematodes infesting black pepper (*Piper nigrum* L.).In:Abstracts-National Congress on Centenary of Nematology in India.Indian Agricultural Research Institute, New Delhi, India, pp.143.

Beena, B., Eapen, S.J. and Ramana, K.V. 2003. Native rhizobacteria for the biological suppression of *Radopholus similis* infesting black pepper (*Piper nigrum* L.). In: Abstracts and Short Papers-Sixth International PGPR Workshop, Calicut, India.

Bridge, J. 1984. Coffee nematode survey of Tanzania. Report on a visit to examine plant parasitic nematodes of coffee in Tanzania, February 1984. Commonwealth Institute of Parasitology, 22 p.

Bridge, J. 1988. Plant parasitic nematode problems in the Pacific islands. *J.Nematol,* 20:173-183.

Campos, V.P., Sivapalan, P. and Gnanapragasam, C. 1990. Nematode parasites of coffee, cocoa and tea. Plant parasitic nematodes in subtropical and tropical agriculture. Luc, M., Sikora, R.A. and Bridge, J. (Eds.). CAB International, UK, p. 387-430.

Central Plantation Crops Research Institute (CPCRI) 1974. *Annual Report,* 1973, Kasargod, 62 pp.

Charles, J.S. 1978. Studies on the nematode diseases of ginger. M.Sc (Ag) Thesis, Kerala Agricultural University, Thrissur, p.

Charles, J.S. and Kuriyan, K.J. 1982. Studies on nematode incidence in ginger. In: Venkatram (ed.) Proc.PLACROSYM -II 1979. Indian Society for Plantation Crops, Kasargod, Kerala, India, pp. 50-57

Chen, C.M., Li,H.Y.and Lii,O.Y.1986.The study on root knot nematodes of common turmeric (*Cucurma domestica* Valet.). *Herald. agric.Sci.* 1: 16-22

D'Souza, G.I., Kumar, A.C., Viswanathan, P.R.K. and Shamanna, H.V. 1970. Relative distribution and prevalence of plant parasitic nematodes in coffee tracts of South Western India. *Indian Coffee* 34: 329-330, 342

Dasgupta, M.K. and Rama, K. 1987. Plant parasitic nematodes associated with plantation crops. *Review. Trop. Pl. Pathol.* 4: 289-304

Eapen, S.J. 1995a. A note on the incidence of Rotylenchulus reniformis in cardamom plants. *Indian J. Nematol.* 25: 213

Eapen, S.J. 1995b. Investigations on Plant Parasitic Nematodes Associated with Cardamom. Final Report, Indian Institute of Spices Research, Calicut, Kerala, India

Eapen, S.J.,Beena,B.and Ramana,K.V. 2000. Preliminary screening of some hyphomycetous fungi for the biological control of root-knot nematodes (*Meloidogyne incognita*) infesting spice crops. In: Abstract of Papers - National Nematology Symposium on Integrated Nematode Management, Orissa University of Agriculture and Technology, Bhubaneshwar, Orissa. p. 18

Eapen, S.J., Chempakam,B.and Ramana, K.V. 1999. Role of defense related enzymes in the resistance of black pepper to Meloidogyne incognita. In: Abstract of Papers - National Seminar on Nematological Research in India, C.S. Azad University of Agriculture and Technology, Kanpur.

Eapen, S.J., Ramana, K.V. and Sarma, Y.R. 1997. Evaluation of Pseudomonas fluorescens isolates for control of *Meloidogyne incognita* in black pepper (*Piper nigrum* L.). In: Biotechnology of Spices, Medicinal and Aromatic Plant - Proceedings of the National Seminar on Biotechnology of Spices and Aromatic Plants, (eds.) Edison, S., Ramana, K.V., Sasikumar, B., Nirmal, B.K. and Eapen, S.J. Indian Society for Spices, Calicut, India, pp. 129-133

Ekanayake, U.B.M.(1964).Annual report of the crop protection officer, Coconut Research Institute for 1963.*Ceylon Coconut Quarterly*,15:5057

Geetha, S.M. 1991. Studies on the Biology, pathogenicity and biocontrol of different populations of *Radopholus similis*. Ph. D. Thesis, Kerala Agricultural University, Thrissur.

Goncalves, W., Thomaziello, R.A., Moraes, M. V., Fernandes, J.A.E., Costa, A.M., Corsi, T., Jungueira, C.A. and Lacerda, L.A.O. (1978). Estimativas de danos ocasionados pelos nematoides do cafeeiro. In: *6th Congresso Beasileiro de Pesquisas Cafeeiras de Ribeirao Preto,* Sao Paulo. Brazil-IBC/GERCA : 182-186.

Govindankutty, M. P. and Koshy, P. K. (1979). Nematodes associated with the coconut palm (*Cocos nucifera* L.) In: *Technical Bulletin 2 - Nematodes, fungi insects and mites associated with the coconut palm.* Central Plantation Crops Research Institute, Kasaragodu, India, 1-50

Gunasekharan, C.R., Vadivelu, S. and Jayaraj, S. 1987. Experiments on Nematodes of Turmeric - A Review.Proc. Third Group Discussions on the Nematological Problems of Plantation Crops.October 29-30,1987.Sugarcane Breeding Institute,Coimbatore,Tamilnadu,India. Abstract: 45

Habeebullah, A., Reghupathy, A., Rethinam, P. and Koshy, P. K. (1983). Residues of aldicarb in coconut. *Seminar on pesticides and environment, 4-5 August,1983.* Coimbatore, Tamil Nadu, India: 21 [Abstr.]

Haider, M.G., Jha, R.N. and Nath, R.P. 1998a. Studies on the nematodes of spices. Pathogenic effect of root knot (Meloidogyne incognita) and reniform nematodes (Rotylenchulus reniformis) alone and in combination on turmeric (*Cucurma longa* L.). *Indian J. Nematol.* 28: 52-55

Haider,M.G.,Jha,R.N.and Nath,R.P.1998b.Efficacy of some nematicides and organic soil amendments on nematode population and growth characters of turmeric (*Cucurma longa* L.). *J. Res. Birsa Agric. Univ.* 10: 211-213

Hedge, L., Krishnappa, K. and Bojappa, K.m. 1993. Recation of different cultivars of cardamom (*Elettaria cardamomum* Maton) to root knot nematode (*Meloidogyne incognita*). *Indian J. Nematol.* 23: 127-128.

Jacob, A. and Kuriyan, J. 1979a. Screening of pepper varieties for resistance against root-knot nematode (*Meloidogyne incognita*). *Agric. Res. J. Kerala* 17: 90

Jairajpuri, M.S. 1966. On *Basiotyleptus caudatus* and a redescription on *Thornenema thionemanii. Proc. Helminthol. Soc. Wash.* 33(1): 30-33

Jasy, T. and Koshy, P.K. 1992. Effect of certain leaf extracts and leaves of *Glyricidia maculate,* (H.B and K.) Steud. As green manure on *Radopholous similis. Indian J. Nematol.* 22:117-121.

Kaur, K.J. 1987.Studies on nematodes associated with ginger (*Zingiber officinale* Rosc.) in Himachal Pradesh. Thesis submitted to Himachal Praesh University, Shimla, India, 198 pp.

Khan, E. and Nanjappa, C.K. 1972. Four new species of Criconematoidea (Nematoda) from India. *Indian J. Nematol.* 2: 59-68

Koshy, P. K. (1998). Biological control of burrowing nematode on palms and black pepper. Final Tech. Report, USIF Project No. 1 N -AES-779 C.P.C.R.I., Kayangulam, Kerala, India, 690533

Koshy, P. K. and Banu, J. G. (2002). Nematode diversity in plantation crops-The World scenario.In: *Nematode Diversity* (ed.) Jairajpuri, M. S. Silverline printers pp. 390- 425.

Koshy, P. K., Raski, D. J. and. Sosamma, V. K (1981). *Brachydorus swarupi* sp. n.(Nemata Dolichodorinae) from soil about roots of arecanut palms in Kerala state. *Indian Journal of Nematology* 13: 401- 404.

Koshy, P. K., Sosamma, V. K. and Nair, C. P. R. (1975). Preliminary studies on *Radopholus similis* (Cobb, 1893) Thorne, 1949) infesting coconut and arecanut palms in South India. *Indian J. Nematol.* 5: 26-35.

Koshy, P. K., Sosamma, V. K. and Samuel, R. (1998). Interactive effect of Vesicular Arbuscular Mycorrhizae (VAM) with *Radopholus similis* on coconut seedlings. Proceedings of the Third International Symposium of AFRO ASIAN Society of Nematologists (TISAAN) Nematology- Challenges and Oppurtunities in 21st century held at Sugarcane Breeding Institute, Coimbatore during April 16-19,1998. pp.106-110

Koshy, P. K., Sosamma, V. K. and Sundararaju, P. (1976). Yellow leaf disease: Nematological studies. *Arecanut and Spices Bulletin*, 8: 37-41.

Koshy, P. K., Sundararaju, P. and Sosamma, V. K. (1978). Occurrence and distribution of *Radopholus similis* (Cobb, 1893) Thorne, 1949 in South India. *Indian J. Nematol.* 8: 49-58.

Koshy, P. K., Sundararaju, P., Sosamma, V. K. and Ravikumar, K. (1985). Efficacy of four systemic nematicides against *Radopholus similis* in coconut nursery. *Indian J. Nematol.* 15:148-151.

Koshy, P.K. 1987b. Integrated Nematode Management in Plantation Crops. *Proceedings of the Third Group Discussion on the Nematological Problems of Plantation Crops. October, 29th -30th, 1987.* Sugarcane Breeding Institute, Coimbatore, Tamil Nadu, India: 49-51.

Koshy, P.K. 1994. Nematodes in plantation crops. *Advances in Horticulture Vol.10- Plantation and Spice Crops Part 2.* Chadha, K.L. and Rethinam, P. (Eds.), Malhotra Publishing House, New Delhi.

Koshy, P.K. and Nair, C.P.R. 1979. Control of *Radopholous similis* (Cobb 1893) Throne, 1949 in coconut nursery. *Indian J. Nematol.* 9:15-19.

Koshy, P.K. and Sosamma, V.K. 1979. Control of the burrowing nematode *Radopholous similis* on coconut seedlings with DBCP.*Indian J.Nematol.* 12:200-203.

Koshy, P.K. and Sundararaju, P. 1979. Response of seven black pepper cultivars to *Meloidogyne incognita. Nematologia Mediterranea* 7: 123-135

Koshy, P.K., Sosamma, V. K. and Sundararaju, P. (1991 a). Nematode Management in coconut and coconut based cropping system. Invited paper Session I on

Pest Management, *International Symposium on Coconut Research and Development – II,* Kasaragod, Kerala, 26-29 November, 1991.

Koshy,P.K.,Sosamma,V.K.and Sundararaju,P.(1991b).*Radopholus similis,* the burrowing nematode of Coconut.*Journal of PlantationCrops.*19(2):139-152.

Koshy, P.K., Sosamma, V.K. and Banu, J.G. 2002. Milestones of Nematological Research in Kerala. Released during the seminar on Centenary on Nematological Research in Kerala during February 21st -22nd 2002 at CPCRI, Kayangulam, Kerala

Koshy, P.K., Sosamma, V.K. and Sundararaju, P. 1977. Screening of plants used as pepper standards against root-knot nematode. *Indian Phytopathol.* 30: 128-129

Koshy, P.K., Sundararaju, P. and Wilson, K.I. 1979. Efficacy of certain systemic nematicides against Meloidogyne incognita in a cardamom nursery. *Nematologia Medi.* 7: 195-198

Koshy, P.K.1986.The burrowing nematode, Radopholus similis (Cobb, 1893) Thorne,1949.In:Swarup,G. & Dasgupta,D.R.(eds.)Plant parasitic nematodes of India-Problems and Progress,IARI,New Delhi,India.pp. 223-248

Krishna Menon, K. (1949). The survey of pollu and root diseases of pepper. *Indian Journal of Agricultural Science* 19: 89-136.

Kumar, A.C. 1984. Resistance in coffee to *Meloidogyne* spp. and occurrence of intersexes in *M. thamesi. Nematologica,* 30: 108-110

Kumar, A.C., Viswanathan, P.R.K. and D'Souza, G.I. 1971. A study of plant parasitic nematodes of certain commercial crops in coffee tracts of South India. *Indian Coffee* 35: 222-224

Latta,R.(1966).Attempts to relate nematodes to lethal yellowing disease of coconut palms in Jamaica.*Tropical Agriculture Trinidad,* 43: 59-68.

Lima, J.A.A., Menezes, M., Karan, M.Q. and Martins, O. F.G. 1975. Genera of plant parasitic nematodes isolated from the rhizosphere of Cashew *Anacardium occidentale* L. *Fitossanidade* 1(2): 32-35

Lokesh, M.S. and Gangadharappa, P.M. 1995. Management of Phytophthora foot rot and nematode diseases in black pepper (*Piper nigrum L.*). *J. Spices Aromatic Crops* 4: 61-63.

Mammen, K.V. 1973. Root gall nematodes as a serious pest of ginger in Kerala. *Curr. Sci.* 42: 549

Mani, A.and Prakash,K.S.1992.Plant parasitic nematodes associated with turmeric in Andhra Pradesh,India. *Int. Nematol. Network Newsl.* 4: 13-18

Mani, A., Naidu, P.H. and Madavachari, S. 1987. Occurrence and control of *Meloidogyne incognita* on turmeric in Andhra Pradesh, India. *Int. Nematol. Network Newsl.* 4: 13-18

Mohandas, C.and Ramana, K.V. 1983. Effect of different levels of Meloidogyne incognita on plant growth of two cultivars of black pepper (Piper nigrum L.) Abstracts of papers.Third Nematology Symposium Organised by Nematological Society of India,Himachal Pradesh,India:9-10

Mohandas, C. and Ramana, K.V. 1987. Slow wilt disease of black pepper and its control. *Indian Cocoa, Arecanut and Spices J.* 11:10-11

Mohanty, K.C., Ray, S., Mohapatra, S.N., Patnaik, P.R. and Ray, P. 1995. Integrated management of root knot nematode in ginger (*Zingiber officinale Rosc.*). *J. Spices. Aromatic Crops* 4: 70-73

Muniappan, R. 1993. Pests and diseases of Eryhtrina: a review. *J. Coffee Res.* 23: 1-13

Nair,R.B.(1964).Nematodes from arecanut soils.*Arecanut Journal*,15:87-88

Nirula, K.K. and Kumar, R. 1963. Collateral host plant of root knot nematodes. *Curr. Sci* . 32: 221-222

Pegg, K.C., Moffett, M.L. and Colbran, R.C. (1974). Disease of ginger in Queensland. *Queensland Agricultural Journal*, 100:611-618.

Pizarro, A. C. (1969). Nematodes associated with roots of coconut and other palms. *Philippine Journal of Plant Industry*. 34: 155-158.

Rama, K. 1987. *Ecology of plant parasitic nematodes affecting some major perennial crops in West Bengal*. Ph.D. Thesis, University of Visva Bharati, India, 276p.

Rama, K. and Dasgupta, M.K.1985. Physiological ecology of plant pathogenic nematodes associated with ginger in Darjeeling District, West Benghal. Indian J. Nematol. *Indian J. Nematol.* 15: 284-285

Ramana, K.V. 1986. Slow-wilt disease of black pepper and the role of plant parasitic nematodes in it's etiology. Proc. Second Group Discussion on Nematological Problems of Plantation Crops, April 24-25, 1985.Central Coffee Research Station,Balehonnur,Karnataka, India: 17-21

Ramana, K.V. 1992. Role of nematodes in the incidence of slow decline (slow wilt disease) of black pepper (Piper nigrum L.) in Calicut district, Kerala. Abstracts of Papers, Third Nematology Symposium Organised by Nematological Society of India. Himachal Pradesh Krishi Viswa Vidyalaya, Solan, Himachal Pradesh, India: 9

Ramana, K.V. 1994. Efficacy of *Paecilomyces lilacinus* (Thom.) Samson in suppressing nematode infestations in black pepper *(Piper nigrum L.). Journal of Spices and Aromatic Crops*, 3: 130-34.

Ramana, K.V. and Eapen, S.J. 1998. Plant parasitic nematodes associated with spices and condiments. In: Nematode Diseases in Plants. (ed.) Trivedi, P.C. CBS Publishers and Distributors, New Delhi, pp. 217-251.

Ramana, K.V. and Mohandas,C.1986. Reaction of black pepper germplasm to root-knot nematode,*Meloidogyne incognita.Indian J. Nematol.* 16: 138-139

Ramana, K.V. and Mohandas, C. 1987b. Nematological Problems in Black pepper (*Piper nigrum* L.) Proceedings of the Third Group Discusion on Nematological Problems of Plantation Crops. October 29-30, 1987. Sugarcane Breeding Institute, Coimbatore, Tamilnadu. India. pp.43-44

Ramana, K.V., Eapen, S.J. and Beena, B. 2002. Biological Control of Plant Parasitic Nematodes of Major Spices, Final Report of ICAR Ad- hoc Project, Indian Institute of Spices Research, Calicut, Kerala, India

Rao, P.S., Krishna. M.R., Srinivas, C., Meenakumari, K. and Rao, A.M. 1994. Short duration, disease-resistant turmerics for northern Telangana. *Indian Horticulture,* 39: 55-56.

Ravindran, P.N.,Ramana,K.V.,Nair,M.K.,Babu, K.N. and Mohandas, C. 1992. Pournami- a high yielding black pepper selection tolerant to root-knot nematode (*Meloidogyne incognita*). *J. Spices and Aromatic Crops* 4: 67-69

Reddy, D. B. 1978. Insect pests, diseases and nematodes of Areca catechu and betel vine (*Piper betle*) recorded in Asia and the Pacific. *Technical Document No. 114. Plant Protection Committee for the South East Asia, Asia and Pacific region,* FAO, Bangkok, 6p.

Routary, B.N., Sahoo, H. and Das, S.N. 1987b. Nemic associations of ginger and turmeric in Orissa. *Indian J. Nematol.* 17: 122-123

Sheela, M.S. 1991. Control of root-knot nematode *Meilodogyne incognita (Kofoid and White) Chitwood, 1949 infesting black pepper (Piper nigrum L.) by bacterial pathogens.* Ph.D. Thesis, Kerala Agricultural University, Vellanikkara, p. 139.

Sheela, M.S. and Venkitesan, T.S. 1992. Bacterial pathogens associated with root-knot nematode in Kerala, Their occurrence and status. (ed.) Ramachandran Nair, R. *Proc. 4th Kerala Science Congress, Feb 27th- 29th, 1992,* Thrissur. Kerala State Committee on Science, Technology & Environment, Government of Kerala, pp.231-233.

Sheela, M.S. and Venkitesan, T.S. 1994. Potential of the bacterial pathogen *Bacillus macerans* on root-knot nematode infesting black pepper. *Proc. Acad. Environ. Biol.* 6: 97-99.

Sheela,M.S.,Bai,H.,Jiji,T.and Kuriyan,K.J.1995.Nematode associated with ginger rhizosphere and their management in Kerala. *Pest Mgmt. hort. Ecosystems* 1: 43-48

Sheela, M.S., Hebsybai and Sivaprasad, P. 2003. Management of phytonematodes by Bioagents. *Nematode Management in plants.* Scientific Publishers, Jodhpur, India, 185-201.

Sheela, M.S., Venkitesan, T.S. and Mohandas, N. 1993. Status of *Bacillus* spp. as biocontrolagents of root-knot nematode infesting blackpepper. *J. Plantn.Crops*, 21: 218-222.

Sivaprasad, P. and Sheela, M.S. 1998. Distribution of phytonematode management and VAM fungi in the rhizosphere of pepper and its impact on nematode management. *Proc. 3rd Int. Symp. Afro-Asian Society of Nematologists*, Sugarcane Breeding Institute, Coimbatore, pp. 102-105.

Sivaprasad,P.Jacob,A.Nair,S.K.and George,B.1990.Influence of VA mycorrhizal colonization on root knot nematode infestation in Piper nigrum L.Trends in mycorrhizal research.Proc.National Conference on Mycorrhiza,Feb.14-16,1990,Haryana Agricultural University,Hisar,India: 100-101

Sivaprasad, P., Jacob, A., Sulochana, K.K., Visalakshi, A. and George, B. 1992. Growth root-knot nematode infestation and phosphorus nutrition in *Piper nigrum* (L.) as influenced by vesicular arbuscular mycorrhizae. Proc. Third International Conference on Plant Protection in the Tropics, March 20-23, 1990, Genting Highlands, Malaysia. 6: 34-37

Sosamma, V, K. 1998. Nematode parasites of coconut and their management. *Indian coconut J.* 29 (4): 107-112.

Sosamma,V.K.(1994).Interactions between Vesicular Arbuscular Mycorrhizae(VAM) and burrowing nematode on Coconut.Final report ICAR AP cess fundscheme 1988-93. C.P.C.R.I, Kayangulam, Krishnapuram, Kerala, India. pp.121

Sosamma, V. K. (1999). Occurrence of *Pasteuria* sp. a hyperparasite of *Radopholus similis* and *Heterodera oryzicola* from Kerala. Annual meeting and Symposium on Plant disease Management for sustainable Agriculture. 14-16th Dec. 1999 at CPCRI, Kayangulam. Abstr p.50.

Sosamma,V.K.(2000).Nematodes of palms.In:*Plant diseases* (ed.) 393-413. pp. P.C.Trivedi Pointer Publishers, Jaipur 302003, Rajasthan, India.

Sosamma, V. K. (2002) Nematodes of coconut, arecanut and oil palm. Paper presented in the seminar on Centenary of Nematological Research in Kerala. during 21-22nd Feb.2002 at CPCRI, Kayangulam, Kerala.

Sosamma, V. K. 1984. *Studies on the burrowing nematode of coconut*. Ph.D. Thesis. Kerala University, Trivandrum, Kerala, India, 166 p.

Sosamma, V. K., Koshy, P. K. and Bhaskara Rao, E. V. V. (1980). Susceptibility of coconut cultivars and hybrids to *Radopholus similis* in field. *Indian Journal of Nematol.* 10: 250-252.

Sosamma, V.K, Koshy, P.K and Samuel, R.1998. Effect of VAM, *Glomus mosseae* on nematode infesting banana under coconut based high density multispecies cropping system. *Proc. III Int .Symp. AASN*, p.94

Sosamma, V.K.and Koshy, P.K. 1997. Biological control of Meloidogyne incognita on black pepper by *Pasteuria penetrans* and *Paecilomyces lilacinus*. *J. Plantation Crops* 25: 72-76.

Sosamma.V.K,Koshy,P.K.and Samuel,R.(1998).Effects of Vesicular Arbusclar Mycorrhizae,*Glomus mosseae* on nematodes infesting banana under coconut based high density multi species cropping system.*Proc. of the Third International Symposium of AFRO ASIAN Society of Nematologists (TISAAN) Nematology- Challenges and Opportunities in 21st century,April 16th-19th,1998* Sugarcane Breeding Institute,Coimbatore, pp. 94-97.

Sreeja, T.P., Eapen, S.J. and Ramana, K.V. 1996. Occurrence of Verticillium chlamydosporium Goddard in a black pepper garden in Kerala, *Indian J. Spices.Aromatic Crops* 5: 143-147

Sudha, S. and Sudararaju, P. 1998. Effect of neem oil cake and nematicides for the control of Radopholus similis in arecanut based cropping system. In third international symposium of Afro Asian Society of Nematologists held at Coimbatore held during April 16-19, 1998

Sundararaju,P and Sudha,S.1998.Effect of neem oil cake and nematicide for the control of burrowing nematode,*Radopholous similis* in the arecanut based on cropping system.*Proc.Third Internal Symp.Afro-Asian Society of Nematologists,* April 16th-19th,1998,Coimbatore, India.

Sundararaju, P. and Koshy, P. K. (1982a). Distribution of phytoparasitic nematodes on arecanut in South India. *Proceedings of the Silver Jubilee Symposium on Arecanut Research and Development, December 12th -14th 1982.* Vittal, South Kanara, Karnataka, India: 143-146.

Sundararaju, P. and Koshy, P.K. 1982a. Distribution of Phytoparasitic Nematodes on arecanut in South India. Proceedings of the silver jubilee Symposium on Arecanut Research and Development, 12-14 December 1982. Vittal, South Kanara, Karnataka, India, 143-146.

Sundararaju, P. and Koshy, P.K. 1982b Response of *Areca* collections against *Radopholus similis. Indian J. of Nematol.* 12: 404-405.

Sundararaju, P. and Koshy, P.K. 1986. Control of *Radopholus similis* on arecanut seedlings with aldicarb, aldicarb sulfone, carbofuran and fensulfothion. *Indian J. Nematol.* 16: 4-7.

Sundararaju, P., Sosamma, V.K. and Koshy, P.K. (1979). Pathogenicity of *Radopholus similis* on ginger. *Indian Journal of Nematology* 9: 91-94.

Sundararaju,P.,Sosamma,V.K.and Koshy,P.K.1984.Additional host records of root knot nematode, *Meloidogyne incognita.Indian Journal of Nematology*,14: 183-184.

Sundararju, P., Koshy, P.K. and Sosamma, V.K. 1979b.Survey of plant parasitic nematodes associated with spices in Kerala and Karnataka. *Proc. Second*

Symposium on Plantation Crops (Plant Protection), Ootacamund. Tamilnadu, India: 39-44.

Thomas, G.V.,Sundararaju, P., Ali, S.S. and Ghai, S.K. 1989. Individual and interactive effects of VA mycorrhizal fungi and root-knot nematodes, *Meloidogyne incognita* on cardamom. *Tropic. Agric.* 66: 21-24.

Van Weerdt, L. G., Birchfield, W. R., and Esser, R. P. (1959a). Observations on some subtropical plant parasitic nematodes in Florida. *Proceedings of the Soil Crop Science Society of Florida*, 19:443-451.

Van Weerdt, L.G., Martinez, A.P.and Esser, R. P. (1959b). Results of a survey designed to determine the etiology of lethal yellowing of *Cocos nucifera.Proc. of the Florida State Horticultural Society*, 72: 421-430.

Venkitesan, T.S.and Setty, K.G.H.(1978).Control of the burrowing nematode *Radopholous similis* (Cobb) on black pepper.*Pesticides*,13:40-43.

Visser, T. (1959). Observations on the prevalence and control of parasitic eelworms in tea. *Tea Quarterly*, 30: 96-107.

Viswanathan, P.R.K., Kumar, A.C.and D'Souza, G.I. 1974. Nematode association of cardamom in South India. *Cardamom News* 6: 5-19.

Weischer, B. (1967). Report to the Government of India on Plant Parasitic nematodes. PL 24/52 *Report No TA 2332, FAO, Rome*, 15p.

FSC
www.fsc.org
MIX
Papier | Fördert gute Waldnutzung
FSC® C083411

Zeitfracht Medien GmbH
Ferdinand-Jühlke-Straße 7
99095 Erfurt, Deutschland
produktsicherheit@kolibri360.de